T0210826

Economic Issues in
Global Climate Change

Economic Issues in Global Climate Change

Agriculture, Forestry, and Natural Resources

EDITED BY

John M. Reilly
and Margot Anderson

Routledge
Taylor & Francis Group

LONDON AND NEW YORK

First published 1992 by Westview Press

Published 2018 by Routledge
52 Vanderbilt Avenue, New York, NY 10017
2 Park Square, Milton Park, Abingdon, Oxon OX14 4RN

Routledge is an imprint of the Taylor & Francis Group, an informa business

Copyright © 1992 by Taylor & Francis

Library of Congress Cataloging-in-Publication Data
Economic issues in global climate change : agriculture, forestry, and
    natural resources / edited by John M. Reilly and Margot Anderson.
        p. cm.
    Includes bibliographical references and index.
    ISBN 0-8133-8435-4
    1. Climatic changes—Economic aspects. 2. Forests and forestry—
Economic aspects. 3. Agriculture—Economic aspects. I. Reilly,
John M. (John Matthew), 1955–   II. Anderson, Margot.
TJ1075.G53 1992
621.8′9—dc20   92-2510
                                                                          CIP
ISBN 13: 978-0-367-01153-6 (hbk)
ISBN 13: 978-0-367-16140-8 (pbk)

Contents

Acknowledgments

The chapters in this volume were initially presented at a conference in November 1990. We are indebted to Sally Kane of the National Oceanic and Atmospheric Administration (NOAA) of the Department of Commerce, who was instrumental in helping initiate the conference, develop topics, and identify authors. We gratefully acknowledge the Economic Research Service (ERS) of the U.S. Department of Agriculture, the Farm Foundation, and NOAA for their financial support, with particular thanks to Jim Hildreth (Farm Foundation) and Sally Kane and Rodney Weiher (NOAA). Leslee Lowstuter provided outstanding assistance in organizing and arranging the conference. Thanks to Michael Deland, Bruce Gardner, Howard Greunspecht, William Hyde, John Lee, Timothy Mount, Garth Paltridge, Bob Robinson, and Tom Schelling for their numerous contributions toward the success of the conference and for helping set the tone of this volume. A special thanks to John Miranowski for his continued support of the global-change research program at ERS.

Many people contributed to the preparation of this volume. We are much indebted to Cindy Allen, who not only provided superb technical editing but also undertook crucial administrative duties in overseeing the production process from start to finish. In many ways this book is as much Cindy's as it is ours. Marjorie Kingston was responsible for the bulk of the word-processing chores. Additional typing, editorial, and administrative assistance was provided by Larry Bostian, Ann and John Lafferty, Leslie K. Pope, Doug Rager, Shauna M. Roberts, and Tiajuana Sizemore. We thank Barbara J. Barnes, Susan De George, Tom MacDonald, Jim Morrison, and Anne E. Pearl for undertaking the often arduous task of preparing camera-ready graphics. Florence Robinson's efforts in preparing the index are greatly appreciated. We also thank Amy Eisenberg at Westview Press for her patience and support. Finally, our thanks to the authors for their outstanding contributions. Their diligence and timely response to our editorial queries helped keep the publication process on schedule.

John M. Reilly
Margot Anderson

Editors' note: We used the International System of Units (SI) throughout this book; non-SI units were included by author's request. For information about SI units and conversion factors, see Standard Practice for Use of the International System of Units

(SI) (the Modernized Metric System), *by the American Society for Testing and Materials Committee E-43 on Metric Practice, ASTM, Philadelphia, 1989.*

Preface

In the last decade of the 20th century, most recognize the potential for human activities to alter global climate with attendant effects on agriculture, forestry, and natural resources. Attempts to analyze, understand, and explain the sources and effects of global climate change have led to a broader global-change research agenda. This research agenda includes human-induced soil and water degradation, ozone depletion, deforestation, increased greenhouse-gas emissions, reduced biodiversity, and degradation of irrigated land. Each of these environmental considerations is individually a problem that can be analyzed and addressed. The challenge of global-change research is to recognize the interdependence of these problems on a global scale. Economic analysis of global change is increasingly important as the debate moves from one of scientific curiosity to serious consideration of proposals to limit the extent of global change. Despite the growing concerns about global environmental change, empirical economic analysis has been limited largely to estimates of the cost of controlling CO_2 emissions from fossil fuels. Few studies have considered the economic effects of global climate change or have examined the economic forces contributing to biological sources of greenhouse gases. Economic considerations are also paramount for defining appropriate measures of global change, determining the implications for trade and competitiveness, estimating the effects on long-term economic growth, and identifying appropriate technological and policy responses.

The chapters in this volume are an attempt to rectify the scarcity of economic analysis within the global-change debate. Not surprisingly, a variety of economic techniques and perspectives are exhibited. The diversity of analysis and perspective not only reflects the difficulty associated with modeling the economics of global change but also reflects experimentation, which is often inherent in any new avenue of research. The scale of analysis ranges from farm-level models to global models; time is included through comparative static analysis or explicit dynamic paths of adjustment; global change is incorporated as an exogenous shock or enters the model endogenously. Some authors focus on methodological issues, such as uncertainty, discounting, and weighting costs and benefits of trace-gas emissions. In many cases, these chapters are works in progress or initial ideas, not definitive solutions to problems. Nevertheless, the chapters fill an important role in encouraging economic analysis and expanding our knowledge of global change.

Part One presents two overview chapters that introduce readers to the science of global change and highlight several of the prominent economic issues. Daniel Albritton uses a pictorial approach to familiarize readers with global-change issues and technical terminology. In addition to providing a scientific background, he helps isolate research areas where economic analysis is needed. For example, he articulates how economic reasoning can be useful in the global research program: Economics can address discounting issues, weigh tradeoffs among a variety of environmental scenarios, and provide policy options. Even though considerable uncertainty exists in the earth sciences, Albritton advocates research on the social and economic effects of global change, particularly with respect to assessing the implications for extreme global-change scenarios, which are associated with significant economic costs.

Richard Stewart's chapter focuses on two fundamental issues in the economics of global change: how to define an appropriate measure of global climate change and how to design solutions to mitigate greenhouse-gas emissions. He is concerned with the disproportionate emphasis on a single greenhouse gas, CO_2, in current measures of global-warming potential. He advocates the comprehensive approach to measuring greenhouse-gas emissions, that is, one that addresses all the sources and sinks of trace gases and that reflects the relative impact (negative or positive) of each gas. The comprehensive approach is particularly relevant for natural resource–based sectors, such as agriculture and forestry, because these sectors not only contribute to greenhouse-gas emissions but also function as trace-gas sinks. In addition, Stewart underscores the need for a time component in greenhouse-gas indices to account for a gas's atmospheric residence time, which can critically affect the global-warming potential of a given gas. The comprehensive approach can also allow more flexible policy options. Policies designed to control a single gas, such as policies restricting CO_2 emissions, can induce technological change or substitution of equally environmentally damaging activities. By including all gas sources and sinks, tradeoffs in abatement can be more readily recognized and policies can be designed to reflect a country's or region's unique economic characteristics.

Part Two builds on themes developed by Stewart in the preceding section. The first chapter in the section critiques economists' reliance on efficiency criteria to determine optimal allocation of resources, and the subsequent chapters present different perspectives on measuring and weighting greenhouse-gas emissions. The themes raised by Richard Norgaard and Richard Howarth are repeated in various guises throughout the volume: What influence do future generations have in deciding how resources should be allocated today, and can current methods of economic analysis fail to lead us to define policies that ensure sustainable economic development? Because cost-benefit analysis satisfies only efficiency criteria, Norgaard and

Howarth contend that intergenerational use of resources will not be considered. The authors argue that providing sufficient weight to the future is essential in global-change analysis because of the extended time component involved and that classical analysis with lower discount rates and/or nonmarket valuation of resources is an inadequate solution to the intergenerational issue.

Jae Edmonds, John Callaway, and Dave Barns return to the trace-gas-index problem by advocating a comprehensive approach to clarifying the role of agriculture, forestry, and land-use sources (AFL) in greenhouse-gas emissions. They reiterate the need for a comprehensive approach, given that AFL can act as both sources and sinks of greenhouse gases. Depending on the index, they estimate that AFL may account for 25–42% of the global-change problem. The low estimate is based on a simple measurement scheme that only considers AFL contributions to radiative forcing, whereas the higher estimate reflects more sophisticated measures that weight AFL emissions by coefficients that reflect, inter alia, the effect of each gas on the radiative profile in the atmosphere and its atmospheric residence time. John Reilly expands the comprehensive-approach theme by calculating an index of trace gases derived from an optimal-control problem. His trace-gas index, which specifically incorporates economic discounting, includes not only the global-warming potential of gases but also the nonclimatic effects of trace gases and an evaluation of the tradeoff between economic damage due to climate change and to radiative forcing itself.

The seven chapters in Part Three use a variety of methods and perspectives to examine connections between global change, agriculture, and natural resources. Along with highlighting several different methodological approaches, the scale of analysis ranges from a single farm to global agricultural production and from one crop to more complex farming systems. Three chapters focus on methodological issues associated with estimating climate-change impacts. Harry Kaiser, Susan Riha, David Rossiter, and Daniel Wilks use a detailed level of analysis to examine long-term adjustment to climate change for a representative farm in Minnesota. They stress the importance of the interdisciplinary approach by combining atmospheric, agronomic, and economic models to examine farm-level effects of climate change. William Easterling, Pierre Crosson, Norman Rosenberg, Mary McKenney, and Kenneth Frederick present the details of a modeling exercise that concentrates on climate-change effects in a four-state region considered to be highly sensitive to climate change—the American Midwest. By focusing on an entire region, they are able to capture interactions between agricultural, forestry, and water resources to present a more aggregated picture of possible climate-change implications. Noel Gollehon, Michael Moore, Marcel Aillery, Mark Kramer, and Glenn Schaible also focus on a broad region, the Western United States, and examine the implications of climate-induced changes in water availability on Western agriculture.

A broader scale of analysis and scope of climate change is considered by Sally Kane, John Reilly, and James Tobey, who focus on changes in world agricultural supply and demand within an aggregated static model of world production and trade. Their analysis focuses on the welfare implications of changes in the quantity and price of major agricultural commodities, given climate-induced yield changes. As they report, the cumulative global effect of climate change on world agriculture is not large, although some individual countries show net welfare gains and others show net welfare losses.

An important aspect of the global-change problem is the relationship between existing agricultural policies and the ability of the agricultural sector to adapt to environmental change. It is generally accepted that in many countries, farm programs (as well as other domestic and trade policies) can have a significant impact on the quantity and quality of the natural resource base. Frequently, the plethora of price supports, trade policies, and environmental regulations are in conflict. Jan Lewandrowski and Richard Brazee address climate-change adjustment under current U.S. farm programs and indicate that farm programs, by slowing the adjustment to environmental change, could impede adaptation to a drier and/or more variable climate. In addition, they point out that current insurance and disaster programs may need to be reassessed given that the probability of bumper and drought years will increase.

Yohe presents an alternative approach to large econometric and simulation models. He argues that although global change is an extremely complicated process, simple economic models incorporating time, risk, and information are useful tools in extracting the important effects of a changing environment. Using large models or systems to compute aggregate effects often requires ignoring endogenous intertemporal adaptation, which can be a crucial aspect of adapting to global change. He illustrates the benefit of parsimony with a utility-based model that incorporates risk aversion and a variable reflecting the likelihood that a change in climate has occurred. He makes the crucial point that because climate changes slowly and it is difficult to determine when exactly a change has occurred, adapting to change will be based on imperfect information.

Thomas Drennen and Duane Chapman return to the trace-gas-index problem by focusing on one gas, methane. The authors are concerned about proposed weighting schemes that apply weights to methane emissions without accounting for the gas's unique characteristics. The issue is important because methane emissions are associated with agricultural activities such as rice and livestock production, which are often concentrated in developing countries. Improperly attributing the source of greenhouse-gas emissions to these countries can impose inequitable and inefficient mitigation strategies. Drennen and Chapman argue that methane's role in global warming is likely overstated; if factors such as atmospheric residence time, sources of gases, and cycling of gases are considered, then a unit of methane from bovine animals is significantly less than a unit of methane emitted through fossil-fuel combustion.

The interaction between forest resources and global climate change is examined in Part Four. Three central issues emerge in this section: (1) How will climate change affect forest resources? (2) How will forest resources affect the rate and degree of climate change? and (3) How can forest resources be managed, given climate change?

Michael Fosberg, Linda Joyce, and Richard Birdsey raise several issues important to modeling the management of forest resources under climate change. Three forest attributes — sustainability, uncertainty, and interactions — are essential in formulating models that yield admissible policy recommendations.

Michael Bowes and Roger Sedjo use a stochastic model of forest growth to simulate the sensitivity of forest growth to various levels of climate change. Their focus is primarily on Missouri, which may be quite vulnerable to climate change. Under some climate scenarios, Bowes and Sedjo predict that without intervention or adaptation strategies, Missouri's forested area could be significantly decreased with attendant consequences for the local timber industry and wildlife populations and ecological impacts associated with increased fire risk. Intervention could take the form of technological responses that improve timber yields under reduced moisture conditions. However, in low-productivity forests such as Missouri's, it is the authors' judgment that active adaptation intervention will not be economically viable.

Richard Adams, Ching-Cheng Chang, Bruce McCarl, and John Callaway focus on evaluating the economic costs and benefits of strategies designed to mitigate agricultural trace-gas emissions and sequester carbon through tree planting. Agriculture and forestry's role in trace-gas emissions is attributed to methane from rice production and livestock, nitrous oxide from nitrogenous fertilizer use and deforestation, and CO_2, primarily from deforestation. They estimate the costs associated with reducing agricultural emissions through more efficient feed rations and through decreased nitrogen fertilizer use. They also investigate costs of tree planting on agricultural land to sequester carbon.

Several proposals are advocated to mitigate the buildup of greenhouse-gas emissions. Sequestering carbon in forests is one suggested approach for limiting the buildup of CO_2 in the atmosphere. The efficacy of such a plan depends on several factors, including tree-planting practices, amount and type of land available, land-rental rates, yield, and carbon uptake. Using detailed data, Kenneth Richards calculates the incremental amount of carbon that could be sequestered through a rural tree-planting and forest-management program in the United States. He indicates that tree planting is an economically viable scheme that may be particularly beneficial in tropical countries.

The effects of global change are expected to vary widely across individual countries. For many developing countries, agricultural production is already at risk because of low-quality soils and restricted access to modern inputs. Farmers are often less able to adjust to change because proper information or alternative cropping practices are not available. In addition, because many countries are producing only at a subsistence level, changes in the quality or availability of agricultural resources may have debilitating effects. On the other hand, countries in the developed world may experience fewer adverse effects because they tend to have higher-quality agricultural resources, a diverse agricultural base, better access to modern inputs, and national policies that can provide adaptation assistance. Unfortunately, in many countries adaptation to climate change may involve more intensive farming methods, which can pose environmental risks. The chapters in Part Five develop several of these themes in more detail by focusing on problems unique to particular countries.

Using a general equilibrium approach, David Godden and Philip Adams examine the implications of global change on the Australian economy. Because their analysis includes supply-and-demand effects and the effects of greenhouse- induced policies, they are able to capture a variety of interactions that cannot be captured in single-sector models. For example, Godden and Adams show that although some agricultural sectors may exhibit welfare losses, agriculture as a whole could gain due to offsetting effects created by changed foreign demand for Australian agricultural products.

Diana Liverman presents a perspective from a middle-income, resource-vulnerable country—Mexico. She indicates that under climate change, Mexico is likely to experience a warmer and drier climate. Liverman's estimates indicate that corn yields will seriously decline unless steps are undertaken to expand irrigation or to improve access to modern seed varieties and chemical inputs.

Gennady Menzhulin predicts the effect of climate change on agriculture in Europe, the USSR, and North America by focusing on two variables: the likely level of crop productivity and the variability of yields, which he estimates by using agronomic models. His likely scenario for future climate change is based on paleoclimate data instead of scenarios generated by circulation models. Menzhulin finds that previous predictions of future agricultural disasters cannot be substantiated, in the long run, for the regions examined.

The problems facing Japan, an increasingly input-intensive agricultural system beginning to experience resource depletion, are explored by Ryohei

Kada. He traces Japan's agricultural history from a fairly diverse farming sector to one reliant on monoculture and modern inputs. Kada raises questions about the long-term sustainability of the current Japanese system and calls for a reevaluation of policies and production methods.

Part Six critically assesses several chapters in this volume. Timothy Mount appraises the models used to estimate the impact of climate change on agriculture. He sees a comparison between the evolution of models used to analyze the energy crisis and models used to estimate climate-change impacts. Over time, the level of analysis evolves from simple models, with exogenous shocks representing changes in resource availability, to sophisticated models that endogenize the resource change and allow economic agents to adjust accordingly. Mount allows that climate- change research is currently at a middle stage of development; the tools are not as sophisticated as they should be for complete analysis but are likely to continue to evolve and become more useful for policy analysis. Because of the gradual nature of climate change, he advocates the use of long-run models, with adjustment, and models that incorporate agricultural resources, such as soil and water.

Steven Sonka focuses on chapters that provide quantitative assessment of the effects of global change on the agricultural sector. He finds that, in general, the analyses fall short of what is needed by those formulating domestic and international policy. He is particularly concerned with the overemphasis on economic change that is precipitated by an exogenous shift in climate conditions and the lack of nonclimatic considerations. Sonka advocates research that can address the climate-change-induced consequences on food security, the economics of food production and consumption, and the implications for investment in the agricultural sector.

W. Kip Viscusi tackles issues critical to economic analysis of global change. Although he covers several pertinent topics, such as discounting and his faith in the efficiency criteria, he focuses primarily on issues surrounding uncertainty. He points to uses of uncertainty that include estimating future climate conditions, assessing climate-change impacts, and modeling the effect of uncertainty on the actions of individuals. Because uncertainty is pervasive in global-change research, he perceives a need to clarify the various definitions of uncertainty and how to best include uncertainty in economic models.

Part Seven jointly addresses data availability and needs and research priorities for economic analysis of global change. Timothy Mount argues that economic analysis will likely become increasingly empirical, putting large demands on data collection. Because it is unlikely that any single organization can provide all the necessary data, cooperation and coordination among agencies will be required.

John Antle creates a useful taxonomy that helps elucidate future research needs. For him, research addressing the economics of global change is either a reducing uncertainty (research directed at reducing the uncertainty surrounding the effects of global change) or technology and policy development (research that increases welfare through intervention given uncertainty). Additional uncertainty-reducing research could focus on distributional impacts. As various chapters in the volume show, even though aggregate impacts of global change may be small, regional impacts could be significant. Along the same theme, additional understanding of the role developing countries will play in the greenhouse-gas debate is of particular concern. In terms of technology- and policy-development research, Antle suggests further research on the tradeoffs among the plethora of mitigation and adaptation interventions. At the same time, Antle argues that we need to seriously consider not only the likelihood of technological change but its effect on the rate and degree of global change.

Michel Potier and Tom Jones stress the need for research in several areas: (1) construction of general equilibrium models to assess economy-wide global-change effects; (2) assessment of the benefits of reducing global change; (3) evaluation of the tradeoffs among economic policy instruments, such as emissions taxes and tradable permits; and (4) development of practical policy advice given the continuing interest in international greenhouse-gas agreements.

In conclusion, this volume represents a broad cross section of current research on the economics of global change. Despite, or perhaps because of, the preliminary nature of many chapters, several authors challenge conventional wisdom and question accepted empirical estimates. Perhaps more importantly, the suggestions for future research may provide guidance and motivation for those who might be beginning research in the area. The small group of researchers that has been focusing on global change over the past decade must expand because the information needs for policy analysis are likely to grow significantly in the future. This volume provides a snapshot of current economic thinking about global change and provides one starting point for researchers who are evaluating the economics of global change within the context of agriculture, forestry, and resource issues.

Margot Anderson
John Reilly[1]

---

[1] From the Resources and Technology Division, Economic Research Service, U.S. Department of Agriculture, Washington, D.C. The views expressed in this preface are the authors' and do not necessarily express the views or policies of the U.S. Department of Agriculture.

PART ONE

Overviews

1

The Science of Global
Change: An Illustrated Overview

Daniel L. Albritton[1]

Introduction

This summary addresses four points: (1) the scientific scope of global change, (2) the characteristics of the three major components of the global system, (3) the status of the current scientific understanding of global change, and (4) what appear to be the most fruitful interactions between science and economics. The comments regarding the last point are the personal viewpoint of a scientist. Other chapters in this volume elaborate on the science-economics interaction from the economist's perspective.

The Global System: What is the Nature of the Science-Economics Arena?

It is useful to think of the global system in terms of forcings, physical processes, physical responses, biological processes, and ecosystem responses (Figure 1.1). A variety of forcing agents activate numerous physical processes that move the global system to a new physical state. The latter, in turn, induces numerous biological processes that cause changes in the world's ecosystems.

The aim of global-change scientific research is to understand the role of humans in the forcing agents and to build a predictive understanding of how the planet will respond to these forcings. Humans and their economic affairs enter into public-welfare decisions associated with the forcing agents and into similar decisions associated with the physical and biological responses. Specifically, there are costs associated with reducing our forcing

---

[1] From the Aeronomy Laboratory, National Oceanic and Atmospheric Administration, Department of Commerce, Boulder, Colo.

of global change, and there are costs associated with coping with the impacts of global change.

Ideally, we want to know enough—both about the global system and the economic systems—to optimize the relation between those two costs. This should be the joint aim of the partnership of the physical sciences and economics, toward which this volume is an excellent start.

Components of the Global System: Three Types

Global-Change Forcing Agents—Human-Influenced and Natural

Radiatively Important Gases. Several trace gases capture part of the surface heat energy that is radiated outward toward space and then radiate it back to the surface, which is the greenhouse effect (Figure 1.2). It is a natural part of the global system. However, over the past century, human activities have increased the atmospheric abundances of gases such as CO_2, methane, chlorofluorocarbons, nitrous oxide, and lower-atmospheric ozone. Thus, the trace-gas emissions associated with human activities have perturbed the natural greenhouse balance of the planet, leading to the possibility of global warming.

Sun-Earth Interactions. The sun is, of course, the driver of the climate system of the Earth. In short, we are a "solar-powered" planet. The variation of sunlight causes daily, week-to-month, seasonal, and long-term variations. The first three variations are clearly discernible. However, it is by no means clear how sun-atmosphere couplings can introduce changes occurring over one to two decades, and the topic is one of lively current debate. The effort to find such mechanisms is driven not only by the need to forecast such natural swings of the climate, but also to aid the search for and the understanding of the human-induced changes that are superimposed on this natural variation.

Physical Response Processes: The Climate "Machine"

The physical climate system is, in many ways, analogous to many of our machines (Figure 1.3).

Physical Air-Sea Coupling (a "Part"). Wind stress on the ocean surface influences evaporation, which influences one of the key greenhouse gases—water vapor. Such a process must be characterized if we are to build a working picture of the greenhouse effect.

Water Vapor Feedback ("Linkages"). Many of the parts are coupled cyclically. For example, a surface warming will cause more evaporation, leading to higher concentrations of water vapor. This, in turn, will increase the greenhouse trapping of radiation, which would then would lead to even higher surface warming. The cycle illustrates "positive feedback," i.e., a reinforcement that amplifies the warming phenomenon.

El Niño-Southern Oscillation (a "Subassembly"). In the large expanses of the tropical Pacific, a major subsystem of the planet marches to its own drummer. The atmospheric circulation pattern (upward motion in the western Pacific, matched by downward motion in the eastern Pacific) waxes and wanes on approximately a 26-month cycle. During the high peaks and deep valleys of this variation, U.S. weather and habitation and fishing along the eastern coasts of the Pacific are severely affected. Hence, considerable research is directed toward building a predictive capability for this major subsystem.

Oceanic Thermal Inertia (a "Warm-up" Period). The main delay in a warming of the Earth's surface is the long time that it takes to warm the oceans. The nature of that delay depends largely on how warm surface water is carried into the colder deep ocean and vice versa. Thus, the large-scale circulation patterns of the ocean, which are difficult to observe, are a key factor in the timing of greenhouse warming.

Biological-Response Processes

Biotic systems have three scales of responses (Figure 1.4).

Microscale. A key example of very-small-scale processes is the uptake of CO_2 via the stomata of leaves. Carbon dioxide can stimulate plant growth, showing that there is an "upside" to what is the large "downside" of this major greenhouse gas. However, there is high variation in the responses of different plants, making it difficult to assess the overall benefit.

Middle Scale. An important example of local responses to temperature change is forest composition. Several climatic processes influence the makeup of forest-tree types: rainfall—regeneration; moisture levels—forest fires; and warmer winters—pest mortality. Forest responses via such mechanisms can be large, even to small changes in physical variables.

Macroscale. Several ecosystems demonstrate that long-term climatic changes can produce large-scale changes in the landscape: drier climes and forest migration. The understanding of these processes is, however, largely based on sparse and empirical data.

Knowns and Unknowns:
How Well Do We Currently Understand this System?

The few examples above illustrate the complexity of the Earth system, with its variety of forcing agents, diversity of the "wheels" and "cogs" that are the processes that make up the planetary machine, and the new physical and biological states that are reached. Although we have learned much, there is still much to learn (Figure 1.5). The status of the science is varied, ranging from "certainties" to "unknowns," which are arranged here in that order. There are policy and economic implications about what we do know and about what we do not know.

The Natural Greenhouse Effect (a "Certainty")

In terms of basic physics, if a body is bathed in visible radiation, it warms up and radiates infrared energy (heat) (Figure 1.6). In terms of our planet Earth, it works the same way, except the atmosphere introduces a "blanket" that traps part of that outbound radiation. It is not the common atmospheric gases—nitrogen and oxygen—that are the wool in the blanket; it is the minute concentrations of gases such as water vapor and CO_2.

There are several key points regarding the greenhouse effect (Figure 1.7): (1) it is a natural part of the Earth; (2) water vapor and CO_2 have been part of the atmosphere for millions of years; and (3) their presence has produced an average surface temperature of ≈ 15 °C. Without them, the average temperature would be ≈ -15 °C, and our planet would be shrouded in ice.

Thus, there is no doubt that the greenhouse effect is real. We understand its basic principles. So, what is the problem and issue regarding the greenhouse effect? It is this: just recently (geologically speaking), we have begun to alter it.

Trace-Gas Forcing of the Radiation Balance (a "Confident Calculation")

In the 1950s, the major gas causing radiative forcing was CO_2, with the combined effect of all the of the other gases amounting to only one-third as much. However, by the 1980s, not only had the total radiative forcing increased fourfold, but also CO_2 was then only about half of it (Figure 1.8). This demonstrates that, from a scientific perspective, policy and economics should consider all of the greenhouse gases and their relative contributions.

Predictions of Future Planetary Responses (Application of Our "Best Tools")

To predict future changes as a result of current forcings requires a "working replica" of the global system (Figure 1.9). These are the global models that reside in large computers and are intended to be the best "replica models" of the complex system itself. While admittedly not perfect miniatures of the system, the best attempts have obviously been made to incorporate the known major processes (e.g., both the warming and cooling roles of clouds and the time delay introduced by ocean circulation).

Several future forcing scenarios are entered into to these models, e.g., "business as usual" (increasing greenhouse gases) and "bite the bullet" (decreasing greenhouse gases). Global models, as our best tools, then yield predictions for each of the scenarios. The results of those predictions include values for variables that are of human interest, e.g., temperature, rainfall, and sea level. The range of these values can be compared with past natural variation—worse or less? They can also be used to search for the first signs that these predictions are actually borne out.

Some predictions from the recent Intergovernmental Panel on Climate Change (IPCC) report are given here (Figure 1.10). The business-as-usual scenario predicts that global-average temperatures will increase 1 °C by 2025 and 3 °C by 2100. Sea level will increase 0.2 m by 2030 and 0.65 m by 2100. If severe cuts in greenhouse-gas emissions were to be made, the above values would be reduced by factors of 2 to 5. Generally, such changes would not be accomplished smoothly because of superimposed natural variation. The continents would warm faster and more than would the oceans. It is important to note that it is currently believed that some warming is very likely.

The View of the Past ("Judgment Calls")

The geological record contains the climate history of the planet. Proxy indicators such as tree rings and fossils give estimates of past temperature variations. They show that, over the past 10,000 years since the last Ice Age, the planet's average temperature has varied 1–3 °C, thereby providing a measure of the natural swings in the planet's surface temperature (Figure 1.11). For the past several hundred years, we have been "rebounding" from the Little Ice Age, which was the most recent minimum in temperature.

Direct temperature measurements have been made for the past century and a half. These data show the details of the most recent end of the warming trend—an increase of ≈0.45 °C over that span. It has occurred largely in two upward jumps, the first in the 1920s and the second in the 1980s. Do we know why these temperature changes have occurred?

What Cannot Be Said ("Unknowns")

The recent 0.45 °C warming can neither confirm nor deny whether a human-influenced warming has occurred. Because the predicted greenhouse warming (≈0.7 °C) is so similar in magnitude to unexplained natural variation, the signal does not stand out clearly from the noise (Figure 1.12). Thus, the jury is still out on whether a greenhouse warming has or has not occurred.

The current models are not accurate enough to predict regional climate changes. The greatest uncertainties arise from an imperfect understanding of sources and sinks of greenhouse gases (which influence the predicted forcing), clouds (which influence the predicted magnitude of the warming), oceans (which influence the predicted timing and patterns of the warming), and polar ice sheets (which influence the predicted sea-level increases).

Science and Economics: Tending the Interface
(Some Personal Impressions and Examples)

The Next Steps Vis-à-Vis the Long Haul

A key example of how science can guide economic research (and vice versa) is in regional changes (Figure 1.13). Science is probably 5–10 years away from being able to predict regional changes reliably. Yet, the majority of the economic impacts are defined by the nature of regional changes. What is to be done to make progress in economic analyses over the next few years?

It is suggested that economics research focus on the upper extreme of the scientific uncertainty. Although the upper extreme of uncertainty is no more likely than the lower one, the decisions regarding the larger impacts associated with the upper extreme (e.g., the highest sea-level rise) are the ones that need the most economic input.

Global Changes—Multiple Issues

A human perturbation can often cause more than one type of environmental issue. Examples are the chlorofluorocarbons, which contribute to stratospheric ozone depletion and greenhouse warming, and nitrogen oxides and volatile organic compounds, which cause acid rain, smog, and global warming (the latter via the production of lower-atmospheric ozone). Science cannot give relative weights to environmental issues, but economics can. Thus, it is suggested that economic studies focus on the relative cost indices of the effects of different environmental issues.

Trace-Gas Sources and Sinks

The science-economics interaction is particularly critical for the greenhouse gases (Figure 1.14). The knowledge of the different sources and sinks of the variety of greenhouse gases ranges from excellent (industrial sources of chlorofluorocarbons) to abysmal (the role of iron in oceanic uptake of CO_2). The studies needed to refine this knowledge are difficult.

Hence, it is suggested that economic advice regarding the high-payoff areas of the source-sink research endeavors would be a distinct aid in research planning. The reverse is also true. Namely, the advice of science regarding the magnitude of the uncertainty range for particular emissions could avoid economic studies that would be of minimum utility.

Discounting the Future

It is clear that there are areas where large uncertainties exist (e.g., regional climate change). It is also clear that there are phenomena that are, for practical purposes, irreversible. An example is the long lifetime of the chlorofluorocarbons, which implies that, even with our most stringent reductions in emissions, the Antarctic ozone "hole" is likely to be with us for a century.

Consideration of action or inaction in the face of uncertainty must weigh the degree of reversibility associated with the phenomenon, i.e., whether we can indeed quit the game later, after we learn that we are being dealt losing hands. It is suggested that economic discounting consider building in a factor that accounts for such irreversibility.

Partnerships

It is clear that both science and economics are valuable tools in the tool kits of today's decision makers. Not only is it expected that each of the two tools be honed to a sharp cutting edge, but it is also clear that the complexity of the environmental issues faced today demands that they act in concert, providing a full picture of options for the decisions that lie before us. Books like this one are links in those interactions. I appreciate the chance to contribute.

10

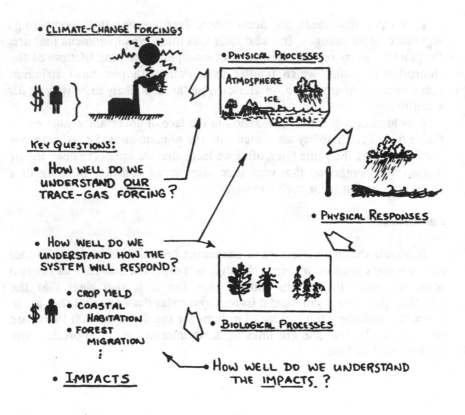

FIGURE 1.1

Ⓑ COMPONENTS OF THE GLOBAL SYSTEM

(3 TYPES)

1. GLOBAL-CHANGE FORCINGS: (HUMAN-INFLUENCED & NATURAL)

• RADIATIVELY IMPORTANT GASES

    ○ CO_2 IS A "GREENHOUSE" GAS. CO_2

    ○ FOSSIL FUEL SOURCE.

    ○ ITS ATMOSPHERIC ABUNDANCE IS INCREASING.

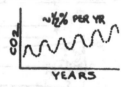

             ~½% PER YR   YEARS

• SOLAR IRRADIANCE   (WE ARE A "SOLAR-POWERED" PLANET!)

—DRIVER→ { • DAILY • WEATHER • SEASONS • CLIMATE ← BUT THE MECHANISMS ARE NOT CLEAR.

FIGURE 1.2

2. PHYSICAL-RESPONSE PROCESSES: ("THE CLIMATE MACHINE")

- **PHYSICAL AIR/SEA COUPLING...** (A "PART")

  AIR ——WIND STRESS——→ ⌂ EVAPORATION
  ────────────────
  OCEAN

- **WATER VAPOR FEEDBACK...** ("LINKAGES")

  ⟳ GREENHOUSE → MORE H_2O → MORE IR → GETS
  WARMING   VAPOR   TRAPPED   EVEN
                                        WARMER ⟲

- **EL NIÑO/SOUTHERN OSCILLATION...** ("SUBASSEMBLY")

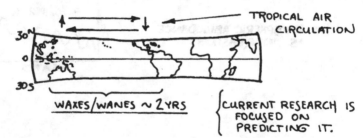

  TROPICAL AIR
  CIRCULATION

  WAXES/WANES ~ 2 YRS
  { CURRENT RESEARCH IS
    FOCUSED ON
    PREDICTING IT.

- **OCEANIC THERMAL INERTIA...** ("WARM-UP PERIOD")

  RADIATIVE FORCING
  ↓↓

  ?(WARM)?   } RESULT: DELAYED
     COLD   WARMING
              BUT HOW LONG?

FIGURE 1.3

3. <u>BIOLOGICAL-RESPONSE PROCESSES:</u>   (EXAMPLES)

• <u>MICROSCALE</u>...   <u>PRODUCTIVITY:</u> FORESTS/CROPS

LEAF STOMATA

<u>MECHANISMS</u>: • CO_2 STIMULATES
                        GROWTH

                    • WARMER TEMPERATURES
                        AID GROWTH

▶ BUT HIGH VARIATION IN RESPONSES!

• <u>MIDDLE SCALE</u>...   <u>FOREST COMPOSITION</u>

<u>MECHANISMS</u>: • RAINFALL/TEMPERATURE
                        AND REGENERATION

                    • MOISTURE LEVEL AND
                        FOREST FIRES

                    • WARMER WINTERS AND
                        PEST MORTALITY

▶ LARGE-MAGNITUDE RESPONSES TO SMALL
    CLIMATE CHANGES

• <u>MACROSCALE</u> ...   <u>AREAL EXTENT OF FORESTS</u>

<u>MECHANISMS</u>: • TEMPERATURE CHANGE
                        AND MIGRATION

                    • LATITUDE/ALTITUDE
                        COUPLING

▶ "UNDERSTANDING" IS LARGELY
    EMPIRICAL & BASED ON SPARSE
    DATA.

FIGURE 1.4

© KNOWNS AND UNKNOWNS

(How well do we understand the workings of the system?)

A Scientific Status Report , arranged
here by decreasing level of confidence:

    (a) CERTAINTIES   ("!")

    (b) CALCULATIONS WITH HIGH CONFIDENCE

    (c) RESULTS OF BEST TOOLS

    (d) JUDGMENT CALLS

    (e) NO STATEMENT CAN BE MADE   ("?")

THERE ARE POLICY AND ECONOMIC
IMPLICATIONS OF EACH

DETAILS FOLLOW...

FIGURE 1.5

(a) THE NATURAL GREENHOUSE EFFECT
(A "CERTAINTY")

● **IN TERMS OF BASIC PHYSICS:**

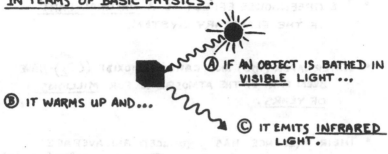

Ⓐ IF AN OBJECT IS BATHED IN **VISIBLE** LIGHT...

Ⓑ IT WARMS UP AND...

Ⓒ IT EMITS **INFRARED** LIGHT.

● **IN TERMS OF OUR PLANET EARTH:**

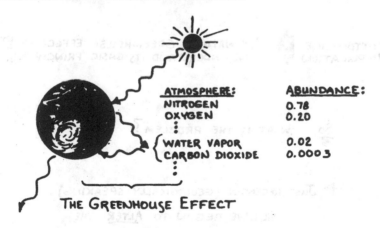

ATMOSPHERE:	ABUNDANCE:
NITROGEN	0.78
OXYGEN	0.20
WATER VAPOR	0.02
CARBON DIOXIDE	0.0003

THE GREENHOUSE EFFECT

FIGURE 1.6

- **KEY ASPECTS:**

    - THE GREENHOUSE EFFECT IS A <u>NATURAL PART</u> OF THE PLANETARY SYSTEM.

    - WATER VAPOR (H_2O) AND CARBON DIOXIDE (CO_2) HAVE BEEN PART OF THE ATMOSPHERE FOR <u>MILLIONS OF YEARS</u>.

    - THEIR PRESENCE HAS PRODUCED AN AVERAGE SURFACE TEMPERATURE ($\overline{T}_S$) OF ~15°C (~60°F)

    - <u>WITHOUT THEM</u>, T_S WOULD BE −15°C (5°F)

BOTTOM-LINE IMPLICATION ➡ THE NATURAL GREENHOUSE EFFECT IS <u>REAL</u>. WE UNDERSTAND ITS BASIC PRINCIPLES.

$\dfrac{So...}{}$ WHAT IS THE PROBLEM?

➡ JUST RECENTLY (GEOLOGICALLY SPEAKING), WE HAVE BEGUN TO <u>ALTER</u> THE GREENHOUSE EFFECT.

FIGURE 1.7

(b) Trace-Gas Forcing of the Radiative Balance

("High-Trust Calculations")

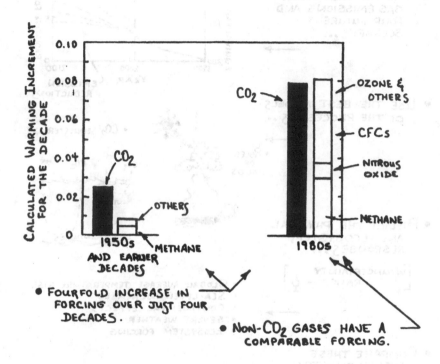

- Fourfold increase in forcing over just four decades.

- Non-CO_2 gases have a comparable forcing.

 • CO_2 is not the whole story.

• Radiative forcing is increasing rapidly.

FIGURE 1.8

(c) Application of the "Best Tools"

The Approach...

- Take the past trace-gas emissions and four future scenarios...

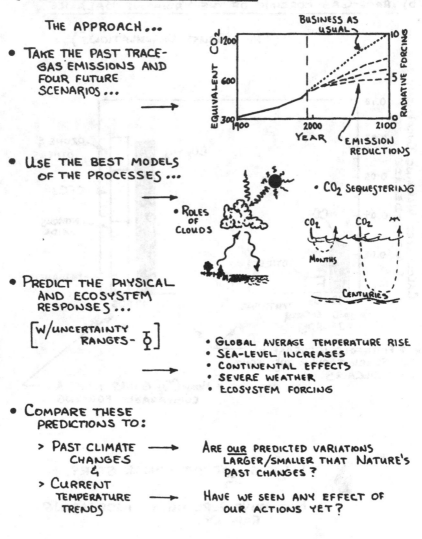

- Use the best models of the processes...

- Predict the physical and ecosystem responses...

$$\left[\text{w/uncertainty ranges} - \phi \right]$$

- Global average temperature rise
- Sea-level increases
- Continental effects
- Severe weather
- Ecosystem forcing

- Compare these predictions to:

  > Past climate changes → Are <u>our</u> predicted variations larger/smaller that Nature's past changes?
  &
  > Current temperature trends → Have we seen any effect of our actions yet?

FIGURE 1.9

THE PREDICTIONS...

- IF THE "BUSINESS-AS-USUAL" SCENARIO WERE TO INDEED OCCUR:

    GLOBAL-AVERAGE TEMPERATURES:

    UP 0.5 - 1.2°C BY 2025   (~1°C)

    2.1 - 4.7°C BY 2100   (~3°C)

    GLOBAL MEAN SEA LEVEL :

    UP   8 - 30 cm BY 2300   (~20cm)

    UP   30 - 110 cm BY 2100   (~65cm)

- IF THE "SEVERE CUTS" SCENARIO WERE TO INDEED OCCUR:

    THESE NUMBERS ARE REDUCED BY FACTORS
    OF 2 - 5.

- GENERALLY, THE INCREASES WILL NOT BE SMOOTH,
    BECAUSE OF SUPERIMPOSED NATURAL VARIATION.

- CONTINENTS WARM FASTER & MORE THAN OCEANS.

 NONE OF THE UNCERTIANTY RANGES FOR
THE "BUSINESS AS USUAL" SCENARIO
INCLUDE ZERO.

(I.E., IT IS CONSIDERED EXTREMELY
UNLIKELY THAT THE INCREASED FORCING
WOULD YIELD NO AVERAGE WARMING.)

FIGURE 1.10

(d) THE BEST JUDGMENT INDICATES THAT...

- A 3°C WARMING WOULD BE LARGER THAN NATURAL VARIATIONS OVER THE PAST 10,000 YEARS:

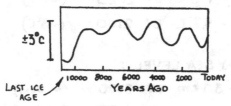

- OVER THE PAST 150 YEARS, THERE HAS BEEN A GLOBAL-AVERAGE TEMPERATURE INCREASE OF 0.45±0.15°C.

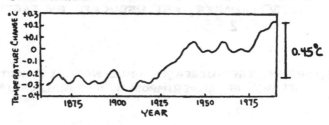

- OVER THE SAME PERIOD, GLOBAL-AVERAGE SEA LEVEL HAS INCREASED 15 ± 5 CM.

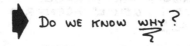 DO WE KNOW WHY?

FIGURE 1.11

(e) WHAT CANNOT BE SAID...

- THE ~0.5°C "RECENT" TEMPERATURE RISE CAN NEITHER CONFIRM NOR DENY WHETHER A HUMAN-INFLUENCED GREENHOUSE WARMING HAS OCCURRED.

  SINCE THE PREDICTED CURRENT ENHANCED GREENHOUSE WARMING 0.5-1.9°C IS COMPARABLE TO NATURAL VARIATION...

  - NATURAL CHANGES MAY BE MASKING A TRACE-GAS WARMING, OR

  - THE OBSERVED CHANGES MAY BE LARGELY NATURAL.

- REGIONAL CLIMATE CHANGES CANNOT YET BE PREDICTED RELIABLY.

- THE GREATEST CURRENT UNCERTAINTIES REGARD THE TIMING, MAGNITUDE, & REGIONAL PATTERNS OF THE PREDICTED CHANGES.

- THESE UNCERTAINTIES ARISE FROM INCOMPLETE UNDERSTANDING OF:
  - SOURCES/SINKS OF GASES (PREDICTED FORCING)
  - CLOUDS (PREDICTED MAGNITUDE)
  - OCEANS (PREDICTED TIMING & PATTERNS)
  - POLAR ICE SHEETS (PREDICTED SEA LEVEL RISE)

FIGURE 1.12

ⓓ SCIENCE & ECONOMICS: THE INTERFACE

(SOME PERSONAL IMPRESSIONS & EXAMPLES)

● THE NEXT STEPS VIS-A'-VIS THE LONG HAUL

    EXAMPLE: PREDICTING REGIONAL CHANGES

        SCIENCE — SUBSTANTIALLY IMPROVED PREDICTIONS
                    WILL BE 5 - 10 YEARS AWAY.

SUGGESTION⟩ ECONOMICS — FOCUS ON THE "UPPER"
                          RANGE OF THE   E.G.,
                          UNCERTAINTY   PREDICTED
                          RANGE ΔT

    WHY? BECAUSE — • BOTH ENDS ARE EQUALLY
                        LIKELY.
                 — • DECISIONS REGARDING LARGEST IMPACTS
                        NEED MOST ADVICE.

● GLOBAL CHANGES — MULTIPLE ISSUES

    EXAMPLES:

        SCIENCE — (1)   CFCs   { OZONE DEPLETION
                                  GREENHOUSE WARMING

             (2)   • NO_x { SMOG - ACID RAIN
                    • HYDRO-   GREENHOUSE WARMING
                      CARBONS   (VIA OZONE PRODUCTION)

SUGGESTION⟩ ECONOMICS — HOW TO WEIGHT ISSUE A vs. B IN
                      POLICY DECISIONS IS NOT A SCIENTIFIC
                      QUESTION.

        BUT ECONOMICS COULD PROVIDE THE COMMON
                BASIS.

FIGURE 1.13

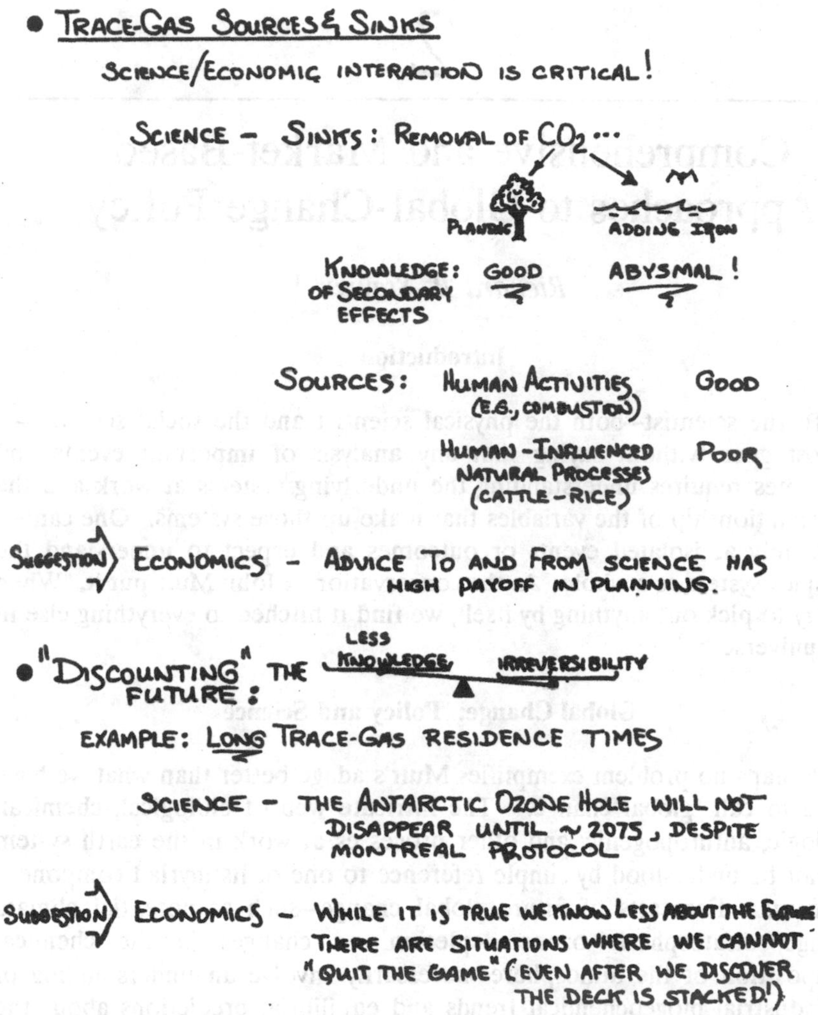

- <u>TRACE-GAS SOURCES & SINKS</u>

  SCIENCE/ECONOMIC INTERACTION IS CRITICAL!

  SCIENCE — SINKS: REMOVAL OF CO_2 ...

  PLANTING   ADDING IRON

  KNOWLEDGE: GOOD   ABYSMAL!
  OF SECONDARY
  EFFECTS

  SOURCES: HUMAN ACTIVITIES   GOOD
  (E.G., COMBUSTION)

  HUMAN-INFLUENCED   POOR
  NATURAL PROCESSES
  (CATTLE-RICE)

  SUGGESTION⟩ ECONOMICS — ADVICE TO AND FROM SCIENCE HAS
  HIGH PAYOFF IN PLANNING.

- "DISCOUNTING" THE   LESS
  FUTURE: KNOWLEDGE ——▲—— IRREVERSIBILITY

  EXAMPLE: LONG TRACE-GAS RESIDENCE TIMES

  SCIENCE — THE ANTARCTIC OZONE HOLE WILL NOT
  DISAPPEAR UNTIL ~2075, DESPITE
  MONTREAL PROTOCOL

  SUGGESTION⟩ ECONOMICS — WHILE IT IS TRUE WE KNOW LESS ABOUT THE FUTURE
  THERE ARE SITUATIONS WHERE WE CANNOT
  "QUIT THE GAME" (EVEN AFTER WE DISCOVER
  THE DECK IS STACKED!)

FIGURE 1.14

2

Comprehensive and Market-Based Approaches to Global-Change Policy

Richard B. Stewart [1]

Introduction

To the scientist—both the physical scientist and the social scientist—it almost goes without saying that any analysis of important events and outcomes requires understanding the underlying systems at work and the interrelationship of the variables that make up those systems. One cannot look only at isolated events or outcomes and expect to understand the complex system as a whole. As the conservationist John Muir put it, "When we try to pick out anything by itself, we find it hitched to everything else in the universe." [2]

Global Change: Policy and Science

Perhaps no problem exemplifies Muir's adage better than what we have come to call "global change." The intricate web of biological, chemical, geologic, anthropogenic, and other processes at work in the earth system cannot be understood by simple reference to one of its myriad component elements. Forecasts of future global change—such as potential climate change, stratospheric ozone depletion, or changes in the chemical composition of the atmosphere—necessarily involve an understanding of preindustrial biogeochemical trends and equilibria, predictions about the

---

[1] From Georgetown University Law Center, Washington, D.C.; between August 1989 and July 1991, at the Environment and Natural Resources Division, U.S. Department of Justice. The author's views are his own and do not necessarily represent the views of the U.S. government or the Department of Justice. The assistance of Jonathan Wiener in the preparation of this paper is much appreciated.
[2] John Muir (1911); diary entry for July 27, 1869, page 110.

enormously diverse socioeconomic activities that might perturb these equilibria, models of the likely resultant physical changes in the earth system, estimates of any eventual impacts on societies and ecosystems, and appreciation of interwoven feedbacks and synergisms.[3]

Policy formulations need to match the underlying ecological systems they address. Yet all too often, policy discussions zero in hastily on only one of many variables, despite our experience that narrow policies addressing only one attribute of a complex system typically provide little or no environmental benefit yet incur significant cost. In global-change discussions, policy commentators initially gravitated toward a narrow focus on one aspect of the issue: the potential global-warming effects of CO_2 emissions from energy-sector activities. Ignored were all the other aspects of "global change," including the several other radiatively active trace gases, their diverse sources and sinks in all sectors of human activity, and their multiple environmental attributes. The result was a policy thrust that omits most of the global-change science and neglects the foundations of effective and efficient policy.

Happily, the policy perspective seems to be broadening to match the ecological reality. Integration of physical and social science is making clear that a comprehensive approach to the complex global system is essential, addressing all the relevant trace gases and their sources and sinks. This volume is itself evidence of such a comprehensive view, because it emphasizes the important role of agriculture, forestry, and natural resources in global-change processes, including both their inputs to global change and the likely effects of global change on them. Briefly, the set of inputs contains emissions of not just CO_2 but a variety of gases, including methane (CH_4), nitrous oxide (N_2O), chlorofluorocarbons (CFCs), and others. Among the anthropogenic trace gases, CO_2 is the least potent radiative-forcing agent,[4] and, despite its abundance, it constitutes only about half of total current anthropogenic contributions to global radiative forcing.[5] From a sectoral standpoint, energy too is only about half the game; agriculture and forestry together account for one-fourth or more of current contributions to radiative forcing.[6] Among the agricultural contributions to radiative forcing, the gases other than CO_2 predominate. Through rice cultivation, animal husbandry, land clearing, and other critical activities practiced worldwide, agriculture and forestry yield major amounts of CH_4, N_2O, and CO_2—though it is in these sectors that the precision of our estimate of net emissions is

---

[3] For example, *see* J. Smith and D. Tirpak, eds. (1989).
[4] From Shine et al. (1990), Table 2.8.
[5] From Houghton et al. (1990), page xx, Figure 7.
[6] From IPCC Working Group III (1990), page 4, Figure 1.

perhaps the lowest, and the need for better emissions characterization the greatest.

Moving from the inputs to the effects of global change, it is clear that the trace gases have environmental attributes—both beneficial and adverse—in addition their potential for radiative forcing. For example, CFCs deplete the stratospheric ozone layer, allowing greater ultraviolet irradiance to affect crops and other life forms, whereas CO_2 in the atmosphere may enrich plants' photosynthetic productivity and aid the efficiency of plants' water use (Rosenberg et al., 1990). These effects, as well as the impacts of radiative forcing, are all of obvious concern in the agriculture and forestry sectors.

Global-Change Policy: Comprehensive and Market-Based Approaches

This brief and incomplete summary, with which you are no doubt familiar, highlights the need to take a system-wide approach to global change. A "comprehensive approach" would address all the sources and sinks of the trace gases, and employ a measure of the relative impacts of the various gases. One such measure, the "radiative-forcing" or "global-warming-potential" index, has already proved helpful in addressing global-change issues; further improvements in such measures, especially to take account of new scientific information, the economics of intertemporal discounting, and environmental impacts beyond radiative forcing, should be a high priority. Calculation of these indices depends in turn on the estimates of sources and sinks, whose refinement must also be a high priority.

At the same time, the complex system of socioeconomic activities that contribute to net emissions of trace gases and that are potentially affected by global change indicates that any potential efforts to limit emissions that contribute to global change could be quite costly, at least for some strategies and in some societies.   The need to minimize costs while achieving environmental goals suggests the utility of employing market-based incentives that (1) ensure that the most is achieved by those who can do so at least cost and (2) encourage innovation, rather than using "command-and-control" tactics that mandate uniform adoption of centrally selected techniques.

But before marching headlong into policy prescriptions, one must confront at least three basic kinds of questions:

1. To what extent and when will global change occur? What would be the impacts of global change, and their costs and benefits? What further scientific research is needed to resolve remaining uncertainties?

2. What are the costs and benefits of measures to limit or adapt to global change? In light of these costs and benefits, what actions, if any, are

warranted now? What is the appropriate combination of measures to limit net emissions of trace gases contributing to global change and measures to adapt to any adverse effects of global change?

3. If limitation efforts are warranted—a big "if" that can only be decided based on a careful look at the costs, benefits, and uncertainties—how should they be designed? Should they be narrowly focused on specific activities, or comprehensive to match the global system? Should they employ traditional command-and-control regulations, or make use of market-based economic-incentive tools?

The answers to these questions are not simple. But experience with past environmental policy and an understanding of the nature of global change suggest quite strongly that the answer to all three questions should be shaped by a comprehensive approach.[7]

A comprehensive approach is essential to shaping future inquiry into the scientific understanding of global change. The comprehensive approach helps to identify needed future scientific research by clarifying the areas of current uncertainty that are most significant for global-change forecasts and global-change policy. If CO_2 from the energy sector is only half the story, what is the rest? And how do the several aspects of global-change science create a coherent understanding of the socioecological systems that determine anthropogenic inputs to global change and the impacts of global change on society?

In considering limitation or adaptation measures, the comprehensive approach suggests the range of options and the likely costliness and effectiveness of different policy designs. If the task is to enumerate current actions that affect net emissions, or to enumerate technologies with the potential to limit net emissions, the comprehensive approach gives a measure for calculating the overall environmental value of each action, and thus for identifying the actions of most value. If limitations on emissions are agreed upon or enacted, the comprehensive approach would provide the environmental guidance and economic flexibility to shape a cost-effective response strategy. By defining the policy goal in terms of net emissions of index-weighted trace gases, the comprehensive approach would give emitters the incentive to achieve their limitations goals through the least-cost mix of strategies addressing the various gases, sources, and sinks.

In addition, any effort to design policies responding to the second and third questions just posed—measures to adapt to or limit global change—should employ market-based approaches. Market-based economic

---

[7] Further elaboration of the ideas in this paper can be found in Stewart and Wiener (1990) and U.S. Interagency Task Force (1991).

incentives are likely to yield results at substantially less cost than would a command-and-control approach.

Comprehensive Coverage or Piecemeal Peril

Those proposing immediate trace-gas reductions typically focus on limits, often through adoption of specific "best-available" technology, on CO_2 emissions from fossil-fuel combustion. Other proposals also tend to be narrowly targeted by gas and sector, such as stand-alone policies aimed at reducing CH_4 from rice cultivation. But such a narrow focus is neither warranted by the factual information about greenhouse gases, nor by sound policy.

First, aiming at CO_2 (or another gas) alone would omit other gases with potentially greater adverse impact on the ecological system. For example, CO_2 is, molecule for molecule, the least potent radiative-forcing agent of the major anthropogenic trace gases. Because any limitation policies must necessarily address future increments of net emissions, it is the comparative impact of additional amounts of each gas that must be addressed. To provide a sound guide to policy choices, a measure of the relative environmental impacts of the gases should be used to set policy priorities; as mentioned earlier, the current radiative-forcing index needs to be used and improved by greater attention to new research findings, discounting, and environmental effects beyond radiative forcing. The problem with piecemeal proposals is not just that all the bluster they entail generates "more heat than light," but that we do not even know how much heat! Rather than taking piecemeal stabs in what might be called the proverbial radiative darkness, good policy choices demand information on the relative effects of the various emissions.

Meanwhile, CO_2 may provide significant environmental benefits that the other trace gases do not: CO_2 is the grist of photosynthesis, and can improve plants' water-use efficiency as well. The other greenhouse gases confer no such benefits, and some pose serious threats beyond radiative forcing; CFCs, for example, deplete the stratospheric ozone layer. Current work on a radiative-forcing index could be expanded to incorporate the full environmental impacts of each trace gas: CO_2 would receive a credit for enriching plant growth while CFCs receive a debit for ozone depletion. To make the point more concrete: if one is interested, say, in promoting global welfare, or just global agricultural output, policies aimed at restricting CO_2 alone are not ideal; for any given environmental result, and assuming for the moment equal costs of abatement across gases, one might prefer to have as much of the emissions in low-forcing, plant-enriching CO_2 as possible, and less due to each of the other trace gases according to their relative environmental impacts.

Second, the comprehensive approach gives sinks the serious attention they deserve. For greenhouse gases, it is *net* emissions that would be of ecological concern, the result of both emissions from surface sources and removal by surface sinks, including oceanic phytoplankton, trees, grasses, soil biota, crops, and tropospheric chemical reactions. Preserving and properly managing forests and other vegetation, or protecting phytoplankton from anthropogenic injury such as from toxic-waste disposal at sea, could help sequester carbon released from surface sources. By addressing net emissions, the comprehensive approach would give incentives for actions that carry with them significant side benefits in increased biodiversity, improved oceanic food webs, reduced soil erosion, and better timber management.

Third, the assumption I made earlier—equal costs of abatement across gases, sources, and sinks—is almost assuredly incorrect. The comprehensive approach would therefore deliver important economic benefits as compared with piecemeal approaches. If applied to any international emissions limitations, it would allow each nation the flexibility to devise its own cost-effective policy mix. Of course, any extensive measures to limit net emissions—even under a comprehensive approach—would probably mean some efforts to limit emissions of CO_2, the most prevalent anthropogenic trace gas. But the extent of any limits on each gas ought to be free to vary according to their environmental impacts and the local costs of abatement. Because the marginal costs of abatement will vary by gas, source, sink, and technique and across nations, this flexibility will enable nations to choose among the diverse responses in a way that ensures an overall least-cost response. For example, the least-cost policy option for limiting net emissions in one nation may be switching from coal to natural gas, while for another nation it may be changing agricultural practices to reduce CH_4 and N_2O emissions, and for another it may be reducing deforestation and ensuring sustainable forest management. Put another way, reducing emissions of CO_2 from fossil-fuel combustion (or stopping deforestation, or any other single tactic) might be the cheapest way to limit overall net emissions in one nation, but the most expensive in another. Global costs of limiting emissions would be significantly reduced by allowing the flexibility afforded under the comprehensive approach.

Fourth, the comprehensive approach would also cure the bane of piecemeal approaches: unwanted shifts or displacements of emissions to unregulated activities that continue to produce environmental degradation. The most oft-mentioned example of such a shift is the natural-gas-fuel-switching case: under a CO_2-only approach, utilities would likely be encouraged to undertake fuel switching from coal to natural gas, because with current combustion techniques, coal burning produces almost twice as much CO_2 per joule (or, per Btu) as does burning natural gas. But use of natural gas means CH_4 leakage from natural gas mining and transportation systems. One recent study (Rodhe, 1990) estimates that a 3–6% rate of CH_4 leakage from natural-gas transport would fully offset all the CO_2 savings from switching from coal to natural gas. Such leakage rates are probably at the high end of the average in many advanced industrialized countries, but may be typical elsewhere. And swift expansion of natural gas transport capacity to comply with a stiff CO_2-reduction strategy could well mean use of hastily designed new facilities, or older facilities in disrepair, with leakage rates higher than today's. The net result could be greater net radiative forcing than that found in the absence of the CO_2-reduction effort. Even if the CH_4-leakage rate only offset, say, 50% of the CO_2 savings, the CO_2-only policy would be severely undermined. The comprehensive approach, on the other hand, would encompass CH_4 emissions and thereby ensure that CH_4 leakage is included in a nation's inventory of net trace-gas emiss2ions and in the incentives and efforts to limit net emissions.

Shifts among gases are possible in the agriculture and forestry sectors as well. Some proposals to reduce CH_4 emissions from wet rice cultivation involve increased application of nitrogenous fertilizers, which yield N_2O emissions (Lashof and Tirpak, 1990). Certain proposals to reduce CH_4 emissions from livestock also depend on increased use of feed grains grown with nitrogenous fertilizers (Lashof and Tirpak, 1990). Given that the radiative-forcing potential of N_2O over a midrange period (100 years) is over 10 times larger than that of CH_4 (Houghton et al., 1990), small trade-offs could be significant enough to offset or outweigh the savings in CH_4-equivalent emissions. Similar risks are present in the forestry sector, where calls to increase the productivity of trees as CO_2 sinks might be pursued by applying nitrogenous fertilizers, which emit N_2O.

Shifts could occur across sectors as well. If a CO_2-only policy induced the substitution of ethanol made from corn for carbon-containing gasoline, CO_2 emissions would decline, assuming the corn farm did not displace CO_2 sinks. But one would have to account as well for increased N_2O emissions: corn cultivation is one of the most nitrogenous-fertilizer-intensive crops (Lashof and Tirpak, 1990), and N_2O has a midrange radiative-forcing index some 200 times larger than CO_2's (Houghton et al., 1990). To take another example, policies to block deforestation and thereby preserve CO_2 sinks could inadvertently increase CO_2 emissions as communities that formerly burned harvested wood for fuel now turned to coal.

These shifts may not be inevitable; technologies and practices could perhaps be chosen that take account of all relevant trace-gas emissions. But without the system-wide outlook and the incentives provided by a comprehensive approach, choosing optimal technologies is highly unlikely. Under the piecemeal proposals being floated on the international scene, there is no reason to think that an overall optimal emissions outcome would emerge, and there is every reason to think that narrow efforts will yield narrow results. Piecemeal policies in the United States have regularly produced inadvertent shifts of emissions: policies focused on limiting one form of emissions from the "end of the pipe" have remade water effluent into solid sludge, transformed air emissions into ash, and routed solid waste to incinerators or ocean dumping. Policies requiring scrubbers to remove one gas, sulphur dioxide (SO_2), had the effect of impairing fuel efficiency and slightly raising emissions of another gas, CO_2. Trace-gas emissions and global change are far more complex than these environmental problems and, we can be sure, offer a much more complex catacomb of unseen passageways.

Building the Comprehensive Approach

The major objection that has been raised to the comprehensive approach is that the current science is not up to monitoring certain sources and sinks, such as nonpoint sources of CH_4 and N_2O. The objectors say that we should "do what we can now" and wait until later to design a comprehensive approach. Yet focusing on CO_2 from energy emissions merely because that is the source most easily measured at present bespeaks a certain complacency about the current stock of research. Many current studies are confined to energy policy alone simply because, in the words of one of the more candid analysts, "[t]his focus suggests itself because the necessary quantitative data for a least-cost analysis are far more developed in the case of energy than for other major sources of greenhouse gases."[8] And unless comprehensive research is carried out, a narrow and incomplete knowledge base will yield incomplete and flawed piecemeal policy responses.

The more inquisitive comprehensive approach thirsts for research into the key unknowns. Measuring many such emissions will not be easy. But it is not beyond our reach, if we focus current research efforts to support a comprehensive approach. Indeed, it is particularly in agriculture and forestry that the research toward improved monitoring is most needed. The net-emissions database is perhaps least well-developed for the diffuse, nonpoint sources and sinks of greenhouse gases that are typical in those sectors. For example, recent studies are advancing understanding of carbon sinks, both the total size of the oceanic vs. terrestrial sinks (Tans et al. 1990), as well as the more localized effectiveness of different types of forestry in sequestering carbon (Harmon et al., 1990). However, the uncertainties surrounding these processes are still large. We are beginning to understand CH_4's diverse sources in rice cultivation, livestock, the energy sector, and waste disposal (IPCC Response Strategies Working Group, 1990)

---

[8] From Krause et al. (1989), volume 1, page I.1-3.

and its sink in tropospheric chemical reactions, but again the uncertainties in these estimates remain significant. The same holds for N_2O. We will need better estimates of these sources and sinks if we are to forecast future concentrations of the gases, or to fashion reliable greenhouse-gas indices. If we are to calculate baseline and future net emissions for each nation or sector, the measurements will need to be more precise.

Moreover, the pertinent question is not what is immediately "feasible," but whether the costs of proceeding with a flawed piecemeal policy design are less than the costs of doing the necessary groundwork to develop a comprehensive approach. One need not wait for perfection; in the interim, proxy-based estimates of difficult-to-measure emissions could be used. And publications like the present one, which bring together physical and social scientists with expertise in the critical areas of uncertainty about emissions, namely agriculture and forestry, are indispensable. Looking toward the not-so-distant future, international efforts should be undertaken to build cooperative networks to track net emissions.

These scientific building blocks are urgently needed if sound policy is to emerge. In addition, any international agreement employing the comprehensive approach would itself be constructed out of institutional building blocks—the several international accords and national actions contemplated or already in place, each of which addresses a discrete term in the global-change equation. For example, the international agreement to phase out ozone-depleting substances, and the upcoming global forestry agreement, would be woven into the fabric of an international comprehensive approach to global change. Related national actions would similarly be recognized, to avoid giving disincentives to useful measures that nations wish to take for other reasons.

Market-Based Approaches

We have learned a great deal about the drawbacks of traditional command-and-control regulatory approaches: by mandating uniform adoption of centrally chosen abatement techniques, these approaches raise costs, discourage innovation and resource-use efficiency, and raise administrative burdens. The virtues of market-based economic incentives for environmental protection are now increasingly well-recognized. The common feature of the new tools is that they respond to market failure—such as excessive pollution—by redirecting and harnessing market forces to correct the problem. They allow flexibility among market actors, promote decentralized decision making about response tactics, further least-cost solutions by allowing those who can fix the problem most cheaply to do so most often, and stimulate efficient resource use and innovation in technologies and practices.

The first step in a market-based approach to devising new policies should be to examine current government policies and their effects on markets that affect and are affected by global change. What are their effects on the inputs to global change and on the ability to adapt to any global change? On the input side, it is often government policies that subsidize activities in the agricultural and forestry sectors and thereby increase net emissions of trace gases. These include counterproductive agricultural price supports and other policies that induce excess crop planting and unduly intensive use of nitrogenous fertilizers, adding to N_2O emissions as well as erecting trade barriers. And they include rules that needlessly encourage even below-cost forest clearing, reducing carbon sinks.[9] On the adaptation side, policies that prevent the development of efficient markets for natural resources—such as water—may undermine the incentives that would be provided to induce conservation if global change put pressure on supplies. Better operation of private markets, on both the input and adaptation sides, could help address global change without major costs to society.

Market-based incentives could also be used in new policies to address global change. Where market failures demand policy interventions, market-based tools can provide the best response options. Fees, tradable allowances, and deposit-refund programs have demonstrated success in several important environmental applications, including the tradable-credits program used to phase out lead in gasoline—achieved at about half the cost of a traditional regulatory program (amounting to savings of hundreds of millions of dollars). Both fees and tradable allowances are now being used in the U.S. program (U.S. EPA, 1988a, 1988b) to phase out CFCs under the Montreal Protocol (UNEP, 1987). And tradable allowances will be employed in the acid-rain-reduction provisions of the new Clean Air Act, with projected national savings of $1 billion annually compared with a command-and-control program.

If the threat of global change is a market failure worthy of policy intervention, then market-based techniques are especially well-suited to implementing limitation measures for trace gases contributing to global change. Because trace-gas emissions arise from so many diverse and pervasive sources, the costs of abatement are bound to vary widely among emitters. Market-based mechanisms use that variation to social advantage by imposing some restraint on total emissions—a limit on the net quantity emitted or a fee on each unit emitted—but then letting the market allocate the burden of mitigative measures to those who can most easily shoulder it.

---

[9] For example, *see* Repetto (1990).

Because the trace gases mix essentially globally and have essentially only global impacts, the possibility under market-based incentives that the spatial distribution of emissions may become uneven—called "hot spots" in the context of toxic substances—would not likely be of much concern.

In the context of limiting inputs to global change, two main economic instruments have been suggested: tradable emissions allowances and emissions fees. Both instruments hold the potential for achieving environmental goals at least cost. Tradable allowances set a total limit on net emissions, issue that sum of allowances to emitters, and let emitters trade them. Those for whom emissions reductions or sink expansions are relatively more expensive will buy allowances, whereas those who can achieve them cheaply will sell allowances. This gives an incentive to each emitter to develop new means of limiting emissions at less cost than its competitors, so that it can sell its allowances at a profit. The choice of response tactics—emissions controls, efficient use of fuels and other inputs, and innovation of new emissions-limitations techniques—is left to the emitter. The market allocates abatement actions to those who do so at least cost, reducing the overall cost to society.

Trading could be employed domestically by nations taking steps to limit their emissions of trace gases, as we are using trading in the phaseout of CFCs (U.S. EPA, 1988a, 1988b). Reallocations among nations of any agreed international obligations would also be advantageous. Such trades could consist of informal, bilateral reallocations of obligations to limit net emissions. One nation could satisfy its obligations by investing in response actions in another. Given significant international variations in marginal costs of limitation, such trades would likely enable the world economy to realize substantial cost savings. Limited international CFCs trading of this sort is now authorized under the Montreal Protocol, but it remains to be seen whether the Protocol's strictures on trading will stifle market activities.

Such trading would also serve as a market-based, decentralized vehicle for introducing needed technology into the developing world. It would point technology toward those who needed it most, and stimulate innovation by industrialized nations of technologies useful in developing nations. At the same time, this framework could obviate creation of a heavily bureaucratized, centralized regulatory authority and technology-assistance fund.

An emissions fee is another important option. The fee could be calibrated to the environmental-impacts index value of the net emissions activity. Like emissions trading, emissions fees offer a least-cost solution that promotes innovation and efficient resource use. Such a plan could make excellent sense domestically, especially where the focus is on specifying the cost of a limitation program more precisely than the quantity of emissions avoided, or where revenue raising is a major goal. Indeed, the

United States is using fees on CFCs in addition to tradable allowances. International application of a fee would raise many more difficult questions: Would nations cede their sovereignty to an international tax authority? How would the fee be set in light of varying effective marginal tax rates, and diverse taxation structures, across nations? How would the potentially enormous revenues raised be allocated and expended?

Incentive-based policies may be well-suited in the agriculture and forestry sectors, where the diversity of response tactics is great and specifying particular response technologies would be costly or counterproductive. For example, options to limit CH_4 emissions from rice cultivation include, among others, employing new or different varieties of rice, changing nutrient additives to the rice field, genetically engineering new rice plants, or switching to alternative crops such as potatoes. Mandating use of one technique by all farmers would raise costs and discourage innovation of better options. An incentive related to performance—emissions of CH_4—would, in contrast, allow farmers the flexibility to achieve results through the least costly option and would simultaneously encourage innovation. Use of incentives would put a premium on the ability to monitor emissions, but the use of proxies or surrogates, such as emissions factors linked to inputs, could aid in this effort. If the cost of estimating emissions, even with proxies, exceeds the benefits of the flexibility allowed under a performance-incentive approach, then an approach that ties incentives to the observed practices employed in the field[10] might be successful as long as it embraced a sufficiently wide variety of practices from which farmers could choose, welcomed innovations, and gave appropriate incentives for farmers' decisions to replace rice with alternative crops.

Market-based incentives could also be used to encourage efficient adaptation practices. Long-range investments, such as coastal construction or water-use planning, might, because of market failures or other institutional failures, be undertaken without giving appropriate weight to any global-change risks (such as rising sea levels or shifting precipitation). Such failures might be addressed by informational or incentive-based policies, such as by requiring coastal construction firms to purchase subsidence (coastal-erosion) insurance, or by fostering a market in water resources that provides incentives for efficient use and long-range risk management.

---

[10] *See* IPCC Response Strategies Working Group (1990), pages 30–32.

Conclusion: Shaping Global-Change Policy

Global change presents a great challenge and a great opportunity: the chance to gather all that we have learned in the past 20 years of environmental policy, science, and economics and to apply that education to a phenomenon of a scope and ubiquity greater than any we have yet confronted. Integrating physical science and social science perspectives will be crucial in this effort; we must have an ongoing conversation among policy analysts, economists, and earth scientists on such diverse topics as discounting, the adaptability of societies and ecosystems to change, means of measuring emissions, the incentive effects of policy options, and forecasting future emissions scenarios. Throughout all of these discussions, the tools of comprehensive and market-based approaches will both sharpen our insight and broaden our analysis.

References

Harmon, M., W. Ferrell, and J. Franklin. 1990. "Effects on Carbon Storage of Conversion of Old-Growth Forests to Young Forest." *Science* 247:699–702.

Houghton, J., G. Jenkins, and J. Ephraums, eds. 1990. *Climate Change: The IPCC Scientific Assessment, Policymakers' Summary.* Pp. vii–xxxiii. Cambridge, U.K.: Cambridge University Press.

IPCC Working Group III. 1991. *Climate Change: The IPCC Response Strategies, Policymakers' Summary.* Pp. xix–lxii. Washington, D.C.: Island Press.

IPCC Response Strategies Working Group. 1990. "Methane Emissions and Opportunities for Control." Coordinated by the Japan Environment Agency and the U.S. EPA. Washington, D.C.: U.S. Environmental Protection Agency.

Krause, F., W. Bach, and J. Koomey. 1989. Energy Policy in the Greenhouse. El Cerrito, Calif.: International Project for Sustainable Energy Paths (IPSEP) and Dutch Ministry of Housing, Physical Planning and Environment.

Lashof, D., and D. Tirpak, eds. 1990. *Policy Options for Stabilizing Global Climate.* Washington, D.C.: Office of Policy, Planning and Evaluation (PM 221), U.S. Environmental Protection Agency. (21P-2003.1.)

Muir, J. 1911. *My first Summer in the Sierra.* San Francisco: Sierra Club Books (published 1988).

Repetto, R. 1990. "Deforestation in the Tropics." *Scientific American* 262: 36–42.

Rodhe, H. 1990. "A Comparison of the Contribution of Various Gases to the Greenhouse Effect." *Scientific American* 248: 1217–19.

Rosenberg, N., B. Kimball, P. Martin, and C. Cooper. 1990. "From Climate and CO_2 Enrichment to Evapotranspiration," in P. Waggoner, ed., *Climate Change and U.S. Water Resources.* Pp. 151–75. New York: John Wiley & Sons.

Shine, K., R. Derwent, D. Wuebbles, and J.-J. Morcrette. 1990. "Radiative Forcing of Climate," in J. Houghton, G. Jenkins, and J. Ephraums, eds. *Climate Change: The IPCC Scientific Assessment.* Pp. 41–68. Cambridge, U.K.: Cambridge University Press.

Smith, J., and D. Tirpak, eds. 1989. *The Potential Effects of Global Climate Change on the United States*. Washington, D.C.: Office of Policy, Planning and Evaluation (PM 221), U.S. Environmental Protection Agency. (EPA-230-05-89-050.)

Stewart, R. B., and J. B. Wiener. 1990. "A Comprehensive Approach to Climate Change." *American Enterprise* 1(6): 75–80.

Tans, P., I. Fung, and T. Takahashi. 1990. "Observational Constraints on the Global Atmospheric CO_2 Budget." *Science* 247: 1431–8.

United Nations Environmental Programme (UNEP). 1987. *Montreal Protocol on Substances that Deplete the Ozone Layer, Final Act*. Nairobi: UNEP. (Updated in London, 1990.)

U.S. Environmental Protection Agency. 1988a. "Protection of Stratospheric Ozone." 53 Federal Register 30,506 (to be codified at 40 C.F.R. pt. 82) (Final Rule) (capping CFC production with marketable permits).

U.S. Environmental Protection Agency. 1988b. "Protection of Stratospheric Ozone." 53 Federal Register 30,604 (Advance Notice of Proposed Rulemaking) (proposing allocation of permits).

U.S. Interagency Task Force. 1991. *A Comprehensive Approach to Addressing Potential Climate Change*. Washington, D.C.: U.S. Department of Justice.

PART TWO

Broader Perspectives

PART TWO

Broader Perspectives

3

Sustainability and Intergenerational Environmental Rights: Implications for Benefit-Cost Analysis

Richard B. Norgaard and Richard B. Howarth[1]

Introduction

Development and environmental economists are asking new questions about future generations and economic theory in response to the widespread concern that, to be sustainable, development must be ensured, and it has not been (WCED, 1987). Is sustainability a constraint on economic optimization, the outcome of economic optimization done correctly, or one path of growth among many that an economy might find by chance? Will economic investment criteria lead to the conditions for sustainable development if nonmarket aspects of resources and environmental services are fully valued or a social rate of discount is used (Markandya and Pearce, 1988)? Will internalizing environmental externalities result in economies operating in a sustainable manner (Pearce and Turner, 1990)? Will economies operating at higher rates of stock-resource use invest in more research and technological change, facilitating access to new stock resources or an earlier and smoother transition to renewable resources (Pezzey, 1989)? Or must the use of resources and environmental services be constrained to ensure sustainability (Daly and Cobb, 1989)?

These are well-reasoned questions. They result, however, from a line of reasoning that is at odds both with the moral questions driving the political discourse on sustainability and with economic theory made whole again. The moral question is with respect to the rights of future generations to

---

[1] From the Energy and Resources Group, University of California, Berkeley (R.B.N.), and the Applied Science Division, Lawrence Berkeley Laboratory, Berkeley, Calif. (R.B.H.).

stock resources and environmental services. Basic economic theory informs us that if resource and environmental rights are redistributed from present generations to future generations, their efficient allocation between uses, both as capital and consumable goods, changes. Thinking of sustainability as a matter of intergenerational transfers reformulates what questions are considered reasonable. Rather than simply taking the existing distribution of rights as given, one can return to general equilibrium theory. This does not reduce the complexities of determining the effects of alternative distributions of rights or the uncertainties of the interplay of resources with technological change. It does, however, clarify the nature of valuation and the discount rate in benefit-cost analysis. Furthermore, it suggests a substantial reframing of how benefit-cost analysis fits into the policy process.

In this paper we (1) develop an explanation for why economists have been formulating the question of sustainability so incompletely, (2) document how the relationships between sustainability and intergenerational transfers affect the discount rate and resource valuation, and (3) investigate the significance of this reformulation to the role of benefit-cost analysis in policy formulation.

Coevolution of Economics with Technocratic Progressivism

Economics coevolved during the past two centuries with Western beliefs in technological progress and progressive social organization. Both the theoretical elaboration of economics and its practice reflect a historical faith in progress. Economists, along with the public at large, have consistently assumed that new technology will continually provide access to new resources, such that the question of whether future generations should have rights to resources can be ignored. Historic beliefs in progress also incorporated the idea that all peoples would eventually rise to a common, more rational, perfectly informed set of understandings about the world. This would eliminate the obviously irrational differences responsible for conflict among peoples, cultures, and nations. Natural resource and environmental economics, even more so than the other subdisciplines of economics, coevolved with the progressive vision of scientifically informed experts making superior technical decisions on the public's behalf, avoiding the irrationalities of politics. The idea of progress and progressive beliefs (Nisbet, 1980) have had as much influence on the selection of assumptions used by economists and the role economists play in public decision making as Newton had on how economists formulated their basic model.

The end of the 20th century represents an important watershed in public beliefs. The widespread concern that the sustainability of development must be ensured marks the demise of our faith that technical and social progress unfolds naturally after the removal of the barriers erected by traditional cultures. Global unity is threatened as partially Westernized nations individually reformulate how they wish to chart their progress while the idea of nations themselves—Canada, India, and the Soviet Union being obvious examples—is threatened by a revitalization of cultural differences within national boundaries. Our inability to foresee and correct in a timely manner the environmental and social consequences of development in the industrialized countries and of removing the barriers of traditional cultures elsewhere is discrediting the superiority of Western knowledge and progressive social organization. These shifts in our beliefs, understandings, and expectations and the dramatic transitions in national and international politics are obviously tightly interrelated.

Given this significant decline in public acceptance of the beliefs with which economic practice and theorizing coevolved, Norgaard (1990) argued that one might predict major changes for economic thought and practice. In the present paper, we follow a similar train of thought and argue that (1) publicly held beliefs about progress and progressive social organization selected against fully using the economic model; (2) for decades, this has affected the teaching of economics, the expectations of students, and the environments in which economists work; (3) hence, the discipline today has lost much of its ability to fully use its own theory; and (4) the theory's full use is sorely needed—especially now that global social beliefs and values are in a state of flux—to attain sustainable development.

Progressivism as a philosophy of social organization, most explicitly formulated by Auguste Comte (1848) during the second quarter of the 19th century, was firmly incorporated into both Marxist and non-Marxist social theory during the third quarter of that century. As a well-established system of beliefs, it became firmly embedded in the design of social institutions as the collaboration between science, technology, and government took on increasing importance in the final quarter of that century. Comte and his followers thought that not only could natural scientists actually know natural reality but that social systems could likewise be understood positively as phenomena apart from human values and the patterns by which people think, design, and act. Comte envisioned that science, by serving the process of social decision making, would free society from the irrationalities of established religions and the tyrannies of arbitrary political powers.

Economists joined the federal and state agricultural agencies established on the progressive model after the Civil War, the water and forestry agencies established at the turn of the century, the soil- and land-management agencies established during the depression, the international-development agencies established after World War II, and the environmental and energy agencies set up during the 1970s. Economists filled a critical niche in these agencies. Their estimates of the "true" costs of inputs and of the value of the nonmarket products of environmental and resource management and their rules for weighing one alternative against another allowed the agencies to go beyond mere technical design and actually make efficient choices on the public's behalf without having to constantly return to legislative bodies for advice. Indeed, governments increasingly mandated that decisions be made through economic reasoning (Nelson, 1987).

To fulfill this progressive mandate, economists have retained positivism and other "modern" assumptions about the nature of science long after other social sciences abandoned them (McCloskey, 1985; Norgaard, in press). The problem, we argue, is that when economists assumed the role within governments of determining which plans and projects were socially efficient, they had to ignore the fact that their theoretical model indicates that under different distributions of income, power, or property rights, there are different efficient solutions (Bator, 1957). To determine but one solution, economists made their efficiency determinations on the implicit assumption that the existing distribution was given. Through this assumption, economists became a conservative force in political contests over who should be allowed to do what. To be sure, as economists assumed this special role, they developed extensive rationales, otherwise known as welfare economics, for when such decisions were justified (Just et al., 1982). While the rationales are weakly convincing, few decisions have fit the assumptions on which the rationales depend.

Bromley (1990) aptly characterized the pragmatic and academic defense of benefit-cost analysis as being rooted in an ideology of efficiency that arose with economists' participation in technocratic progressivism. It has become commonplace for economists, on the one hand, to acknowledge that equity factors are insufficiently incorporated into benefit-cost analysis and, on the other hand, to argue that at least benefit-cost analysis identifies the efficient solution. This has left the impression that there is necessarily a trade-off between equity and efficiency, and indeed many economists commonly refer to just such a trade-off. In fact, economists have presented the public with a false dichotomy for so long that they have begun to believe it themselves. For every distribution of rights to resources among individuals, there is an efficient allocation of resources to intermediate and end uses. Benefit-cost analysis has traditionally only been used to identify the efficient allocation of resources given the existing distribution of rights.

Although most environmental and resource issues are resolved in part through the reassignment of rights, achieving sustainable development will probably require even more explicit and more significant redistributions of rights to resources, environmental services, and other productive assets. Although we emphasize in this paper the intergenerational transfers that are under political consideration, others have argued that sustainable development will not be possible without international transfers of rights as well as reassignments among peoples within nations (Blaikie and Brookfield,

1987; Guha, 1990; Redclift, 1987; WCED, 1987). Economists, armed with only part of their theoretical framework or working in organizational environments where they can use only part of their theory, are having difficulty addressing the issues of sustainable development. They will not meet the policy challenges of sustainable development without using general equilibrium models to explore the implications of alternative assignments of rights.

Intergenerational Rights, Resources, and Efficiency

Economists have developed an extensive literature on the "optimal" depletion of stock resources over time based on the model elaborated by Hotelling (1931). In this now classic paper, he explored how resource producers would behave in a world of perfect competition and perfect knowledge of future market conditions. Profit-maximizing producers would equate the return realized through holding units of the resource for future extraction to the return available from extracting the resource and investing the net revenue earned from the sale in the capital market. Accordingly, the resource royalty—the difference between price and marginal extraction cost—would increase over time at the rate of interest (Dasgupta and Heal, 1979). This basic outcome for the case where a resource can be extracted at constant cost is illustrated in Figure 3.1 under the conditions where a backstop technology of known cost replaces the stock resource after depletion.

Since the energy crisis of 1973–1974 alerted the profession to the importance of stock resources, economists have written over 1000 articles elaborating how this basic result differs under various combinations of about a dozen different assumptions, mostly addressing what resource allocators might really know and when. Within this extensive formal literature, no one ever developed a model in which future generations had explicit resource rights. Well-respected economists pondered the equity implications of this line of reasoning but never formally showed that there are different efficient solutions for different intergenerational distributions of resources and that the equity implications of alternative efficient allocations can be significant (Howarth, 1990; Howarth and Norgaard, 1990).

One does not have to go into the intricacies of our intergenerational model to understand the effect of assigning rights to future generations on the use of the resource. Imagine that the stock resource was divided up evenly among people who are born every 25 years all together on quarter-century marks. Imagine, furthermore, that no resource trade occurs between generations. Each generation would then allocate its share of the resource according to Hotelling's reasoning, and the price curve would come out as in Figure 3.2. From here, it is easy to imagine that if there

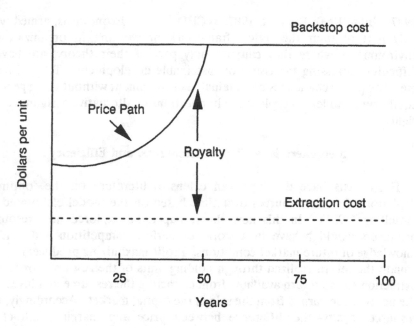

FIGURE 3.1 Efficient resource use with the rights assigned to the current generation.

were trade between people in the quarter-century groups, the sawtooth shape would even out, perhaps such that the price of the resource would always be the same as that of the backstop. In any case, the efficient use of the resource would clearly be different from when all of the rights to the resource are assigned to those in the first quarter-century period.

In part, economists implicitly assumed that progress would take care of the interests of future generations by ensuring them access to resources through new technology; in part, they have been trained not to think about distributional equity. In any case, the economics of intertemporal resource use was reduced to questions of efficient allocation as if the present generation had all of the rights to resources. Economists have repeatedly argued, for example, that energy-resource markets work *socially optimally* as long as externalities are internalized. Similarly, they have argued that the extinction of some species may be optimal without considering that the policy decision under consideration is whether future generations might have rights to biological diversity. Clearly, environmentalists see these as equity issues. We suspect that future generations, if they could speak for themselves, would express the same concern.

Within solely an efficiency framework, which implicitly assumes that the current generation holds all of the environmental rights, economists have explored two approaches that would favor sustainability. They have first

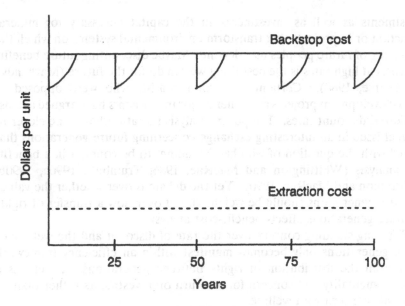

FIGURE 3.2 Efficient resource use with the rights distributed across generations.

pondered whether a lower rate of interest would not be more appropriate when determining net present value, and second, they have argued that valuing nonmarket benefits and costs would improve the situation for future generations. Each of these approaches may be appropriate, but a full theoretical framing is necessary to carry them out correctly.

Sustainability and the Rate of Discount

Economists have long recognized the perversity of discounting the benefits received and costs borne by future generations in efficiency analyses (Ramsey, 1928). Sustainability has once again highlighted the problem. How can one discount the values of benefits received and costs borne by future generations while being concerned with their welfare? Are not lower discount rates more consistent with sustainability than higher ones, since a zero discount rate treats generations equally? Low discount rates, for example, favor both letting trees grow longer and the planting of trees that take longer to grow. Perhaps a lower, social rate of discount would lead to sustainable development. Perhaps, in effect, economic practice should be a little less perverse.

The net effect of lower discount rates, however, has led to a tangled thicket of reasoning within the efficiency half of economic theory. When interest rates are low, many investments are made, including poor investments as well as investments in the capital necessary for mineral extraction or in projects that transform environmental systems on which the well-being of future peoples may depend. Maybe discounting future benefits and costs at high rates is the best thing we can do for the future (Markandya and Pearce, 1988). Certainly, environmentalists who were opposed to water-development projects and other major investments have argued against subsidized discount rates. The policy-analysis literature has come closer to the real issue in an interesting exchange concerning future generations that started with the question of who has "standing" to be counted in a benefit-cost analysis (Whittington and MacRae, 1986; Trumbull, 1990a, 1990b; Whittington and MacRae, 1990). Yet the debate is over whether the values of future generations should be weighted, not over how a transfer of rights to future generations affects benefit-cost analysis.

This long-standing concern over the rate of discount and the welfare of future generations only becomes manifest within an efficiency framework apart from the distribution of rights between generations. It reflects a banker's mentality and concern for a return on investments rather than an economic concern with welfare.

Howarth (1990) showed through the exploration of an intergenerational model with capital and resources that the rate of interest changes as income is redistributed between generations. If rights to the services of resources and environmental systems are transferred to future generations, both the supply of savings and the demand for investment funds in the present change, determining a new rate of interest (Norgaard, 1986; Howarth, 1990; Norgaard and Howarth, 1991). Although the discount rate has been treated as an instrumental variable, efficiency dictates that it equal the rate of interest. It is inappropriate to think of the rate of interest as anything other than simply another price, as something that equilibrates savings and investment on the margin. What is important is the extent to which current and future generations have rights to resources, environmental services, and human and physical capital. These rights over generations determine the kinds and levels of investments made within generations. The emphasis must be on the kinds and levels of investments and hence on what is passed between generations, not on the level of the equilibrating mechanism. The discount rate should still be important to bankers, but not to economists and policy makers.

Sustainability and Environmental Valuation

For the same reasons that the rate of interest or discount is a function of how rights to environmental services are distributed across generations, the values of environmental services are also a function of how they are distributed. Clearly, if the present generation has complete rights to the environment without any responsibilities to protect it because future generations have no rights, then the marginal value of environmental services will be lower than it would be if the present generation had more restricted access to environmental services because the present generation is responsible to future generations.

Although this is theoretically straightforward, economists have put their emphasis elsewhere, in a rather roundabout argument. They have argued, in part rightly, that environments are misused and degraded, reducing the future level of potential services, because their full value is not reflected in markets. Too many old-growth forests are cut because there is not a market for biological diversity. By indirectly determining what people are willing to pay for biodiversity, economists argue that better environmental-management decisions can be made. Nonmarket valuation makes considerable sense with respect to making better use of the resource for the current generation. However, economists have advocated nonmarket valuation as a means of protecting the rights of future generations as well. There is something contradictory about trying to protect the rights of future generations through more fully measuring how the current generation values the resource. The valuations, furthermore, are either based on behavior that is guided by market signals generated with the current generation holding all of the rights, or they are based on preferences revealed through interviews in the context of the current distribution of rights.

The valuation of nonmarket goods and services is based on the assumption that the absence of these goods and services from the market is a marginal problem. If their absence significantly distorts how people perceive opportunities and behave, then we can neither ask nor observe people to determine how they value things. The fact that attendance at national parks is not rationed by the price mechanism may not distort the economy and people's perceptions and actions much, but if the price of gasoline does not reflect its true market cost, the value of national parks cannot be determined through the travel-cost method. If nonmarket goods and services make up a small portion of the total economy, then the inclusion of these goods and services in the market would not significantly change behavior or the relative prices between existing goods and services. However, if they make up a significant portion, the problem of nonmarket goods is significant and the need for economists is more than trivial, yet valuation becomes illogical (Norgaard, 1989). If sustainability were a marginal problem, it would probably not be on the political agenda.

With respect to protecting future generations, environmental valuation as it is now undertaken, to make the point strikingly clear, resembles an effort to determine how men really value women in a society that condones rape. Valuation can be analogous to trying to determine how untouchables value dung collecting when the caste system gives them few other alternatives. The analogies help emphasize the point that rights are a prior issue, that rights affect the structure of society, and that, most importantly, what people do, not the level of a hypothetical market-equilibrating mechanism, is the issue. To use another analogy, if income were more evenly distributed in the United States, the price of Cadillacs would be lower. The price, however, is just an equilibrating mechanism. The important point is that there would be fewer people driving around in Cadillacs.

The obsession economists have had with the equilibrating mechanism, both in the case of the discount rate and in the case of valuation, rather than being concerned with the nature of the equilibrium reached, identifies a related contradiction. Neoclassical economists espouse the use of the market mechanism because it systematically links complex phenomena. Demand and supply curves for each good are a function of the prices of numerous other goods. Prices are constantly changing, most markedly in response to key substitutes and complements and in response to key inputs to goods production, but, ultimately, in response to any change in tastes, technology, and resource availability with respect to any good in the economy. It is this property of the price system, of everything being connected to everything else by an invisible hand, that central planners have had such difficulty matching, hence the success of the market economies. Of course, if the relative values of goods were simply fluctuating randomly, planners could use an average and perhaps even improve on the performance of a market economy. Or if relative values were moving systematically in accord with phenomena that the central planners could identify, they might be able to speed the adjustment and avoid overshooting the new equilibrium. The problem, of course, is that the movements are neither random nor knowable. This explanation of the success of the price system gained widespread acceptance during the Reagan-Thatcher era. And the success of the price system is the miracle now sought in the Soviet Union and Eastern Europe.

Neoclassical economists also present quite a different story. They argue that one thing can be weighed against another according to their economic values. The values of benefits and costs can be determined, summed, and compared. The problem is that when they undertake benefit-cost analysis, they forget their other story with respect to the constantly changing nature of economic forces. If demand curves are constantly changing, values are constantly changing. And if they are changing in ways that are neither random nor predictable, values over the period of economic decisions cannot be determined. When engaged in benefit-cost analysis, neoclassical economists assume the burdens of the central planners they chide.

Implications for the Policy Process

The historic standoff between the effectiveness of progressive technocratic decision making and the social desirability of democratic decision making may be heading toward a rout as the environmental and social consequences of Western technologies undercut the unique authority science has assumed within government bureaucracies. During the past decade only, there has been a phenomenal rise in nongovernmental organizations staffed with scientists, an acceleration of the politicization of science, and the near breakdown of progressive technocratic decision making due to the inability of agencies to maintain, in the public's understanding, the agencies' definitions of problems and solutions. Rational positivism has been undermined, no longer simply by the ingenuousness of thinking that values can be kept separate from fact, but also by the inability of technocracies to defend their syntheses of the contributions from the separate environmental and social sciences. Economists have taken their share of criticism for their role in the process. They may be the subject of more derision when the public understands that economists have been basing their arguments about the efficiency of alternative plans to protect the rights of future generations or to transfer rights to them on arguments that implicitly assume that this generation has the rights to fully exploit all resources, the environment, and other assets.

Though the policy environment is changing rapidly, there is little indication that economists have begun to notice. Economists were asked by legislatures to simplify their agenda by helping the agencies weed out inefficient plans and projects (Nelson, 1987; Heineman et al., 1990). They were, in effect, told, and have since been trained to think, that policy making should consist of determining values and weighing the benefits and costs of projects and plans prepared by technicians "below" them. Policy making has never followed economists' expectations that they should do the final sifting. Yet economists have affected the policy process significantly during the past 30 years. To the extent that economists have captured the process, they have reduced the range of policy discourse, and probably increased the risk of error (Hoos, 1983; Mishan, 1986a). Legislators and the public still wish—to a considerable extent—that more decisions could be made by the rules economists claim to provide.

Others have pointed out that economists have been treating distributional questions inadequately (Bromley, 1989; Hoos, 1983; Mishan, 1986b). Our analysis demonstrates that the problem is more serious than a matter of slighting the questions of equity. It is not simply that economists are not looking at the full set of options by excluding distributional considerations. Where distribution is the critical issue, and in most questions of public policy it surely is, the incomplete use of economic theory results in illogical information being fed into the policy process. In the case of sustainable development, the problem is that economists are questioning whether plans of action that redistribute rights are efficient by using values based on the implicit assumption that the redistribution does not take place (Nordhaus, 1990).

In critical cases, as in rape and caste, Western European peoples have put the question of rights before efficiency. The people of the United States, for example, are not about to let economists weigh the benefits and costs of the First Amendment. A quarter-century ago, both the administration and Congress had the good sense not to ask economists to weigh the benefits and costs of civil rights legislation. To be undertaken properly by economists, such analyses would require a social-welfare function to weigh the array of efficient solutions generated by alternative specifications of rights to speech or civil liberties. In fact, while the rights to speech and civil liberties are constantly being modified, their modification has been undertaken directly with respect to other rights rather than through a complex analysis of alternative efficient outcomes. We have methods of adjudicating rights quite apart from economics. And while economic considerations no doubt motivate different interest groups to favor or not favor changes in basic rights, economic arguments are typically recognized as unethical in these cases and not used openly in either political or legal discourse.

Less-critical cases of changes in rights are probably the majority, and here the assignment of rights and their economic implications need to be considered together. Economists, however, were mistakenly asked by society to simplify the politics of assigning rights by appealing to economic efficiency alone. Economists mistakenly accepted this request but could only carry it out by ignoring half of their theory. The solution is to join full economic reasoning with the political and policy process, letting economists describe the economic implications of alternative assignments of rights, rather than having partial economic reasoning dominate the process or act as a separate filter through which political decisions must pass.

Whether sustainability is critical or not, it clearly hinges on the distribution of rights. The ever widening and deepening discourse among European Greens is over politics, not technique. Green politics is rooted in social theory rather than natural science. Third World environmental movements are addressing issues of human rights, cultural survival, social structure, and the global economic order all at once. And while the environmental movements in the United States and Canada initiated the interpretation of environmental problems, they are now rapidly trying to catch up with the broader and deeper understandings that are developing globally. Within this changing milieu, the partial economic reasoning and the structural and procedural environment in which economists work in the United States and in the multilateral assistance agencies are temporarily obfuscating the policy process.

What constitutes proper benefit-cost analysis and its role in the policy process has long been debated (Bromley, 1990). The institutionalization of particular procedures, both through legislative mandates and through the regulations and conventions of diverse agencies, make reform extremely difficult. Healy and Ascher (1990) document how the Congress and the U.S. Forest Service established ever more complex planning procedures in an effort to respond to competing interests. The procedures attempted to incorporate ever more information in the vain hope that the facts would bring the competing interests to a consensus. In fact, excessive information distorted the policy process along an ever more diligent pursuit of rationality, which missed the critical issues, the very issues of who *should* be allowed to do what. In our judgment, the problem starts with the false hope that economic analyses emphasizing efficiency could resolve distributional issues in the first place. Our concern with the welfare of future generations shows that this problem runs deeper than we have heretofore understood.

References

Bator, Francis. 1957. "The Simple Analytics of Welfare Maximization." *American Economic Review* 47: 22–59.

Blaikie, Piers, and Harold Brookfield. 1987. *Land Degradation and Society*. London: Methuen.

Bromley, Daniel W. 1989. *Economic Interests and Institutions: The Conceptual Foundations of Public Policy*. Oxford: Blackwell.

Bromley, Daniel W. 1990. "The Ideology of Efficiency: Searching for a Theory of Policy Analysis." *Journal of Environmental Economics and Management* 19(1): 86–107.

Comte, Auguste. 1848. *A General View of Positivism*. (Translated into English by J. H. Briggs in 1865.) Dubuque, Iowa: Brown Reprints.

Daly, Herman E., and John B. Cobb, Jr. 1989. *For the Common Good: Redirecting the Economy Toward Community, the Environment, and a Sustainable Future*. Boston: Beacon Press.

Dasgupta, Partha S., and Geoffrey M. Heal. 1979. *Economic Theory and Exhaustible Resources*. Cambridge, U.K.: Cambridge University Press.

Guha, Ramadchandra. 1990. *The Unquiet Woods: Ecological Change and Peasant Resistance in the Himalaya*. Berkeley: University of California Press.

Healy, Robert, and William Ascher. 1990. "Incorporating Environmental Information in Natural Resources Policymaking: A Policy Process Approach." Paper presented at the Association for Public Policy Analysis and Management meeting, Oct. 19, San Francisco.

Heineman, Robert A., William T. Bluhm, Steven A. Peterson, and Edward N. Kearny. 1990. *The World of the Policy Analyst: Rationality, Values, and Politics*. Chatham, N.J.: Chatham House Publishers.

Hoos, Ida R. 1983. *Systems Analysis in Public Policy: A Critique*. Revised edition. Berkeley: University of California Press.

Hotelling, Harold. 1931. "The Economics of Exhaustible Resources." *Journal of Political Economy* 39: 137–75.

Howarth, Richard B. 1990. Economic Theory, Natural Resources, and Intergenerational Equity. Ph.D. Thesis, Energy and Resources Program, University of California at Berkeley.

Howarth, Richard B., and Richard B. Norgaard 1990. "Intergenerational Resource Rights, Efficiency, and Social Optimality." *Land Economics* 66(1): 1–11.

Just, Richard E., Darrell L. Hueth, and Andrew Schmitz. 1982. *Applied Welfare Economics.* Englewood Cliffs, N.J.: Prentice-Hall.

McCloskey, Donald N. 1985. *The Rhetoric of Economics.* Madison: University of Wisconsin Press.

Markandya, Anil, and David Pearce. 1988. "Environmental Considerations and the Choice of the Discount Rate in Developing Countries." Environment Department Working Paper 3. Washington, D.C.: World Bank.

Mishan, Ezra J. 1986a. "The Future Is Worse than It Was," in Ezra J. Mishan, ed., *Economic Myths and the Mythology of Economics.* Pp. 169–93. Atlantic Highlands, N.J.: Humanities Press International.

Mishan, Ezra J. 1986b. "The Mystique of Economic Expertise," in Ezra J. Mishan, ed., *Economic Myths and the Mythology of Economics.* Pp. 78–107. Atlantic Highlands. N.J.: Humanities Press International.

Nelson, Robert H. 1987. "The Economics Profession and Public Policy." *Journal of Economic Literature* XXV(1): 49–91.

Nisbet, Robert. 1980. *History of the Idea of Progress.* New York: Basic Books.

Nordhaus, William. 1990. "Greenhouse Economics: Count Before You Leap." *The Economist* 316(7662): 21–24.

Norgaard, Richard B. 1986. "Economic Theory, Natural Resources, and Intergenerational Equity." Paper presented at the Western Economic Association Meetings, July, San Francisco (earlier version presented at the American Economic Association Meetings, December 1985, New York).

Norgaard, Richard B. 1989. "Three Dilemmas of Environmental Accounting." *Ecological Economics* 1: 303–14.

Norgaard, Richard B. 1990. "Sustainable Development and the Possibilities for the Coevolution of a Participatory Economics." Paper presented at the Western Economics Association Meetings, July, San Diego.

Norgaard, Richard B. In press. "Environmental Science as a Social Process." *Environmental Monitoring and Assessment.*

Norgaard, Richard B., and Richard B. Howarth. 1991. "Sustainability and Discounting the Future," in R. Costanzo, ed., *Ecological Economics: The Science and Management of Sustainability.* Pp. 88–101. New York: Columbia University Press.

Pearce, David W., and R. Kerry Turner. 1990. *Economics of Natural Resources and the Environment.* Baltimore: Johns Hopkins University Press.

Pezzey, John. 1989. "Economic Analysis of Sustainable Growth and Sustainable Development." Environment Department Working Paper 15. Washington D.C.: World Bank.

Ramsey, F. P. 1928. "A Mathematical Theory of Saving." *Economic Journal* 38: 543–59.

Redclift, Michael. 1987. *Sustainable Development: Exploring the Contradictions*. London: Methuen.

Trumbull, William N. 1990a. "Reply to Whittington and MacRae." *Journal of Policy Analysis and Management* 9(4): 548–50.

Trumbull, William N. 1990b. "Who Has Standing in Cost-Benefit Analysis." *Journal of Policy Analysis and Management* 9(2): 201–18.

Whittington, Dale, and Duncan MacRae. 1986. "The Issue of Standing in Cost-Benefit Analysis." *Journal of Policy Analysis and Management* 5(4): 665–82.

Whittington, Dale, and Duncan MacRae. 1990. "Comment: Judgments about Who Has Standing in Cost-Benefit Analysis." *Journal of Policy Analysis and Management* 9(4): 536–47.

World Commission on Environment and Development (WCED). 1987. *Our Common Future*. Oxford: Oxford University Press.

4

Agriculture in a Comprehensive Trace-Gas Strategy

Jae Edmonds, John M. Callaway, and Dave Barns[1]

Introduction

Although agriculture is usually thought of as potentially harmed by changes in CO_2 levels and/or climate, it is itself a significant source of greenhouse-gas emissions to the atmosphere. Just as CO_2 is an unavoidable combustion byproduct in the production of energy from fossil fuel, the emission of greenhouse gases is the unavoidable consequence of specific large-scale agricultural activities, including raising ruminant livestock, cultivating rice, tilling soil, and altering land use (IPCC, 1990; Lashof and Tirpak, 1989; Reilly and Bucklin, 1989; Wuebbles and Edmonds, 1991; WMO, 1985). Just as there are specific energy-production activities that generate no greenhouse-gas emissions, there are agricultural activities that emit no greenhouse gases. In fact, there are agricultural activities that enhance greenhouse-gas removal from the atmosphere.

The purpose of this paper is to review the current understanding of the role of the human activities associated with raising crops and livestock, silviculture, and land-use change in relation to the emission of greenhouse gases. We do not attempt to assess the potential future emissions of greenhouse gases associated with agriculture, forestry, and land-use change.

Direct Emissions of Greenhouse Gases from Agriculture

Carbon Dioxide Emissions

Anthropogenic emissions of CO_2 are released principally from two human activities: fossil-fuel use [≈ 5.4 petagrams (Pg) C/year in 1986 (Marland et

---

[1] From Pacific Northwest Laboratory, Washington, D.C.

al., 1989)] and land-use changes (deforestation) [≈1.3 Pg C/year (Trabalka, 1985)], with lesser amounts released by industrial processes such as cement manufacture [≈0.1 Pg C/year in 1986 (Marland et al., 1989)] (Table 4.1). Land-use change is currently thought to be a junior partner in the net emission of CO_2 to the atmosphere. This was not always the case. In the late 19th century, agricultural emissions were the dominant source of net anthropogenic release of CO_2 to the atmosphere (Trabalka, 1985; Bolin et al., 1986). The demand placed on land use by expanding human activities, such as cattle raising, forestry, and cultivation, are clearly important motivations for land-use change. The process, of course, is not that simple. In many parts of the world, the demand for traditional biomass fuels is a locally important determinant of land-use change.

Unlike the fossil-fuel resource base, which provides essentially unlimited quantities of carbon that could be vented to the atmosphere, the carbon stored in above-ground terrestrial biomass is limited. There are ≈560 Pg C in the form of above-ground terrestrial biomass, principally stored in forests. This is estimated to be ≈15-20% (≈120 Pg C) less than was present in the mid-19th century. Carbon stored in soils provides a larger, though still limited, stock of carbon that can be released to the atmosphere. Current estimates place the stock of carbon in soils at ≈1500 Pg C (MacCracken et al., 1990).

TABLE 4.1 Historical Carbon Dioxide Emissions

Sources	Quality of Emissions Data	Uncertainty	Notes
Fossil-fuel use	Good	±10-20%	Global emissions in 1988 = 5.9 Pg C/year. Emissions estimates available for 1850-1986, by country and by fuel
Cement manufacture	Good	±10-20%	Minor importance
Land-use change	Fair/poor	±50-100%	Global emissions in 1980 = 0.4-2.6 Pg C/year, primarily from tropical deforestation

Source: Wuebbles and Edmonds (1991).

Knowledge of the net annual emissions of carbon from land-use changes is far less certain than are emissions estimates for fossil-fuel use. Estimates of net annual CO_2 release as emissions from land-use changes have been made for the year 1980 by various researchers. Net release is calculated as the difference between annual gross harvests of biomass, plus releases of carbon from soils, less biomass carbon whose oxidation is long delayed (e.g., stored in forest products such as telephone poles, furniture, and housing) and less additions to the stock of standing biomass. Houghton et al. (1987) estimated that 1980 emissions were between 1.0 and 2.6 Pg C/year. Detwiler and Hall (1988) estimated that 1980 emissions from the tropics were 0.4–1.6 Pg C/year. Net emissions from land-use change are dominated by tropical deforestation. Houghton et al. (1987) estimated that all but 0.1 Pg C/year of net release is from tropical forests.

Estimates of deforestation in 1980 for all the countries in the world are greatest for Brazil, Columbia, the Ivory Coast, Indonesia, Laos, and Thailand. Estimates of net CO_2 emissions from land-use change have increased over recent decades. Before 1950, significant deforestation is estimated to have occurred in the temperate latitudes as well as in the tropics (Houghton et al., 1987; Lashof and Tirpak, 1989).

Conventional estimates of net CO_2 release from land-use change do not take the possibility of a CO_2-fertilization effect into account. Although it is a matter of heated debate, it has been suggested that increases in the atmospheric concentration of CO_2 could act to accelerate the rate at which the terrestrial biosphere stores carbon. The conventional wisdom (see Bolin, 1986) is that the net flux from the biosphere must be small. Recent papers, including Tans et al. (1990), Goudriaan (1991, 1989), Goudriaan and Ketner (1984), and Esser (1991), have estimated that the terrestrial biosphere may be a large net sink for carbon. Tans et al. (1990) inferred that the northern midlatitudes are a large net sink for carbon. Goudriaan (1991) estimates that the CO_2-fertilization effect could more than compensate for carbon releases from land-use changes and still provide a net sink for carbon in the range of 0.5–1.5 Pg C/year. Esser (1991) finds that the CO_2-fertilization effect may have resulted in a net carbon uptake of 70 Pg over the period 1860–1980.

Methane Emissions

Annual observations of emissions sources are not generally available for methane (CH_4). Source-strength uncertainties are so high that emissions budgets are typically referenced to a decade rather than to an individual year. Whereas CO_2 emissions from fossil fuels and from land-use changes are developed for specific years on the basis of databases for a small number of human activities, the sources and sinks for CH_4 are developed by using an observed, globally averaged atmospheric burden of CH_4 [≈4800 teragrams (Tg) CH_4], an average annual rate of increase (≈1%/year), and an atmospheric lifetime (≈10–12 years), derived from atmospheric-chemistry models to calculate a global-emissions-budget constraint of ≈540 (400–640)

Tg/year. Information about the changing isotopic ratio of CH_4 (i.e., $^{14}CH_4/^{12}CH_4$) is used to partition the constraint by broad period of origin and, by inference, to bound the contribution of fossil fuels to total emissions [*see*, for example, Wahlen et al. (1989), Cicerone and Oremland (1988), or Ehhalt (1985)]. A summary of the best current understanding of the sources and sinks of CH_4 is given in Table 4.2.

At this point, it is not clear that all of the major sources of CH_4 have been identified, and the emissions rates of those that have been identified are subject to significant uncertainty. Anthropogenic activities are currently thought to contribute approximately half of all CH_4 emissions to the

TABLE 4.2 Historical Methane Emissions

Sources	Quality of Emissions Data	Uncertainty	Notes
Natural sources	Poor	> ±75%	Global emissions = 4.5 Pg CH_4/year. Natural sources are ≈50% of global emissions. It is not clear that all sources and sinks for CH_4 have been identified. Total emissions are derived from atmospheric observations and calculations from atmospheric-chemistry models. Currently known sources are (in Tg CH_4/year) enteric fermentation (wild animals), 4±3; wetlands, 110±50; lakes, 4±2; tundra, 3±2; oceans, 10±10; termites, 25±20; methane hydrates, 5±?; other, 40±40.
Agriculture	Fair–poor	±40–50%	25–40% of global emissions. Current source estimates are (in Tg CH_4/year) rice cultivation, 70±30; ruminant digestive systems of domesticated animals (cattle), 77±35; slash-and-burn agriculture and/or land clearing, 55±30.
Energy	Fair–poor	±40–50%	20–25% of global emissions. Current source estimates are (in Tg CH_4/year) deep coal mining, 25±20; natural-gas production, transport, and distribution, 40±15; incomplete combustion (e.g., automotive exhaust, fuel-wood burning), 15±8; landfills, 30±30.

Source: Wuebbles and Edmonds (1991).

atmosphere. The three principal human activities that have been identified as emissions sources are cattle raising, rice production, and energy production and use. Although human activities have been identified as major sources of atmospheric emissions, there remains great uncertainty surrounding emissions-source estimates and the time profile of those emissions.

Roughly 40% of total atmospheric CH_4 emissions are estimated to come from agriculture and land-use change. An approximately equal total amount can be attributed to natural phenomena, with the residual accounted for by energy-related activities. The principal agriculture and land-use-related sources of emissions are the production of rice and ruminant livestock ($\approx$15% of total emissions each) and land-use change ($\approx$10% of total emissions). Whereas energy production and consumption is dominated by developed-country activities, the production of rice is generally associated with developing nations. Eighty percent of total area under rice cultivation can be found in seven Asian developing nations: India, China, Bangladesh, Indonesia, Thailand, Vietnam, and Burma (Table 4.3). Livestock emissions of CH_4 come from every country on the planet, but five countries—India, the Soviet Union, Brazil, the United States, and China—account for half of the total (Table 4.4).

Nitrous Oxide Emissions

The sources of nitrous oxide (N_2O) emissions are poorly documented, as indicated in Table 4.5. Emissions rates are small relative to atmospheric stocks. Although the atmospheric burden and annual rate of increase are known with some confidence, the atmospheric lifetime is uncertain within the range 100–175 years. This leads to significant uncertainties in the estimates of sources and sinks, which can be derived from atmospheric-chemistry models. Individual source terms are subject to even greater uncertainty. Total annual emissions are estimated to be between 7 and 21 Tg N. Emissions studies are inconsistent in their categorization of emission-producing activities. The chief sources of emissions are currently thought to be biogeochemical activities in soils and combustion activities. The biogeochemical activities include N_2O releases from uncultivated and fertilized and unfertilized cultivated soils. Combustion activities include savanna burning, forest clearing, fuel-wood use, and fossil-fuel combustion. Other sources of emissions include oceans and contaminated aquifers (IPCC, 1990; Lashof and Tirpak, 1989).

TABLE 4.3 Rice Area Harvested in 1984

Country	Rice Area ($\times 10^9\ m^2$)	Percent of Total (%)
India	426	29
China	343	23
Bangladesh	106	7
Indonesia	97	7
Thailand	97	7
Vietnam	66	4
Burma	47	3
Japan	23	2
Others	279	19
Total	1484	

Source: U.S. EPA (1990).

TABLE 4.4 Domestic-Animal Methane Emissions, by Country

Country	CH_4 (Tg/year)	Percent of Total (%)
India	10.27	20
USSR	8.05	16
Brazil	7.46	15
United States	6.99	14
China	4.37	9
Argentina	3.11	6
Australia	1.90	4
Pakistan	1.54	3
France	1.52	3
Mexico	1.51	3
Bangladesh	1.43	3
Ethiopia	1.20	2
Sudan	1.00	2
Others	25.48	34
Total	75.83	

Source: Lerner et al. (1988).

TABLE 4.5   Historical Nitrous Oxide Emissions

Sources	Quality of Emissions Data	Uncertainty	Notes
Natural sources	Poor	Great	Global emissions = 10–19 Tg N_2O/year. Natural sources are ≈45–65% of global emissions. Natural sources include oceans and estuaries; natural soils; and wildfires, lightning, and volcanoes. It is not clear that all sources and sinks for N_2O have been identified. Although N_2O is a chemically highly stable gas in the atmosphere, the average lifetime uncertainty ranges from 100 to 175 years, leading to a 100% uncertainty in the global emission rate.
Nonenergy	Poor	±50–100%	35–55% of global emissions. Sources include soil cultivation, nitrogen-fertilizer application, slash-and-burn agriculture, and land clearing.
Energy	Poor	Great	10–40% of global emissions. The source is fossil-fuel combustion. Combustor studies originally showed high rates of release of N_2O from fossil-fuel use in high-temperature combustion. Later analysis revealed a sampling artifact, which, when removed, greatly reduces the emission coefficient.

Source: Wuebbles and Edmonds (1991).

The dominant human activities associated with N_2O emissions are related to agriculture (savanna burning, soil cultivation, and fertilizer application) and energy production (wood burning and fossil-fuel use). Although only ≈20% of total emissions are currently thought to be associated with agriculture and land-use change, those emissions are 75% of total anthropogenic emissions. The single most important anthropogenic N_2O emission source is thought to be the release of nitrogen in the process of cultivation. Next in importance is the application of nitrogen fertilizer. As Table 4.6 indicates, the use of nitrogen fertilizer is not distributed uniformly throughout the world. The "Big Three," the United States, the Soviet Union, and China, together account for approximately half of the total.

TABLE 4.6 Nitrogen-Fertilizer Consumption 1984–1985

Country	Nitrogen (Tg)	Percent of Total (%)
China	13.7	20
USSR	10.9	16
United States	9.5	14
India	5.7	8
France	2.4	3
United Kingdom	1.6	2
Federal Republic of Germany	1.5	2
Canada	1.3	2
Indonesia	1.3	2
Poland	1.2	2
Mexico	1.2	2
Others	18.9	27
Total	69.2	

Source: U.S. EPA (1990).

Until recently, the dominant man-made emissions source was thought to be fossil-fuel combustion. This conclusion was based on flask samples taken from combustion experiments. This research has been shown to be subject to a sampling artifact that produced N_2O in the flask between the time the sample was taken and the time the flask was analyzed (Muzio and Kramlich, 1987; Linak et al., 1989). It is possible that fossil-fuel emissions are a relatively minor source of N_2O emissions, but this is by no means certain. It is also possible that the chemistry that occurred in the flask may also occur in nature.

Carbon Monoxide Emissions

Carbon monoxide (CO) is not an important radiatively active gas, but it is extremely important because it reacts with the hydroxyl radical, OH, and thereby affects the residence time of radiatively active gases such as CH_4. Carbon monoxide has a relatively short lifetime, on the order of a few months, before it is transformed to CO_2; because of the relatively high efficiency of combustion processes, this source of CO_2 is relatively minor.

Even greater uncertainty surrounds the atmospheric CO budget than the CH_4 budget. As indicated in Table 4.7, CO is generated by incomplete combustion processes (complete combustion yields CO_2 rather than CO), including savanna burning, slash-and-burn agriculture, and deforestation; anthropogenic hydrocarbon oxidation; CH_4 decomposition; and other minor processes.   Because CO is highly reactive, it has a relatively short atmospheric lifetime (0.4 years) and is poorly mixed in the global atmosphere.

TABLE 4.7 Historical Carbon Monoxide Emissions

Sources	Quality of Emissions Data	Uncertainty	Notes
Natural sources	Poor	Great	Global emissions = 1.5–4.0 Pg CO/year. Natural sources are ≈50% of global emissions. Natural sources include plants, wildfires, oceans, and oxidation of CH_4. It is not clear that all sources and sinks for CO have been identified. CO is highly reactive and, therefore, global source-sink relationships are highly uncertain even in aggregate.
Nonenergy	Poor	±50–100%	40–60% of anthropogenic emissions. Sources include slash-and-burn agriculture and/or land clearing.
Energy	Fair–poor	±40–50%	40–60% of anthropogenic emissions. Sources include incomplete combustion (e.g., automotive exhaust, fuel-wood use) and oxidation of anthropogenic hydrocarbons. U.S. energy-related CO emissions have been estimated within the National Acid Precipitation Assessment Program (NAPAP).

Source: Wuebbles and Edmonds (1991).

Annual CO emissions estimates range from 600 to 1700 Tg C/year. Roughly equal amounts of emissions come from natural and anthropogenic sources. Anthropogenic emissions in turn are approximately evenly divided between those associated with incomplete combustion in the production and use of energy and incomplete combustion associated with agriculture and land use, i.e., slash-and-burn agriculture and deforestation.

Overall Contribution of Agriculture, Forestry, and Land-Use Change to Greenhouse Warming

The role of agriculture, forestry, and land-use change in the current pattern of emissions of the major greenhouse gases varies from gas to gas. For some gases, CFCs for example, these activities play no role at all. As we have seen, for CH_4, current estimates indicate that these sectors may contribute as much as two-thirds of total anthropogenic emissions. To assess the overall contribution of agriculture, forestry, and land-use change to greenhouse warming, several methods are available. The simplest method available for estimating agriculture's contribution is to multiply the percentage change in radiative forcing over a recent period by the fraction of anthropogenic emissions associated with agriculture, forestry, and land-use-change activities. This approach yields an estimated contribution to radiative forcing of ≈25% by agriculture, forestry, and land-use-change activities (Table 4.8).

TABLE 4.8   Estimating the Contribution of Agriculture, Forestry, and Land-Use Change to Radiative Forcing During the 1980s

Gas	Total Forcing in 1980s[a] (%)	Emissions from AFL Activities[b] (%)	AFL Forcing in 1980s (%)
CO_2	49	19	9
CH_4	19	65	12
N_2O	5	75	4
CFCs	27	0	0
Total	100	—	25

[a] Percent change in radiative forcing during the 1980s.

[b] AFL, agriculture, forestry, and land-use change.

Sources: For percent total forcing in the 1980s, Hansen et al. (1988); for percent emissions from AFL activities, the tables in this paper.

A more sophisticated method has been developed by researchers such as Lashof and Ahuja (1990), Derwent (1990), Derwent et al. (1990), Victor (1990), and the Intergovernmental Panel on Climate Change (IPCC)(1990). The greenhouse-warming-potential (GWP) coefficient was developed by these researchers to provide a common metric for comparing the emissions of various greenhouse gases. The GWP coefficient is a relative measure, stated in terms of 1 kg of CO_2 released into the atmosphere. Values for the GWP coefficient have been computed by the IPCC. These values take into account the effect of each gas on the radiative profile of the atmosphere, its atmospheric residence time, its atmospheric chemical reactivity, and the effect of all chemical-reaction byproducts on the radiative profile of the atmosphere over the gas's atmospheric residence time.

GWP coefficients are computed by following the history of 1 kg of a greenhouse gas released into the atmosphere for a specific period of time and then comparing its *cumulative* effect on the overall radiative profile of the atmosphere for a specific period of time. Time periods of 20, 100, and 500 years have been examined. For example, CH_4 has a GWP coefficient of 63 when a 20-year history is examined. This means that 1 kg of CH_4 released into the atmosphere today would have 63 times the impact on radiative forcing of 1 kg of CO_2 released into the atmosphere today. But CH_4 is removed relatively quickly from the atmosphere, compared with CO_2, and as time passes, its GWP coefficient declines. For a 100-year history, the value declines to 21, and for a 500-year trace, the value declines to 9. Although CH_4 has a greater GWP coefficient than does CO_2, it is released into the atmosphere in much smaller quantities, so the 1990 contribution of global CH_4 release accounted for only 15% of the total GWP. GWP coefficients for core radiatively important gases are given in Table 4.9.

TABLE 4.9 Greenhouse-Warming-Potential Coefficients

Gas	20-Year	100-Year	500-Year
CO_2	1	1	1
CO	7	3	2
CH_4	63	21	9
N_2O	270	290	190
CFC-11	4500	3500	1500
CFC-12	7100	7300	4500
CFC-113	4500	4200	2100
CFC-114	6000	6900	6500
CFC-115	5500	6900	7400

Source: IPCC (1990).

Total and percentage contributions to GWP by activity and gas are given in Table 4.10. Note that the relative importance of agriculture, forestry, and land-use change depends on the time frame of the GWP coefficients. When the time frame is 20 years, the contribution of agriculture, forestry, and land-use change to radiative forcing is 40% of anthropogenic GWP, but declines to 31% of anthropogenic GWP if 100-year coefficients are used and declines still further—to 28% of anthropogenic GWP coefficient—if 500-year coefficients are used. The role of agriculture appears more important when any GWP measure is used than when a simpler attribution mechanism, or calculation, is used, such as the one for Table 4.8 (i.e., adding to the total the gas's emission rate times the associated GWP coefficient).

An alternative approach was suggested by Reilly (1992), who dropped the implicit assumption that damage is proportional only to the change in radiative forcing and adopted a discount rate as a mechanism for dealing with the time dimension of the problem. He explored the implication of a quadratic damage function and an allowance for a CO_2-fertilization effect and obtained a trace-gas index (TGI), which, like the GWP coefficient, is indexed relative to CO_2. Emission-coefficient values of the TGI are generally, but not always, higher than for the GWP coefficient. Applying the TGI values yields a relative contribution to CO_2 and climate change by agriculture, forestry, and land-use change of 42% of anthropogenic emissions (Table 4.11). This is the highest value encountered. The reason this approach yields a relatively high assessment of the role of agriculture, forestry, and land-use change is that the role of CO_2 is diminished by adding a credit for the CO_2-fertilization effect, and the roles of the other gases are enhanced. This makes CH_4, a gas whose anthropogenic emissions are predominantly determined by agriculture, forestry, and land-use change, the largest single contributor to the index, 43% of the total TGI. This result is extremely important, because it implies that CO_2 and CH_4 are roughly equal partners in the greenhouse issue, and that energy and agriculture, forestry, and land-use change also provide roughly equal emissions contributions!

We note, of course, that all of these measures are highly uncertain and based on present understandings of a wide variety of phenomena. In addition, we assumed that the change in radiative forcing is due entirely to changes in human activities and that if human activities were excluded, no changes in the atmosphere would be observed.

Summary

In this paper we examined the role agriculture and land-use change plays in greenhouse-gas emissions. Our analysis indicates that when a comprehensive approach is used that includes all emissions from all activities weighted by GWP coefficients, the role of agriculture and land-use change

TABLE 4.10 Agriculture, Forestry, and Land-Use-Change Contributions to GWP

Gas	Agriculture and Land Use (Pg CO_2)[a]	Other Anthropogenic Activities (Pg CO_2)	Percent Gas Contribution to Total GWP (%)
20-year IPCC coefficients[b]			
CO_2	4,767	19,800	41
CO	4,410	3,920	14
CH_4	12,726	6,930	32
N_2O	2,376	849	5
CFC-11		1,328	2
CFC-12		2,673	4
CFC-113		819	1
CFC-114		18	0
CFC-115		17	0
100-year IPCC coefficients[c]			
CO_2	4,767	19,800	57
CO	1,890	1,680	8
CH_4	4,242	2,310	15
N_2O	2,552	911	8
CFC-11		1,033	2
CFC-12		2,748	6
CFC-113		764	2
CFC-114		21	0
CFC-115		21	0
500-year IPCC coefficients[d]			
CO_2	4,767	19,800	71
CO	1,260	1,120	7
CH_4	1,818	990	8
N_2O	1,672	597	7
CFC-11		443	1
CFC-12		1,694	5
CFC-113		382	1
CFC-114		17	0
CFC-115		22	0

[a] Pg CO_2 equivalent.

[b] Percent of anthropogenic GWP from agriculture and land use, 40%; from other anthropogenic activities, 60%.

[c] Percent of anthropogenic GWP from agriculture and land use, 31%; from other anthropogenic activities, 69%.

[d] Percent of anthropogenic GWP from agriculture and land use, 28%; from other anthropogenic activities, 72%.

TABLE 4.11 Agriculture, Forestry, and Land-Use-Change Contributions to the Trace-Gas Index (TGI): Quadratic Climate Effects Plus CO_2 Fertilization[a]

Gas	Agriculture and Land Use (Pg CO_2)	Other Anthropogenic Activities (Pg CO_2)	Percent Gas Contribution to Total Anthropogenic Emissions (%)
CO_2	4,767	19,800	37
CO	2,331	2,072	7
CH_4	18,584	10,120	43
N_2O	2,288	817	5
CFC-11		443	1
CFC-12		1,872	5
CFC-113		3,433	1
CFC-114[b]		17	0
CFC-115[b]		22	0

[a] TGI coefficients from Reilly (1992). Percent of anthropogenic GWP from agriculture and land use, 42%; from other anthropogenic activities, 58%.

[b] Values taken from Reilly (1990), Table 10, "After 500 Year."

is greater than would be estimated by simpler methods. Depending on the specific integration used (20, 100, or 500 years), the relative contribution of agriculture and land-use change varies between 28% and 40% of the total GWP-weighted emissions. An interesting experiment for deriving an alternative emissions-weighting index by Reilly (1992) yielded a value for the percent of total weighted emissions from agriculture and land-use change of 43%.

References

Bolin, B. 1986. "How Much CO_2 Will Remain in the Atmosphere?" in B. Bolin, B. R. Doos, J. Jaeger, and R. A. Warrick, eds., *The Greenhouse Effect, Climatic Change, and Ecosystem*, Scientific Committee on Problems of the Environment (SCOPE) 29. Pp. 93–156. New York: John Wiley & Sons.

Bolin, B., B. R. Doos, J. Jaeger, and R. A. Warrick, eds. 1986. *The Greenhouse Effect, Climatic Change, and Ecosystem*, Scientific Committee on Problems of the Environment (SCOPE) 29. New York: John Wiley & Sons.

Cicerone, R., and R. Oremland. 1988. "Biogeochemical Aspects of Atmospheric Methane." *Global Biogeochemical Cycles* 2: 299–328.

Derwent, R. G. 1990. *Trace Gases and Their Relative Contribution to the Greenhouse Effect.* Harwell Laboratory, Harwell, Oxfordshire: U.K. Atomic Energy Authority. (AERE-R 13716.)

Derwent, R., H. Rodhe, and D. Wuebbles. 1990. "Global Warming Potential of Greenhouse Gases," special report for the United Nations Environmental Programme. Harwell Laboratory, Harwell, Oxfordshire: U.K. Atomic Energy Authority.

Detwiler, R. P., and C. A. S. Hall. 1988. "Tropical Forests and the Global Carbon Cycle." *Science* 239:42–74.

Ehhalt, D. H. 1985. "Methane in the Global Atmosphere." *Environment* 27(10): 6–12,30–33.

Esser, G. 1991. "Uncertainties in the Dynamics of Biosphere Developments with the Accent on Deforestation." *Pure and Applied Chemistry* 63(5): 775–8.

Goudriaan, J., and P. Ketner. 1984. "A Simulation Study for the Global Carbon Cycle, Including Man's Impact on the Biosphere." *Climatic Change* 8: 167–92.

Goudriaan, J. 1989. "Modeling Biospheric Control of Carbon Fluxes Between Atmosphere, Ocean and Land in View of Climatic Change," in A. Berger, S. Schneider, and J. C. Duplessy, eds., *Climate and Geo-Sciences.* NATO-ASI Series C, Vol. 285. Dordrecht, The Netherlands: Kluwer Academic Publishers.

Goudriaan, J. 1991. "Uncertainties in Biosphere/Atmosphere Exchanges, CO_2 Enhanced Growth." *Pure and Applied Chemistry* 63(5): 772–5. .

Hansen, J., I. Fung, A. Lacis et al. 1988. "Global Climate Changes as Forecast by Goddard Institute for Space Studies Three-Dimensional Model." *Journal of Geophysical Research* 93: 9341–64.

Houghton, R. A., B. D. Boone, J. R. Fruci et al. 1987. "The Flux of Carbon from Terrestrial Ecosystems to the Atmosphere in 1980 Due to Changes in Land Use: Geographic Distribution of the Global Flux." *Tellus* 39b: 122–39.

Intergovernmental Panel on Climate Change (IPCC). 1990. *Scientific Assessment of Climate Change.* New York: Cambridge University Press. (World Meteorological Organization report.)

Lashof, D. A., and D. R. Ahuja. 1990. "Relative Global Warming Potentials of Greenhouse Gas Emissions." *Nature* 344: 529–31.

Lashof, D. A., and D. A. Tirpak. 1989. *Policy Options for Stabilizing Global Climate,* draft report to Congress. Washington D.C.: U.S. Environmental Protection Agency, Office of Policy, Planning and Evaluation.

Lerner, J., E. Matthews, and I. Fung. 1988. "Methane Emission from Animals: A Global High-Resolution Data Base." *Global Geochemical Cycles* 2: 139–56.

Linak, W. P., J. A. McSorley, E. Hall et al. 1989. "N_2O Emissions from Fossil Fuel Combustion." Presented at the Air and Waste Management Association Conference. (Document 89-4.6).

MacCracken, M. C., E. Aronson, D. Barns, et al. 1990. *Energy and Climate Change: Report of the DOE Multi-Laboratory Climate Change Committee.* Chelsea, Mich.: Lewis Publishers.

Marland, G., T. A. Boden, R. C. Griffin, S. F. Huang, P. Kanciruk, and T. R. Nelson. 1989. *Estimates of CO_2 Emissions from Fossil Fuel Burning and Cement Manufacturing, Based on the United Nations Energy Statistics and the U.S. Bureau* of Mines Cement Manufacturing Data. Oak Ridge, Tenn.: Carbon Dioxide Information Analysis Center, Oak Ridge National Laboratory. (ORNL/CDIAC-25 NDP-030.)

Muzio, L. J., and J. C. Kramlich. 1987. "An Artifact in the Measurement of N_2O from Combustion Sources." *Geophysical Research Letters* 15: 1369–72.

Reilly, J. M. 1990. "Climate Change Damage and the Trace Gas Index Issue." Washington, D.C.: U.S. Department of Agriculture, Economic Research Service. (Draft paper.)

Reilly, J. M. 1992. "Climate-Change Damage and the Trace-Gas-Index Issue," in J. M. Reilly and M. Anderson, eds., *Economic Issues in Global Climate Change: Agriculture, Forestry, and Natural Resources.* Pp. 72–88. Boulder, Colo.: Westview Press.

Reilly, J. M., and R. Bucklin. 1989. "Climate Change and Agriculture." Washington, D.C.: U.S. Department of Agriculture. (Draft paper.)

Tans, P. P., I. F. Fung, and T. Takahashi. 1990. "Observational Constraints on the Global Atmospheric CO_2 Budget." *Science* 247: 1431–8.

Trabalka, J., ed. 1985. *Atmospheric Carbon Dioxide and the Global Carbon Cycle.* Springfield, Va.: National Technical Information Service, U.S. Department of Commerce. (DOE/ER-0239.)

U.S. Environmental Protection Agency (EPA). 1990. *Greenhouse Gas Emissions from Agricultural Systems, Volume 1—Summary Report,* report of the Intergovernmental Panel on Climate Change, Response Strategies Work Group, Subgroup on Agriculture, Forestry, and Other Human Activities. Washington, D.C.: Office of Policy, Planning and Evaluation, U.S. EPA. (PM-221).

Victor, D. J. 1990. "Sensitivity of Global Warming Potentials to Feedback and Carbon Cycle Assumptions: Implications for Policy." Cambridge, Mass.: Massachusetts Institute of Technology. (Draft paper.)

Wahlen, M., N. Takata, R. Henry et al. 1989. "Carbon-14 in Methane Sources and in Atmospheric Methane: The contribution of Fossil Carbon." Science 245: 286–90.

World Meteorological Organization (WMO). 1985. *Atmospheric Ozone 1985.* Washington, D.C.: U.S. National Aeronautics and Space Administration, Earth Science and Applications Division. (WMO, Global Ozone Research and Monitoring Project Report 16.)

Wuebbles, D., and J. Edmonds. 1991. *A Primer on Greenhouse Gases.* Chelsea, Mich.: Lewis Publishers.

5

Climate-Change Damage and the Trace-Gas-Index Issue

John M. Reilly[1]

Introduction

The 1983 U.S. Environmental Protection Agency (EPA) report *Can We Delay A Greenhouse Warming?* somewhat inadvertently illustrated that trace gases other than CO_2 could, despite virtually whatever was done to limit CO_2 emissions, make it impossible to prevent or even significantly delay climate change. In doing so, the report illustrated that if serious efforts were undertaken to limit climate change as induced by human activity, the efforts could not be limited to CO_2 alone. As economists, we are fond of recognizing that the conditions under which a least-cost combination of options occurs are where the marginal costs of a bit more of each action are equal. If it costs $100 to remove a unit of carbon through an afforestation program in Australia but building a nuclear power plant in the United States can be done for the equivalent of $10/unit, then we should pursue the nuclear plant before the afforestation program. On the other hand, we might continue to pursue halts to deforestation in Brazil if the halts can be had for $10/unit of carbon. Although a unit of carbon is a unit of carbon in terms of radiative forcing, comparing CO_2 with methane (CH_4), nitrous oxide (N_2O), and chlorofluorocarbons (CFCs) requires a conversion index.

The index question has received considerable attention over the past year. Lashof and Ahuja (1990) provided a candidate index for computing the

[1] From the Economic Research Service, U.S. Department of Agriculture, Washington, D.C. The views expressed in this document are solely those of the author and do not necessarily reflect the views or policies of the U.S. Department of Agriculture or the U.S. government. The helpful comments and suggestions of Bruce Larson, Dick Brazee, and Jae Edmonds are gratefully acknowledged, while the author takes full responsibility for any remaining errors.

relative contribution of different trace gases to warming. Scientists working on the Intergovernmental Panel on Climate Change (IPCC) *Scientific Assessment of Climate Change* (Houghton et al., 1990) offered significant refinements to this index by improving the representation of physical processes. If one is to use an index of trace gases for political bargains or for identifying a cost-effective trade-off among gases, the problem is considerably more complex and involves both economic and physical considerations. Effective strategies that minimize costs must deal with the trade-off among gases whether or not an index is explicitly used in the process. I plan to sketch out the calculation of a gas index incorporating these considerations because it provides a framework for comparing different trace-gas-limitation strategies. The problem is readily identified as a problem in optimal control, which has become a familiar tool in solving dynamic problems of intertemporal resource use and investment decisions.

Comprehensive Consideration of Trace Gases and Their Effects

The concepts I wish to highlight are (1) the relevant index should be an index of economic effect of climate-environmental change, which requires adjustments to simple measures of radiative forcing because the gases are removed from the atmosphere at different rates and marginal damages are likely to vary over time; (2) gases have non-climate-related economic effects that differ among gases, and these should be counted as credits (e.g., direct CO_2 fertilization of crops) or debits (e.g., CFCs as contributors to ozone depletion); and (3) programs that limit emissions or enhance sinks are not simply described as either a permanent or one-time reduction in emissions (or enhancement of a sink) but rather represent unique dynamic profiles—forestation is an enhanced sink over 30 or 40 years with a neutral effect thereafter (if the forest achieves equilibrium) or a bulge of emissions (if clearing occurs). Similarly, programs of carbon taxes (recognizing that there are rents associated with fossil resources) may be better cast as delays in fossil-fuel consumption than as complete avoidance. I have chosen to represent physical processes simply in order to streamline the mathematical presentation. In general, any one of the simple representations of economic and physical processes are supported by complex multiequation models.

To define the problem, I identify society's objective as that of maximizing the present value V(0) of climate-related resource endowments where net returns, R(t), at time t are discounted by

$$V(0) = \int_0^{\infty} R(t)exp^{-rt}dt \tag{1}$$

and defined as profits, π, from activities using climate-related resources minus investment expenditures, $P_I(t)I(t)$, where

$$\pi(t) = p(t)F[C(t),K(t),N(t),G_i(t),E_i;t] - w(t)N(t) \tag{2}$$

and where p and w are prices of output and variable inputs and F is the production function relevant in time t relating capital (K), other inputs (N), direct effects of the atmospheric concentrations of trace gases on production (G_i, i=1,...n, where n is the number of trace gases considered), climate-related resources (C), and emissions of trace gases, (E_i). F_K, F_N, and F_{Ei} > 0 and F_{KK}, F_{NN}, and F_{EiEi} < 0. These conditions are traditionally assumed for conventional inputs K and N. For trace-gas emissions, the interpretation is that control of emissions could be accomplished by increasing capital and other inputs; thus, relaxing controls would lead to increased output.[2] F_{Gi} < 0 and F_{GiGi} > 0 for gases like CFCs, which have negative consequences for output through effects on stratospheric-ozone depletion, whereas F_{Gi} > 0 and F_{GiGi} < 0 for gases like CO_2, where the direct effects of increased concentrations include increased crop yields. Climate-related resources (C) are determined by the aggregate concentration of trace gases. This transformation of trace-gas concentrations to climate can be described simply as

$$C'(t) = C[\sum_i \alpha_i G_i(t);t] \tag{3}$$

where the function C can be thought of as a climate model and α_i are the instantaneous gas-specific radiative-forcing effects per unit of gas.

For simplicity, the radiatively weighted atmospheric stock

$$C(t) = G(t) = \sum_i \alpha_i G_i(t) \tag{4}$$

can be used to represent an index of climate. IPCC discusses the validity of this representation and suggests mathematical representations that may better represent radiative-forcing comparisons (Houghton et al., 1990). In particular, saturation of radiative effects may occur as concentrations increase in CO_2, for example. A more complex representation could include the specific equations in the IPCC report or, in the general case, the

---

[2] I would like to thank Bruce Larson, Economic Research Service, for suggesting this representation (personal communication, 1990).

radiative-forcing-function dependence on the atmospheric stock of the gas or of other gases.

The equation of motion for capital stock is

$$\dot{K}(t) = \Psi[I(t),K(t);t] - \delta K \tag{5}$$

where $\dot{K}(t)$ is the rate of change of capital stock, Ψ is a function describing the effectiveness of investment, $\Psi_I > 0$, $\Psi_{II} < 0$, $\Psi_K < 0$, $\Psi_{KK} > 0$, and δ is the rate of depreciation of capital.

Equations of motion for changes in atmospheric trace gases are simply the identity

$$\dot{G}_i(t) = E_i(t) - \delta_i G_i(t) \tag{6}$$

where δ are gas-specific dissipation rates. The constant dissipation rate is a relatively simple assumption that is unlikely to represent the atmospheric trace-gas process any better than it does capital depreciation. One way to provide a somewhat richer description is to make dissipation a function of the atmospheric concentration of the gas. For CO_2 in particular, a variety of carbon-cycle models provides a richer description of the fate of carbon once it is injected into the atmosphere *(see* Houghton et al., 1990).

The gas-specific dissipation rates are simply the inverse of the more commonly used concept of atmospheric lifetimes and are comparable to capital-depreciation rates.

To complete the description of the system, we add initial conditions:

$$K(0) = \overline{K} \tag{7}$$

$$G(0) = \overline{G}_i \tag{8}$$

To obtain expressions relating the appropriate index values for each gas, we solve the optimal-control problem. The current-value Hamiltonian, suppressing t, is

$$H = (\pi - P_I I) + \lambda_K[\Psi(I,K;t) - \delta K] + \sum_i \lambda_i(E_i - \delta_i G_i) \tag{9}$$

where λ_K and λ_i are, respectively, the current-value multipliers for capital and trace gases $i = 1,...,n$. Choice variables are E_i, I, and N.

Necessary conditions for a maximum include[3]

$$\pi_N = 0 \tag{10}$$

---

[3] Sufficient conditions for a maximum are that Ψ_K and R(t) are differentiable, concave functions of N, K, and G_i.

$$\lambda_K \Psi_I - P_I = 0 \tag{11}$$

$$\pi_{EI} + \lambda_i = 0 \tag{12}$$

The λ_K and λ_i can be interpreted as the current marginal value of the constraints. Equation 10 indicates that conditions for an optimum require that variable inputs be used to the point where an additional unit yields zero additional profits. Equation 11 indicates that the shadow value of capital is equal to the price of investment goods adjusted for the effectiveness of investment. Equation 12 gives the static optimality condition that one should equate the marginal cost of emissions control in terms of its impact on profits to the shadow value of the emissions constraint.

The necessary conditions for a maximum also require that

$$d\lambda/dt = r\lambda - (\partial H/\partial K) \tag{13}$$

and

$$d\lambda_i/dt = r\lambda_i - (\partial H/\partial G_i) \tag{14}$$

The resulting expressions for trace gases are[4]

$$\frac{\alpha_i p F_C + p F_{Gi} + (d\lambda_i /dt)}{r + \delta_i} = \lambda_i \tag{15}$$

The numerator in equation 15 includes a term that adjusts the economic impact of trace-gas-induced climate change by the radiative-forcing effect (the general presumption is that this term will be negative—that is, a warmer and drier climate will adversely affect economic output). The second term is the direct effect of trace-gas accumulation on economic activity. This effect can be positive or negative depending on the gas. For example, increased CO_2 concentrations have a positive effect on agricultural yields whereas increased CFC emissions contribute to stratospheric-ozone depletion and consequent effects on human health and other biological systems resulting from increased ultraviolet (UV) radiation. The final term is the expected change in the value of the constraint. Its value depends on the future values.

The denominator of the expression is a discount factor (or adjusted discount rate) that involves the interest rate and the rate of dissipation of the gas. The discount rate will be the same across gases. Significant differences among gases will occur due to differences in the dissipation rate.

---

[4] The expression for capital is similar.

To interpret this expression in terms of an index of gases, we can choose gas j as a numeraire and define the denominator in equation 15 as the gas-specific discount factor γ_i. As a result, the index, β_i, is

$$\beta_i = \left(\frac{\alpha_i pF_C + pF_{Gi} + d\lambda_i /dt}{\alpha_j pF_C + pF_{Gj} + d\lambda_j /dt} \right) \frac{\gamma_j}{\gamma_i} \tag{16}$$

Equation 16 demonstrates the need to consider economic damages in any index. The intuition for this result is fairly clear. To add apples and oranges (climatic and nonclimatic effects), one needs a common metric. Monetary value is the (economist's) usual answer to such problems. Note that equation 16 reduces to a simple expression of relative radiative forcing adjusted by the discount factor if nonclimatic effects of trace gases, the pF_G terms, and the $d\lambda/dt$ terms are ignored:

$$\hat{\beta}_i = \frac{\alpha_i \, \gamma_j}{\alpha_j \, \gamma_i} \tag{17}$$

To identify the effect of the $d\lambda/dt$ terms, equation 12 can be totally differentiated with respect to time:

$$- \frac{d\lambda_i}{dt} = \frac{\partial^2 \pi}{\partial E_i \partial E_i} \dot{E}_i + \frac{\partial^2 \pi}{\partial E_i \partial G_i} \dot{G}_i + \frac{\partial^2 \pi}{\partial E_i \partial C} \dot{C} + \frac{\partial^2 \pi}{\partial E_i \partial K} \dot{K} + \frac{\partial^2 \pi}{\partial E_i \partial t} \tag{18}$$

To explore the nature of equation 18, note the conditions under which terms of $d\lambda_i/dt$ vanish. If a steady-state condition is achieved where annual emissions for each gas balance its dissipation rate, then emissions growth will be zero, concentrations will stabilize, and no additional climate forcing will occur. Thus, the first three terms of equation 15 will disappear. In general, there is no reason to believe that complete stabilization of climate is optimal and, therefore, no reason that in the calculation of an optimal index, these terms will vanish. Even if complete stabilization or radiative forcing was to occur, this could be achieved by a combination of continuing growth in emissions of some gases and reductions in others; thus, the second term would remain. Regarding the fourth term, a growing economy will continue to accumulate capital, but if capital is not a substitute for emissions, then this term will vanish. This is a fairly unlikely possibility because energy conservation, alternative fuels, forestation, and other strategies for reducing emissions generally entail the substitution of capital. The final term describes how the production function will change over time with regard to the importance of trace-gas emissions. A possibility is that technological advance will create additional opportunities for reducing emissions. This leaves open the possibility that the fourth and fifth terms could offset one another. There is no guarantee that such a balance will occur and, in fact, technological change could be neutral or biased toward emissions.

Although the conclusion that the all the terms of $d\lambda_i/dt$ vanish is difficult to support, equations 16 and 18 do provide the basis for calculating appropriate trace-gas indices given presumptions about economic effects of climate, other effects of gases, and paths of emissions growth. The index as expressed in equations 16 and 18 describes the change in the shadow value over time in terms of the emissions-control side of the problem. At optimal values, the index calculation from either the emissions or costs side will be equivalent. The goal of this paper is to illustrate how changing the characterization of effects and damages affects the relative value of reducing different gases. For purposes of calculating relative index values that are explicitly dependent on the effects, recognize that equation 15 can be rewritten as

$$\lambda_i = \int_0^\infty [\alpha_i pF_C C(t) + pF_{Gi} G_i(t)] \exp^{-(r+\delta_i)t} dt \qquad (19)$$

Equation 19 also permits calculation of indices by breaking up the problem by sector, assuming that sectoral effects are independent, or by time-period component through

$$\lambda_i = \int_0^T X(t)dt + [\int_T^\infty X(t)dt]/(1+r)^T \qquad (20)$$

where $X(t)$ is the function in equation 19. Such a representation is useful if, for example, the marginal effects of climate change are not representable as a conveniently integrable function. Such might be the case if one suspects that climate change may be beneficial through some amount of change (e.g., up to 2 °C warmer) but damaging at greater amounts of change (e.g., > 2 °C), or that there are damage (benefits) thresholds or ceilings.

Estimated Trace-Gas Indices

Table 5.1 provides the basic physical data necessary to evaluate equation 19. There remains considerable uncertainty about physical parameters, even regarding measurement of current emissions and atmospheric stocks. Parameters describing future emissions, radiative forcing, and dissipation each summarize complex physical and economic systems. The values in Table 5.1 attempt to summarize these complex systems, about which there is uncertainty, with a single parameter.

TABLE 5.1 Characteristics of Important Trace Gases

Gas	Concentration ($\mu mol/mol$, or ppm)	Atmosphere Stock (10^{12} g)	Emissions in 1990 (10^{12} g)	Dissipation Rate [1/gas lifetime (in years)]	Radiative-Forcing Mass[a]	Emissions Growth Rate[b]
CO_2	354	2710000	26000	0.0083	1	0.006
CO	0.09	430	200	0.1	2.3	0.0085
CH_4	1.717	4800	300	0.1	58	0.005
N_2O	0.3097	1400	6	0.0067	206	0.003
NCFC-22	0.0001	1.51	0.1	0.0067	5440	-0.08
CFC-11	0.00028	6.7	0.3	0.0167	3970	-0.08
CFC-12	0.000484	10.19	0.4	0.0077	5750	-0.08
CFC-113	0.00005	1.63	0.15	0.0143	3710	-0.08

[a]The instantaneous-radiative-forcing effect on a mass basis relative to CO_2.
[b]Emissions growth rates are best estimates of Wuebbles and Edmonds (in press), except CO_2, which is the median value from Reilly et al. (1985). CFC emissions growth rates are based on the Montreal Protocol target of a 50% reduction by 1998 and assume that reductions continue beyond 1998 at the same rate.

Sources: Houghton et al. (1990), Lashof and Tirpak (1989), and Wuebbles and Edmonds (1988).

To provide an indication of the effect of incorporating economic considerations into trace-gas-index calculation, characterize emissions of trace gas i as exponential growth:

$$E_i(t) = E_i(0)\exp^{g_i t} \tag{21}$$

This implies that concentrations will exhibit exponential growth as well, described by

$$G_i(t) = TG_i(0)\exp^{g_i t} + (1-T)G_i(0) \tag{22}$$

where

$$T = \frac{E_i(0)}{(g_i + \delta_i)G_i(0)} \tag{23}$$

If g_i has been constant over time and is assumed to be the same constant rate into the future, then $T=1$.

Little evidence of aggregate future damages exists, but the economic effects on agriculture may be illustrative. One estimate is that an effective doubling of trace-gas concentrations would cause a global-welfare loss of $38 billion (1986 dollars) due to lost agricultural productivity. This estimate includes the partially offsetting effects of direct CO_2 fertilization (Kane et al., in press). Direct CO_2 fertilization offsets about one-half of the yield loss in the United States due to climate change when CO_2 contributes about two-thirds of the effective doubling. Assuming that this ratio holds for the rest of the world on average and that it can be directly applied to the welfare change means that CO_2 fertilization may contribute a benefit of $38 billion, with the climate-alone loss at $76 billion. There have not been comprehensive studies of economic damages caused by climate change stemming from changes in water resources, sea level, forests, and fisheries; of changes in recreation opportunities and biodiversity; or of the cost of possible population migration. For illustrative purposes, total damage from all sources is assumed to be six times the estimated global damage from agricultural changes alone. However, the benefits of CO_2 fertilization are assumed to apply only to agriculture.

To illustrate the difference in the index value, a comparison between a linear damage function

$$D = d_1 + d_2 C \tag{24}$$

and a quadratic damage function

$$D' = d_1' + d_2' C + d_3' C^2 \tag{24'}$$

is made. Assume that the benefit function for CO_2 is linear:[5]

$$B = b_1 + b_2 G_{CO_2} \tag{25}$$

with the parameter estimates given in Table 5.2.

Aggregate evidence on climate damages and gas benefits is related directly to the concentration of atmospheric trace gases expressed as effective μmol/mol, or ppm, CO_2, whereas emissions are more typically specified in mass units (i.e., grams). The radiative-forcing effect is denominated in mass units relative to CO_2. Equation 4 thus identifies an aggregate of trace-gas mass adjusted for relative radiative forcing.

TABLE 5.2 Damage-Function Parameters[a]

Linear Damage	Quadratic Damage	Linear Benefit
$d_1 = -543$	$d_1' = -250$	$b_1' = 57.57$
$d_2 = 1.538$	$d_2' = -0.521$	$b_2' = -0.163$
	$d_3' = 0.001954$	

[a]Determined by assuming $0 damages (benefits) in 1990 when the "effective" CO_2 concentration (C) was 441 μmol/mol (or 441 ppm); $76 billion × 6 sectors at 706 μmol/mol; and for the quadratic damage function, $152 billion × 6 sectors at 882.5 μmol/mol. Benefit parameters applied only to a CO_2 assumed benefit of $38 billion at 0.66 × 706 μmol/mol.

To convert this to the climate index—μmol/mol, or ppm, CO_2—equation 4 becomes

$$C(t) = \Theta[\textstyle\sum_i \alpha_i G_i(t)] \tag{26}$$

It is possible to evaluate the integral in equation 19 if it converges. With damages a function of exponential growth in concentrations, discounting will

---

[5] Agronomic evidence suggests that the benefits of CO_2 fertilization may be largely exhausted at 600 μmol/mol (600 ppm) CO_2, suggesting the need for a different functional form. The linear form is, however, likely to provide a reasonably accurate representation because the saturation effect will not occur for 50–100 years under most emissions growth estimates. The discounted difference effects beyond 50–100 years are likely to be quite small.

generate convergence as long as the exponential rate of growth is less than the discount rate plus the rate of gas dissipation. In general, these conditions are easily met by most parameterizations. Emissions growth projections are on the order of 0.003–0.015, whereas discount rates typically used are on the order of 0.03–0.05. Assuming that equation 19 converges, the expression for the climatic-effects part of equation 19 is

$$\lambda_{iC} = \Theta\alpha_i\left(\frac{d_2}{g_i-r-\delta_i}\right) \qquad (27)$$

in the linear-damage case and

$$\lambda_{iC} = \Theta\alpha_i\left\{\frac{d_2'}{g_i-r-\delta_i}+2d_3'\left[\sum_i\Theta\alpha_i\left(\frac{G_i(0)T}{2g_i-r-\delta_i}+\frac{(1-T)G_i(0)}{g_i-r-\delta_i}\right)\right]\right\} \qquad (27')$$

in the quadratic case. These must be summed with the direct-gas-effects component:

$$\lambda_{iG_i} = \Theta\left(\frac{b_2}{g_i-r-\delta_i}\right) \qquad (28)$$

to generate the index described in equation 16.

With the above and the data in Table 5.1, the trace-gas indices can be computed. These are compared with the IPCC indices, which are based only on physical parameters, the direct radiative forcing, and a calculation showing the effect of ignoring the damage calculation but including the discount rate (Table 5.3). The simplest measure of comparison is the instantaneous radiative forcing, which fails to take into account the differences in atmospheric lifetimes.

The IPCC report (Houghton et al., 1990) attempted to address this with indices of global-warming potential (GWP). In recognizing the infinite horizon of the problem, the authors developed indices by truncating the integration time to a finite, arbitrarily defined number of years.[6] The calculations in the present paper that were closest to those in the IPCC report were for linear-damage estimates of climate effects. With constant marginal costs, damage drops from the index, and the only economic parameter is the discount rate. Although comparable to the IPCC calculations, it has the advantage of more smoothly decreasing the importance

---

[6] In addition to the 20- and 500-year calculations, a 100-year calculation was made but is not reported here.

TABLE 5.3 Comparisons of Trace-Gas Indices

	Economic-Based			Physical Effects Only		
	Climate + CO_2 Fertilization (Quadratic)	Climate Effects (Quadratic)	Climate Effects (Linear)	Radiative Forcing	IPCC GWP[a] 20 years	500 years
CO_2	1	1	1	1	1	1
CO	3.7	2.9	0.9	2.3	7	2
CH_4	92	74	21	58	63	9
N_2O	260	208	201	206	270	190
HCFC-22	8950	7174	1427	5440	4100	510
CFC-11	6343	5085	1389	3970	4500	1500
CFC-12	9119	7309	2140	5750	7100	4500
CFC-113	5917	4743	1319	3710	4500	2100

[a]Global-warming potential.

Sources: Economic-based values were calculated. Physical-effects-only values are from a 1990 IPCC report (Houghton et al., 1990).

of future years rather than abruptly delineating years that matter and years that do not.

Although the desirability of discounting is sometimes debated, the IPCC calculations implicitly use a discount rate of 0 for the first 20 or 500 years and a discount rate of infinity for out-years. Decreasing the discount rate is somewhat comparable to increasing the integration period; both changes increase the weight of out-years in the index. In general, calculating GWP from the integrated effect over the lifetimes of the gases reduces the value of cutting the emissions of short-lived gases such as CH_4, carbon monoxide (CO), and HCFC-22 compared with calculating an index based only on instantaneous radiative forcing.

Under the assumption of a quadratic damage function, the index for shorter-lived gases also increases. Finally, the full index includes a positive offset for CO_2. This has the expected effect of increasing the benefit of reducing all other gases relative to CO_2.

While maintaining the relatively simple framework I have sketched out above, I believe it is possible to add somewhat more accurate representations of dissipation and radiative forcing by making these processes functionally dependent on concentrations. In the current formulation, I have characterized future emissions as a rate of growth. Other simple formulations, such as linear or log-linear growth (decay), should be explored to determine the sensitivity of the index to such assumptions. In particular, effort is required to characterize the economic damage associated with climate and the direct effects of trace-gas accumulation. As with emissions growth, effort must be focused on whether such damages increase linearly, exponentially, or log-linearly with increased climate change—that is, whether marginal cost is increasing, decreasing, or constant.

Some studies have begun to address these issues *(see* NCPO, 1989). For example, costs of sea-level rise have been characterized as increasing at an increasing rate with each centimeter rise. Agricultural effects of climate change may be very small through a doubling of trace-gas concentrations but may increase more rapidly thereafter. Although such efforts to quantify the index will require speculation and simplification that will have much less than the usually hoped for scientific foundation, writing down the speculations and comparing them within an index framework would be an improvement from the current state of affairs.

The index calculations illustrate the importance of considering the timing of damages as well as the time profile of gases. In addition, some particular caveats apply. The negative effect of CFCs on stratospheric ozone were not included. These would significantly increase the index values for CFCs relative to CO_2 and other gases. In addition, CO_2 fertilization was only considered as a positive effect on agriculture. It may benefit forestry, the hydrological cycle, and natural ecosystems as well.

Miscellaneous Issues

The appeal to mathematics in the previous section sacrificed the richness of many discussions regarding global-change policy. The framework for broadly comparing different gases also provides perspectives on several issues frequently discussed in regard to climate-change policy.

No-Regrets Policies

A popular notion is that the best response to climate change is to pursue policies that would make sense to pursue even without climate change. The argument is that because we really do not know how earth systems will respond to increasing emissions, nor do we know whether the societal and/or economic effects of such changes will be severe or even negative, an appropriate response is to do things that are justified on nonclimate grounds that also contribute to reducing emissions. Thus, we will not later regret doing them if climate change does not occur, or if it is not as bad as expected or is even good. The logic behind this concept is troubling. If actions made sense anyway, then why have they not been taken? In other words, the fact that an action has not been taken may mean that it does not make sense. Because no one answers or even raises this basic question, the notion has become an excuse for retreading policies that were previously tried and dismissed, or that some segments of society felt were worthwhile but never gained enough public acceptance to put into place.

For example, an entire set of climate-change policies has been proposed as "no-regrets" policies because they reduce fossil-energy dependence, yet the validity of such a goal remains in dispute. Markets may be more successful at allocating fossil resources over time than are public officials. Thus, such a policy has "no regret" only if one objects, in a particular way, to market allocation of resources.

Similarly, the concept opens the door for strategic loopholes in international bargaining. Nations that may be considering actions that make sense anyway have an incentive not to take such actions until they are sure that they will receive credit in an international agreement. Nations that have chosen for other reasons to tax gasoline, regulate fuel efficiency, or pursue nuclear power will argue that the actions they have taken in the past year, decade, or century should generate credit toward emissions reduction. All of this provides wonderful public relations but does nothing to reduce emissions beyond what would have been done if the climate-change effect had never been imagined. In fact, the potential for perverse incentives may lead to increasing emissions.

Scale of Impact—Scale of Action

There seem to be good reasons not to overburden international bargaining with issues that can be handled within single countries or with bilateral or regional agreements. This consideration has direct implications for decisions regarding which nonclimatic factors should be included in an international bargaining index. The ozone hole and direct effects of CO_2 on agriculture are global phenomena. The case for their inclusion in an international-index calculation is strong. Acid rain, nitrate contamination of groundwater, particulates, and direct health effects of gases such as CO are environmental problems of subglobal dimensions. The scale of action on these issues should be consistent with the scale of impact, which will argue against including them as part of international negotiations on global change or as part of a gas-index computation. This should not prevent individual countries from including domestic and/or subglobal environmental goals or other public policy goals in their private calculations.

What Targets, Which Instruments

Various target emissions concentrations have been proposed, and there are efforts to estimate the cost of complying with such targets. There is, however, a variety of different points within the global-climate-change system at which an international agreement could be negotiated, each point providing a different set of monitoring, enforcement, and efficiency considerations. Emissions targets have appeal because emissions are a point in the system where there are existing, or easily imagined, instruments of public control. But, economically efficient negotiated targets must include more than just emissions. Afforestation to enhance carbon withdrawal is an obvious action excluded from an emissions-policy focus. Although proposals to fertilize the ocean to enhance carbon uptake or to send into orbit panels that cut down incoming solar radiation may be farfetched, these actions are among a general class of actions including (1) removal of gases from the atmosphere, (2) climate modification to offset change, and (3) adjustment of production processes that use climate resources that will be underencouraged if international negotiations focus on emissions only.

These considerations suggest that rather than view emissions as a pollutant (the perspective of the mathematical problem in the previous section), international management of the atmosphere and climate to maximize the value of the resource be adopted. This metaphor suggests that we think of "selling" the atmosphere as a sink for emissions and "selling" climatic resources to resource-using sectors, recognizing the interdependencies of these two activities and also recognizing that there is a variety of other actions we can take to increase the amount of atmospheric sink and climate resource we have to sell that ranges from reforestation to climate modification. The value of this metaphor over the pollutant metaphor is that it emphasizes the ongoing management problem that must be faced and it emphasizes that a simple emissions target (with or without trading) will not generate an optimal solution.

Because of the dynamic nature of the jointly managed atmosphere-climate problem, an international bargaining process that aims at an agreement on emissions targets will likely be rendered obsolete by new information and technology before it is put into place. Flexibility to respond to new information will require mechanisms for ongoing management. From this perspective, dealing with the problem in a comprehensive manner is appropriate, but it is difficult to imagine designing a single instrument that can achieve an efficient outcome. Recognizing the need for multiple instruments, even if an ideal gas index could be created, frees us to consider separate instruments for each gas, thus eliminating the need for an explicit numerical index. For example, it may well make sense to have a mixed set of internationally negotiated policies, with the level and mix of policies adjusted through the international bargaining process as the costs and benefits of action become better identified. Such a mix could include taxes on fossil fuels or tradable permits in gross carbon emission, identified reforestation goals, further research on ocean fertilization, and variety development of rice that may reduce CH_4. Calculating an index to support the negotiation process would be beneficial, but, given the uncertainty of such a calculation, institutionalizing a particular value in a trading scheme may be premature.

Bargaining, Buyouts, and Unilateral Action

Unilateral action is not very effective in reducing trace-gas accumulation—hence the need for international bargaining. Whatever the policy instrument, distributional consequences are unavoidable. Recognizing this up front can probably save considerable effort; rights could be distributed on the basis of population, income, the inverse of income, emissions, or geography. One might make an appeal for allocating rights to maintain neutrality with respect to income distribution. A component of neutrality would be that the benefits of climate and trace-gas avoidance minus the cost of emissions reduction are "equal" among signees. But, equality is likely to be judged, for example, in per capita, per unit gross national product (GNP), or per value of natural-resource-stock terms because, given the difference in size among countries (however measured), equal total benefit is unlikely to be perceived as a neutral distribution.

Conclusions

Limiting climate change by reducing trace-gas concentrations will require reductions in emissions of several different gases. Because the relationship of damages to climate may be nonlinear and gases have different lifetimes in the atmosphere, a simple index based on physical parameters will provide an inefficient basis for trading off reductions among gases. An explicit index is needed if an emissions-trading scheme is to be used, but, even if an explicit trading system is not implemented, informed trade-offs among policies will require an understanding of the relative value of decreasing different gases. In addition, gases have nonclimatic effects that should be considered. In particular, CO_2 has a positive effect on plant growth and thus on agricultural output. Failure to include such effects in decisions regarding control of trace gases will lead to overly severe limitations on CO_2 relative to other gases such as CFCs and CH_4.

References

Houghton, J. T., G. J. Jenkins, and J. J. Ephraums. 1990. *Climate Change: The IPCC Scientific Assessment*, a report prepared for the Intergovernmental Panel on Climate Change by Working Group 1. Cambridge, U.K.: Cambridge University Press.

Kane, S., J. Reilly, and J. Tobey. In press. "Climate Change: Implications for World Agriculture." Washington, D.C.: Economic Research Service, U.S. Department of Agriculture.

Lashof, D. A., and D. R. Ahuja. 1990. "Relative Global Warming Potentials of Greenhouse Gas Emissions." *Nature* 344: 529–31.

Lashof, D. A., and D. Tirpak, eds. 1989. *Policy Options for Stabilizing Climate. Vol. I.* Washington, D.C.: Office of Policy, Planning, and Evaluation, U.S. Environmental Protection Agency.

National Climate Program Office (NCPO). 1989. "Climate Impact Response Functions," report of a workshop held at Coolfont, W.Va., Sept. 11–14, 1989.

Reilly, J. M., J. A. Edmonds, R. H. Gardner, and A. L. Brenkert. 1985."Uncertainty Analysis of the IEA/ORAU CO_2 Emissions Model." *The Energy Journal* 8(3): 1–29.

U.S. Environmental Protection Agency (EPA). 1983. *Can We Delay a Greenhouse Warming?* Washington, D.C.: Office of Strategic Studies, U.S. EPA.

Wuebbles, D. J., and J. A. Edmonds. 1988. "A Primer on Greenhouse Gases." Washington, D.C.: U.S. Department of Energy. (DOE/NBB-083.)

Wuebbles, D. J., and J. A. Edmonds. In press. *A Primer on Greenhouse Gases.* Chelsea, Mich.: Lewis Publishers.

PART THREE

Agriculture, Natural Resources, and Global Change

6

Agronomic and Economic Impacts of Gradual Global Warming: A Preliminary Analysis of Midwestern Crop Farming

Harry M. Kaiser, Susan J. Riha,
David G. Rossiter, and Daniel S. Wilks[1]

Introduction

To what extent can farmers adapt to gradual climate change? Although an increasingly popular view is that agriculture can adapt to a changing climate, our understanding of the interactions among the physical, biological, and economic forces that determine the potential for adaptation is very limited. Studies to date on potential climate-change impacts on the farm sector have either not considered important aspects of these interactions or used limited or unrealistic component models of the interaction process, or both.

The purpose of this paper is to report on a multidisciplinary research project currently under way at Cornell University that addresses some of these shortcomings in examining the farm-adaptability question. Our interdisciplinary group at Cornell is developing three models of the relevant component processes, and protocols for their linkage, to address climate-change impacts and agricultural adaptability. The overall goal is to improve the basic framework within which climate impact assessments are conducted, and to improve the state of knowledge of the interactions among the

[1] From the Department of Agricultural Economics (H.M.K) and the Department of Soil, Crop, and Atmospheric Sciences (S.J.R., D.G.R., D.S.W.), Cornell University, Ithaca, N.Y. Funding for this research is provided by a cooperative agreement from the Resources and Technology Division of the Economic Research Service, U.S. Department of Agriculture.

physical, biological, and economic processes that detemine the potential for agricultural adaptation.

Because the project is still in its initial phase, much of this paper is devoted to describing the methodology that will be followed in examining climate-change impacts on the farm sector. The methodology is unique in two ways. First, gradual climate change, rather than the conventionally assumed static "doubled-CO_2" climate, is being simulated. To provide any insight on the farm-adaptability question, it is imperative that climate, crop, and economic models be dynamic, or transient, in nature. Second, this project uses a multidisciplinary approach that focuses on constructing a unifying model from several disciplines to study a broad-reaching issue. With the exception of Adams et al. (1990), most previous research on climate change has not used an interdisciplinary approach. The present study differs from the Adams et al. study in that this is a farm-level analysis (in Minnesota) whereas the Adams et al. study was a sector-level analysis of agriculture for the entire United States.

The sections that follow outline the major features of each component of the unifying model and their interactions. In addition, preliminary results for one location and two climate-change scenarios are presented.

The Model

The unifying model consists of three components: the atmospheric, agronomic, and economic components. Figure 6.1 illustrates the major inputs and outputs of each component and how each component relates to the others. The output of the atmospheric component includes daily values for minimum and maximum temperature, precipitation, and solar radiation. These three outputs become inputs in the agronomic component. On the basis of these climatic variables, the output of the agronomic component includes crop yields, grain moisture content, and field time (i.e., the amount of time permitted by weather conditions that farmers could be performing field operations). The crop yields, moisture content, and field time in turn become inputs in the farm-level economic component.[2] Finally, the output of the economic model includes optimal crop mix, scheduling of field operations, and net farm income. The following section addresses each component separately in more detail.

---

[2] Grain moisture content will vary depending on when the crop is planted and harvested. This parameter is important in the economic model because it influences grain-drying costs. Because grain drying represents a significant cost to farmers, it has a major influence on when crops get planted and harvested.

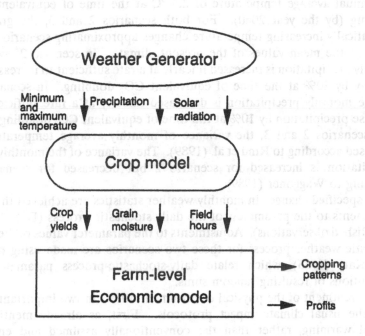

Weather Generator

Minimum and maximum temperature ↓   Precipitation ↓   Solar radiation ↓

Crop model

Crop yields ↓   Grain moisture ↓   Field hours ↓

Farm-level Economic model   → Cropping patterns   → Profits

FIGURE 6.1 The components of the unifying model.

Atmospheric Component

The atmospheric component, which is a stochastic weather generator (Richardson, 1981; Richardson and Wright, 1984), is used to generate the three climatic variables that influence crop yields, field time, and grain moisture. The atmospheric component requires that historical weather data be characterized in terms of the parameters of a particular stochastic process. Several changed climate scenarios are generated to determine a wide range of potential crop and economic responses. Three scenarios are presented here. The first scenario assumes that there is no change in the climate. This scenario is used as a base line for comparison with the other two scenarios. The second scenario assumes a warmer and wetter climate, whereas the third scenario assumes a warmer and drier climate.

The second and third scenarios are developed from plausible, but arbitrary, estimates of climate change that may occur in the medium term (50–100 years). These two scenarios are based on globally average temperature changes portrayed in the transient "scenario B" of Hanson et al.

(1988). This is a relatively mild rate of change, producing increases in global and annual average temperature of 2.5 °C at the time of equivalent CO_2 doubling (by the year 2060). For both scenarios 2 and 3, the gradual, quadratically increasing temperature changes approximating scenario B are added to the mean values of the present climate. In scenario 2, average monthly precipitation is increased linearly at a rate sufficient to increase this quantity by 10% at the time of equivalent CO_2 doubling. In scenario 3, average monthly precipitation is decreased linearly at a rate sufficient to decrease precipitation by 10% at the time of equivalent CO_2 doubling. For both scenarios 2 and 3, the variance of monthly average temperature is decreased according to Rind et al. (1989). The variance of the monthly total precipitation is increased for scenario 2 but decreased for scenario 3 according to Waggoner (1989).

The specified changes in monthly weather statistics are achieved through adjustments to the parameters of the daily stochastic process (D. S. Wilks, unpublished observations). Adjustments to the parameter values of the daily stochastic weather process for these two scenarios are made using results from Katz (1985), which relate daily-stochastic-process parameters to distributions of resulting random sums.

This treatment of the physical environment differs in two important ways from the usual climate-impact protocols. First, as already mentioned, gradual warming, rather than the conventionally assumed and entirely hypothetical doubled-CO_2 climate, is simulated. Although CO_2-doubling studies have provided important contributions to the understanding of the climate system per se, there is no reason to believe that atmospheric concentrations of greenhouse gases will stop increasing when the concentrations equivalent to a doubling of the preindustrial CO_2 concentration are reached. Furthermore, the characteristics of the so-called "transient response" are expected to be qualitatively different from those of a static, equilibrated climate. The second fundamental difference in the treatment of the climate component lies in the construction of the climate-change scenarios. Although the conventional approach of adding assumed changes in temperature or precipitation (typically obtained from a particular run of a general circulation model of the atmosphere) to historical meteorological data is simple and intuitively appealing, it cannot produce changes in variability. Yet changes in climatic variability may have consequences that are as important as changes in means. Consequently, the stochastic weather generator outlined above is used because the underlying site-specific parameters of this model can be adjusted to produce simulated daily meteorological values consistent with specified changes in monthly means and variances of relevant variables.

Agronomic Component

The typical assumption that the same crops will be grown in the same places in the same ways is a serious limitation for examining gradual climate-change impacts on agriculture. Clearly, allowing changes in crop choice and other management decisions at the local level is at the heart of the credible treatment of adaptability and shifts in cropping patterns. Dynamic simulation models of crop-environment interactions, which incorporate crop physiological response to environmental factors including CO_2 concentrations, are used to predict crop response to changing climate. In addition to crop yield, these models are used to predict other important aspects of the cropping system including grain moisture content and field-time availability.

The specific crop-simulation software used, which is called the *General Purpose Atmosphere Plant Soil* model *(GAPS)*, was developed by Buttler and Riha (1989). This model provides for flexible simulation along the lines of similar efforts used by crop physiologists to model crop growth in the field environment [e.g., van Keulen and Wolf (1986)]. It is stronger than previous approaches in its integration of soil, plant, and near-plant atmospheric processes. GAPS also allows different representations of the physical environment to be interfaced with different crop models. Consequently, the degree to which particular results depend on particular crop or environmental components can be analyzed. Corn (maize) and sorghum are currently included in the GAPS modeling framework. The corn model in GAPS was developed by Stockle and Campbell (1985), and the sorghum crop model in GAPS is SORKAM, developed by Rosenthal et al. (1989). Sorghum is included in the present study because it exhibits superior tolerance to heat and drought, and thus has the potential to displace other crops in a warming environment. SORKAM was modified and incorporated into GAPS so that the model corn and sorghum crops could share the same simulated soil, water, and evapotranspiration modules. Although soybeans are not currently part of the GAPS environment, a soybean-crop model (SOYGROW), developed by Wilkerson et al. (1983), is used. Soybeans complement corn production because they can be planted after corn is planted and harvested before corn is harvested. The framework will be expanded later to include wheat.

Economic Component

As the climate changes, farmers will be forced to reevaluate their production decisions—in particular, the choice of which crops to grow. This study attempts to predict these future decisions for each climate scenario by using discrete stochastic sequential programming (DSSP) (Cocks, 1968; Rae,

1971). DSSP is a mathematical programming technique that treats the decision process as a multistage, sequential one. Consequently, decisions made at any particular point in time depend on decisions made in previous stages and outcomes of previous random events, or "states of nature" (Kaiser and Apland, 1989). The advantage of using DSSP is that it models a farmer's decision-making process sequentially. Crop-farm decisions are sequential because plans made at any point in time depend on decisions made at earlier times as well as on outcomes of past states of nature. For example, a farmer's harvest decisions depend on how much land was planted to each crop in the spring, as well as on random events such as yield amounts.

The economic model divides the decision-making process into two stages: stage 1 (preharvest) and stage 2 (harvest). Stage 1 decisions include spring plowing and planting operations, which can take place in four production periods (Table 6.1).[3] Stage 2 decisions include fall plowing and harvesting decisions, which can also take place in four production periods (Table 6.1). The constraining resources for both stages include full- and part-time labor by production period, crop area, and risk.

Four important sources of risk are included in the model: field time, crop yields, grain-drying costs, and output prices. Figure 6.2 presents the decision tree for the economic component. At the beginning of stage 1, the farmer must plan his spring plowing and planting decisions while facing three states of nature for field time. It is assumed that the farmer expects each of the three stage 1 states to be equally likely. At the beginning of stage 2, the farmer must plan his harvest and fall-plowing decisions. At this time, the farmer has perfect knowledge of which stage 1 state has occurred, but only probabilistic knowledge of which stage 2 state will occur. The stage 2 states (10 states conditional for each of the three stage 1 states) consist of random field time, crop yield, grain-drying cost, and crop-price parameters. Hence, 30 joint-net revenue events are possible. Again, it is assumed that the farmer expects each of the 30 joint events to be equally likely. Values for the field-time, yield, and crop-moisture states of nature are generated by GAPS. Grain-drying-cost states of nature are determined by grain moisture content at harvest, yield level, and costs per bushel (or per 0.035 m^3) per percentage point. Specifically, the following equation is used to generate these cost states:

---

[3] Postplant cultivation, fertilization, and pesticide operations are not included as explicit activities in the model because they are not usually constrained by time. Rather, the variable costs associated with these operations are added to the planting activities.

TABLE 6.1 Calendar of Production Operations

Stage	Period	Plow	Plant Corn	Plant Soybeans	Plant Sorghum	Harvest Corn	Harvest Soybeans	Harvest Sorghum
1	1 (Apr. 7–22)	X	X		X			
	2 (Apr. 23–May 11)	X	X	X	X			
	3 (May 1–31)	X	X	X	X			
	4 (June 1–8)	X		X				
2	5 (Sept. 15–30)	X					X	
	6 (Oct. 1–16)	X				X	X	X
	7 (Oct. 17–31)	X				X		X
	8 (Nov. 1–30)	X						

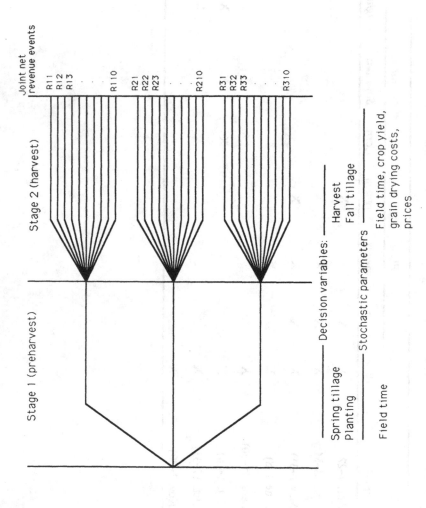

FIGURE 6.2. The decision tree for the discrete stochastic sequential programming (DSSP) farm-level economic component.

$$DC_i = 0.024PT_iY_i$$

where DC_i is drying cost per acre (or, per 4050 m²) for harvest state i; PT_i is the number of percentage points of moisture removed for harvest state i; and Y_i is yield per acre (per 4050 m²) for harvest state i. It is assumed that it costs $0.024/bushel ($0.686/m³) to remove one percentage point of moisture and that grain must be dried only if it has a moisture content of 17% or higher. To generate the price states of nature, each crop price (where all prices are deflated by the consumer price index) is regressed on its yield and price in the previous year. The estimated-price equations for each of the three crops have a negative yield coefficient and a positive lagged-price coefficient, as should be expected. The mathematical formulation of the model is presented in Table 6.2. The objective function 1 in Table 6.2 is to maximize expected net revenue. Constraints 2 through 4 are accounting constraints that define the 30 joint-net revenue events (2), the expected net revenue (3), and the negative deviation from expected net revenue (4). Risk is incorporated into the model by constraint 5, which limits the total negative deviations from expected net revenue to less than or equal to some specified level (D^*). When D^* is set to a number large enough to make this constraint nonbinding, this represents the risk-neutral, or profit-maximizing, solution. On the other hand, when constraint 5 is binding, this represents a risk-averse solution.

Constraint 6 restricts the use of farm labor (full- and part-time) by field operations in the preharvest stage to endowed levels. Note that the right-hand-side parameter in this constraint is stochastic, corresponding to available field hours by production period for the three stage 1 states of nature. Constraint 7 is the land constraint, which limits land planted to endowed levels. Constraint 8 restricts the use of farm labor by field operations in the harvest stage to endowed levels. The right-hand-side parameters for this constraint are also stochastic, corresponding to the available field hours by production period for the 30 stage 2 states of nature. Crop-output constraints are represented by 9, which limits the amount of crop that can be sold to the amount that is harvested for each preharvest state of nature. Finally, constraints 10 through 13 are "sequencing" restrictions, which preserve the proper sequence of field operations in the model. These constraints guarantee that spring plowing occurs before planting (10), that planting activities are matched with harvest activities (11), that harvesting occurs before fall plowing for each stage 1 state (12), and that any land not plowed in the fall is plowed in the spring (13). Constraint 14 is a nonnegativity restriction, which requires that all decision variables in the solution be nonnegative.

TABLE 6.2　The Mathematical Formulation of the DSSP Model

Maximize: E (1)

Subject to:

Accounting Constraints

$$R_{ij} + C_{11}X_{11} + C_{12}X_{12} + C_{21ij}X_{21i} + C_{22}X_{22i} - P_{ij}M_i = 0$$
$$(i=1...3; j=1...10)$$ (2)

$$\sum_{i=1}^{3} \sum_{j=1}^{10} a_i b_j R_{ij} - E = 0$$ (3)

$$R_{ij} - E - d_{ij} \leq 0$$
$$(1=1...3; j=1...10)$$ (4)

Total-Absolute-Deviations Constraint

$$\sum_{i=1}^{3} \sum_{j=1}^{10} d_{ij} \leq D^*$$ (5)

Resource Constraints

$$a_{11}X_{11} + a_{12}X_{12} \leq b_{11}$$ (6)

$$L_1X_{12} < b_2$$ (7)

$$a_{21}X_{21i} + a_{22}X_{22i} \leq b_{3ij}$$
$$(i=1...3; j=1...10)$$ (8)

$$-H_{ij}X_{21i} + M_i \leq 0$$
$$(i=1...3; j=1...10)$$ (9)

Sequencing Constraints

$$-B_1X_{11} + B_2X_{12} \leq 0$$ (10)

$$-IX_{12} + B_3X_{21i} \leq 0$$
$$(i=1...3)$$ (11)

(continues)

TABLE 6.2 (*continued*)

$$-B_4X_{21i} + B_5X_{22i} \leq 0$$
$$(i=1...3)$$
(12)

$$B_7X_{11} - \sum_{i=1}^{3} a_i B_6 X_{22i} \leq 0$$

$$(i=1...3)$$
(13)

$$R_{ij}, d_{ij}, X_{11}, X_{12}, X_{21i}, X_{22i}, M_i \geq 0$$
(14)

where
E = expected net revenue
R_{ij} = total net revenue preharvest state i, harvest state j
C_{11}, C_{12}, C_{22} = variable-cost vectors for spring plowing, planting, and fall plowing
X_{11}, X_{12} = spring plowing and planting vectors, preharvest stage
C_{21ij} = variable-cost vector for harvest, preharvest state i, harvest state j
X_{21i}, X_{22i} = harvest and fall-plowing vectors, preharvest state i
P_{ij} = output price vector, preharvest state i, harvest state j
M_i = marketing-decision vector, preharvest state i
a_i = probability of preharvest state i occurring
b_j = probability of harvest state j occurring, given preharvest state i
d_{ij} = negative deviation from expected net revenue, preharvest state i, harvest state j
D^* = total negative deviations allowable; this is parametrically varied to trace out
   expected net revenue-risk efficient frontier

a_{11} a_{12}
a_{21} a_{22} = matrices of resource requirements for all field operations in stages 1 and 2

b_{1i} = vector of stage 1 resource endowments, preharvest state i
L_1 = vector of ones
b_2 = total crop land endowment
b_{3ij} = vector of stage 2 resource endowments, preharvest state i, harvest state j
H_{ij} = vector of crop yields, preharvest state i, harvest state j
$B_1,...B_7$ = sequence-preserving matrices for field operations
I = identity matrix

Model Simulation Procedures

Actual locations throughout the corn belt will be used to examine gradual climate-change impacts on crop farming. For each location, a representative county will be selected. County agricultural-census data, experiment-station data, and data from the state's land-grant university will be used to construct a hypothetical cash grain farm for each location. Climate data for each location will be collected from agricultural-experiment stations.

The dominant (by land area) soil map unit in the representative county will be used to determine model soil profiles, which will in turn be used to provide soil parameters for GAPS. The profile data will be obtained from the U.S. Department of Agriculture's Soil Conservation Service official soil-service descriptions and map-unit definitions. This simplification implies that the farm has only one soil map unit (or, that other map units do not behave differently from the selected one for the purposes of farm operations and crop growth), and that a single model profile adequately represents the map unit.

Three sites that will initially be examined are (1) the northern limit of corn production in southern Minnesota [major-land-use area (MLRA) 103—central Iowa and Minnesota till prairies], (2) the corn-sorghum transition area of southeastern Nebraska and northern Kansas (MLRA 106—Nebraska and Kansas loess-drift hills), and (3) the corn-soybean area in the heavy, wet lake-plain soils of northwestern Ohio (MLRA 99—Erie-Huron lake plain). Within MLRA 103, the selected county is Redwood, Minn. (Murray, 1985). Within MLRA 106, the selected county is Lancaster, Neb. (Brown et al., 1980). Within MLRA 99, the selected county is Henry, Ohio (Flesher et al., 1974).

For each location, the stochastic weather model will be run in a Monte Carlo sense to generate values for daily weather variables for a changing climate over a period of 100 years, 1980–2079. The resulting daily weather values will then be fed into the crop component to generate annual values for yields, grain moisture, and field time.[4] Finally, the crop yields, grain moisture, and field-time parameters will be fed into the farm-level economic component to generate optimal management strategies and expected net

---

[4] The crop component generates these values disaggregated by time period. For example, for each crop, six yields are generated—one for each combination of planting and harvest periods. Likewise, six values for grain moisture content are generated for each crop—one for each combination of planting and harvesting periods. Field-time output from the crop model is generated in the form of available field hours for each of the eight production periods.

revenue. To generate the 30 joint events for the economic component, the weather component will produce 30 realizations of daily data representative of each decade for the agronomic component. When these climatic realizations are fed into the agronomic component, 30 yield, field-time, and grain-moisture states of nature are generated for the economic model. Because these 30 joint events are equally representative of a decade, the economic model will be solved for each decade in the simulation period, and each solution can be thought of as a representative year within the decade.

To summarize the procedures, the flow of data in this study is from simulated climate, through the agronomic component, to the economic component, ultimately resulting in optimal crop mix, field-operation scheduling, and net revenue. There will initially be three climate scenarios, each producing 100 years of daily data over the period 1980–2079. These data will be grouped into decades that are assumed to have no internal trend. Thus, each decade's weather can be used as the basis of an empirical probability distribution with 30 realizations. Because the first scenario does not have a changing climate, only one decade (1980s) needs to be generated. The DSSP component will then be run for each decade and for two risk-aversion levels at each of the three sites. While price, yield, and field time vary over time, all variable costs are held constant in the economic component. Additional climate-change scenarios will be constructed and analyzed later to obtain a wider range of potential crop and economic responses to gradual climate change.

Some Preliminary Results

The Redwood county farm in southern Minnesota is used here to illustrate the type of analyses that will be conducted for the other regions in the study. The subsections that follow describe the data sources and results for two climate-change scenarios.

Data Sources

Most of the data for the economic component for the hypothetical Minnesota farm that are not generated by GAPS come from Kaiser (1985). These include all resource-requirement parameters and all variable costs for corn and soybean production. Data for sorghum, which currently is not grown in this region, come from national-average statistics. It is assumed that the farm controls 2.4 km² (600 acres) of tillable land on which corn, soybeans, and/or sorghum can be planted. Furthermore, there are two full-time workers and one additional part-time worker who can be hired at a cost of $6.00/hour. The farm uses a conventional tillage system. To account for down time due to machinery repairs, all labor-resource requirements [which are expressed in hours per acre (i.e., per 4050 m^2)] are inflated by 20%.

The yield data for the three crops are generated by GAPS and SOYGROW. Because GAPS and SOYGROW simulate potential rather than actual farm yields, all yields are normalized to the county average for the 1980s. This is done by multiplying each GAPS (or SOYGROW) yield by the ratio of the actual county average yield for the 1980s to the GAPS (or SOYGROW) predicted average potential yield for the 1980s.

Field hours by production period and state of nature are also generated by GAPS. Thirty sets of values for field hours were initially generated to represent the 30 joint states of nature for each decade. All 30 values were used for the four stage 2 periods. However, because stage 1 requires only three states of nature for field hours, three summary statistics of these 30 observations were used: average field hours, average field hours plus 1 standard deviation (SD), and average field hours minus 1 SD. Although GAPS did a reasonable job of predicting field hours for most periods, several adjustments were made after actual observations for the 1980s were compared with predicted observations. First, GAPS significantly overpredicted field hours in the first production period. Hence, all subsequent period 1 field hours generated by GAPS were divided by the ratio of actual to predicted hours based on average observations for the 1980s. The other adjustment that was made was that GAPS-generated field hours for the four stage 1 periods were reduced by 30%, and GAPS-generated field hours for the four stage 2 periods were reduced by 10%. This adjustment was made because the original field-hour parameters, all operations in stage 1, were completed before June, which is somewhat unrealistic for this region. A similar adjustment was required in Kaiser's 1985 study of this region.

On the basis of the 1980 values, the grain moisture content for corn and sorghum predicted by GAPS also needed adjustment. GAPS significantly overpredicted grain moisture, and it was subsequently reduced by 35%. This reduction was determined by comparing actual average drying costs for this region with those calculated when GAPS moisture-content parameters were used.

The price states of nature are based on the following regression equations, which were estimated by using time-series data from 1960 to 1988:

$$RPC_t = 1.65 - 0.0075YC_t + 0.644RPC_{t-1} \quad R^2 = 0.56 \quad DW = 1.43$$
$$(1.70) \quad (-1.12) \qquad (3.94)$$

$$RPSB_t = 5.03 - 0.087YSB_t + 0.648RPSB_{t-1} \quad R^2 = 0.53 \quad DW = 1.49$$
$$(2.53) \quad (-1.73) \qquad (4.60)$$

$$RPSG_t = 1.99 - 0.022YSG_t + 0.679RPSG_{t-1} \quad R^2 = 0.67 \quad DW = 1.12$$
$$(1.99) \quad (-1.64) \qquad (4.68)$$

where RPC is the real price of corn (price adjusted for inflation by dividing nominal price by retail consumer price index for 1980, which was set to 1.0); YC is the county average corn yield; RPSB is the real price of soybeans; YSB is the county average soybean yield; RPSG is the real price of sorghum; YSG is the national average sorghum yield; R^2 is the coefficient of variation; *DW* is the Durbin-Watson statistic; and numbers in parentheses are *t* values. For each decade, the 30 price states for each crop were generated by substituting the yield states and the previous period's price into these equations.

Climate-Change Impacts on Crop Yields

The atmospheric, agronomic, and economic components of the unified model were solved for the three climate scenarios. The yield states of nature, averaged by decade (1980–2079), for climate scenarios 2 and 3 are presented in Figures 6.3 and 6.4.[5] In these figures, the corresponding coefficients of variation for each crop yield are displayed in the bottom graph. Note that because there is no climate change in scenario 1, the average crop yields of the 1980s provide a base-line comparison for the three other scenarios, where the climate is gradually changing from 1990 through 2070.

Climate scenario 2 appears to have no adverse impact on crop yields at this relatively cool location (Figure 6.3). While the climate is gradually warming, the accompanying increase in precipitation prevents the crops from experiencing water stress for the simulation period. In fact, sorghum and soybean yields increase slightly over time, while corn yields remain relatively stable. Interestingly, the variability of yields tends to decrease over time compared with the no-climate-change scenario represented by the 1980s decade. The coefficient of variation of yields for all three crops has a downward trend over time for this scenario, because the variance on precipitation decreases over time under this climate scenario. The robustness of yields is a result of the assumption of a relatively mild change in climate, the relatively cool location, and the introduction of later-maturing varieties later on in the simulation period. Later-maturing

---

[5] For each crop, average yield was calculated by taking a simple average of the yields for the six possible planting-harvesting dates over the 30 states of nature for each decade.

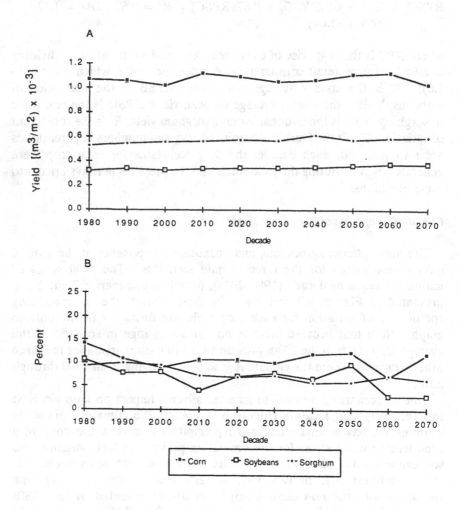

FIGURE 6.3. Average crop yields (A) and coefficients of variation (B) for climate scenario 2, 1980–2070.

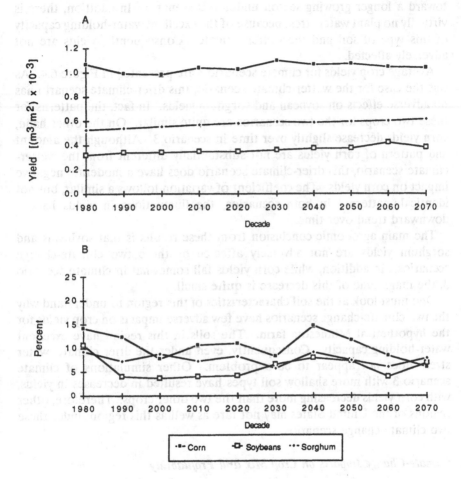

FIGURE 6.4. Average crop yields (A) and coefficients of variation (B) for climate scenario 3, 1980–2070.

varieties have a higher potential yield and take advantage of the trend toward a longer growing season under this scenario. In addition, there is virtually no plant water stress because of the excellent water-holding capacity of this type of soil and the wetter climate. Consequently, yields are not adversely affected.

Average crop yields for climate scenario 3 are presented in Figure 6.4. As was the case for the wetter-climate scenario, this drier-climate scenario has no adverse effects on soybean and sorghum yields. In fact, the patterns for these two crops for the two scenarios are quite similar. On the other hand, corn yields decrease slightly over time in scenario 3. Although the amount and pattern of corn yields are not substantially different from the wetter-climate scenario, this drier-climate scenario does have a moderate negative impact on corn yields. The coefficient of variation follows a similar, but not identical, pattern. In both scenarios, the fluctuations in yields have a downward trend over time.

The main agronomic conclusion from these results is that soybeans and sorghum yields are not adversely affected by these two climate-change scenarios. In addition, while corn yields fall somewhat in climate scenario 3, the magnitude of this decrease is quite small.

One must look at the soil characteristics of this region to understand why the two climate-change scenarios have few adverse impacts on crop yields for the hypothetical Minnesota farm. The soils in this region have excellent water-holding capacity. Consequently, even under the drier climate, water stress does not appear to be a problem. Other simulations of climate scenario 3 with more shallow soil types have resulted in decreases in yields, with corn yields decreasing more than the two other crops. Therefore, other regions of the United States may not fare as well as this region under these two climate-change scenarios.

Climate-Change Impacts on Crop Mix and Profitability

Expected farm net revenue for the three climate scenarios for the risk-neutral case is shown in Figure 6.5. No clear pattern emerges for net farm revenue under scenario 2 compared with no climate change. In this case, net revenue is slightly higher for five decades and slightly lower for four decades relative to no climate change. On the other hand, net farm revenue under scenario 3 is higher for virtually all decades (except 2010) than it is under no climate change. The reason why net revenue is higher in scenario 3 than in the two other scenarios is that there is more field time available in this drier-climate scenario. Because there is more field time, the farmer can plant more corn (which is more profitable than the other crops) relative to the other two scenarios. In addition, although corn yields are

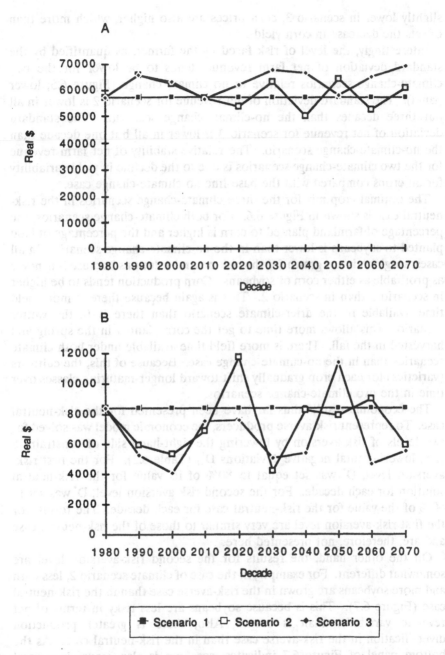

FIGURE 6.5. Average (A) and standard deviation (B) of net revenue by climate scenario for the risk-neutral case, 1980–2070.

slightly lower in scenario 3, corn prices are also higher, which more than offsets the decrease in corn yields.

Interestingly, the level of risk faced by the farmer, as quantified by the standard deviation of net farm revenue, tends to be lower for the two climate-change scenarios relative to no climate change (Figure 6.5, lower panel). The standard deviation of net revenue for scenario 2 is lower in all but three decades than the no-climate-change scenario. The standard deviation of net revenue for scenario 3 is lower in all but one decade than the no-climate-change scenario. The relative stability of net farm revenue for the two climate-change scenarios is due to the decline in yield variability for all crops compared with the base-line no-climate-change case.

The optimal crop mix for the three climate-change scenarios in the risk-neutral case is shown in Figure 6.6. For both climate-change scenarios, the percentage of farmland planted to corn is higher and the percentage of land planted to soybeans is lower than in the no-climate-change scenario. In all cases, sorghum is not grown for the risk-neutral solutions because it is never as profitable as either corn or soybeans. Corn production tends to be higher in scenario 3 than in scenario 2. This is again because there is more field time available in the drier-climate scenario than there is in the wetter scenario. This allows more time to get the corn planted in the spring and harvested in the fall. There is more field time available under both climate scenarios than in the no-climate-change case. Because of this, the cultivars (varieties) for each crop gradually shift toward longer-maturing classes over time in the two climate-change scenarios.

The economic results thus far have been presented for the risk-neutral case. To represent risk-averse producers, the economic model was solved for two levels of risk aversion by lowering the right-hand side of constraint 5 (i.e., lowering total negative deviations D^*) (Table 6.2). For the first risk-aversion level, D^* was set equal to 80% of its value for the risk-neutral solution for each decade. For the second risk-aversion level, D^* was set to 60% of the value for the risk-neutral case for each decade. The results for the first risk-aversion level are very similar to those of the risk-neutral case and are, therefore, not presented here.

On the other hand, the results for the second risk-aversion level are somewhat different. For example, in the case of climate scenario 2, less corn and more soybeans are grown in the risk-averse case than in the risk-neutral case (Figure 6.7). This is because soybeans are less risky in terms of net revenue variability than is corn. Also, there is greater production diversification in the risk-averse case than in the risk-neutral case. As the bottom panel of Figure 6.7 indicates, sorghum is also grown in several decades to reduce risk. This is a classic way of reducing risk in portfolio analysis. Although not shown, expected revenue is lower in the risk-averse case than in the risk-neutral case because of the positive tradeoff between

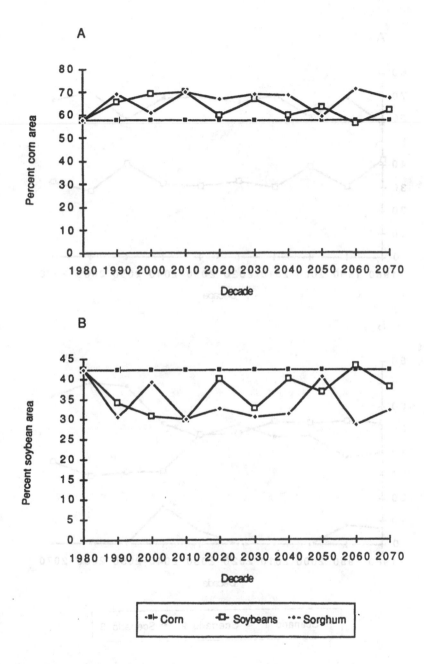

FIGURE 6.6. Choice of corn (A) and soybeans (B) by climate scenario for the risk-neutral case, 1980–2070.

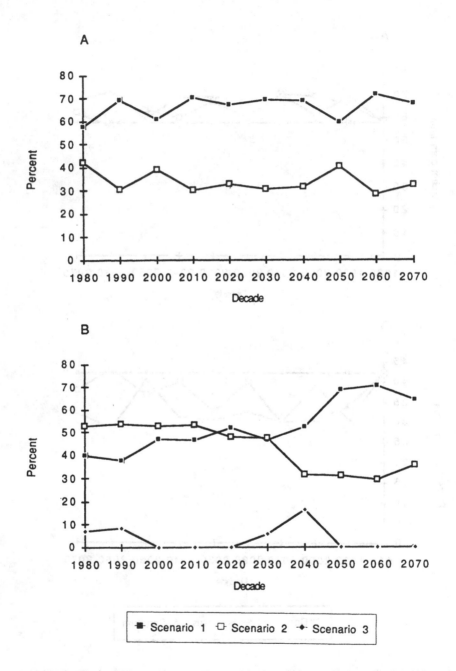

FIGURE 6.7. Choice of crops for climate scenario 2 for the risk-neutral (A) and risk-averse (B) cases, 1980–2070.

risk and income. Similar results with respect to risk hold for climate scenario 3.

Conclusion and Future Work

If climate is warming as rapidly as many projections indicate, agriculture will have to respond. Climate change will affect the agricultural ecosystem in a variety of ways. The course of growth and the yield of crops will change, as will the periods of time available for field operations. These changes will be reflected in decisions made by farmers and, more broadly, in changes in crop production and geography. This paper has reported initial results of an effort to develop an interdisciplinary protocol to address the interactions among climatology, agronomy, and economics and to model the effects of climate change on farm operations and profitability.

The results presented here are preliminary. Some fine-tuning of the unified model is still needed before the results can be finalized. In the future, a wider range of hypothetical climate changes will be investigated. In particular, these will include different rates of warming, greater warming in the winter, and a wider range of possible changes in average precipitation. The physiological effects of gradually increasing CO_2 concentrations will also be simulated in future work. Although preliminary, the results thus far indicate that crop farmers in the southern region of Minnesota can adapt to a mildly changing climate (i.e., warmer and either wetter or drier). This adaptability is due primarily to the relatively cool present climate and the excellent water-holding capacity of the soil in this region. Adaptive strategies for both scenarios include taking advantage of the longer growing season by producing more corn than under the no-climate-change scenario.

To build more realistic estimates of production, and hence of prices, other North American grain-producing areas will have to be simulated. This means that in addition to the crops treated here, wheat production will have to be simulated. A schematic representation of the procedure we plan to follow in future project development is shown in Figure 6.8. For reference, elements treated in the present preliminary work are outlined by the dashed line. In the future, these additions and improvements to the model will be pursued.

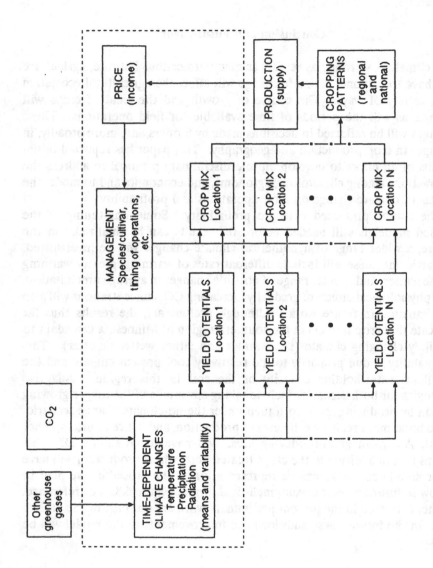

FIGURE 6.8. Schematic representation of future development of the unifying model.

References

Adams, R. M., C. Rosenzweig, R. M. Peart et al. 1990. "Global Climate Change and U.S. Agriculture." *Nature* 345: 219–24.

Brown, L. E., L. Quandt, S. Scheinost, J. Wilson, D. Witte, and S. Hartung. 1980. "Soil Survey of Lancaster County, Nebraska." Washington, D.C.: U.S. Department of Agriculture.

Buttler, I. W., and S.J. Riha. 1989. *GAPS: A General Purpose Simulation Model of the Soil-Plant-Atmosphere System. Version 1.1 User's Manual.* Ithaca, N.Y.: Cornell University Department of Agronomy.

Cocks, K. D. 1968. "Discrete Stochastic Programming." *Management Science* 15: 72–9.

Flesher, E. C. Jr., K. L. Stone, L. K. Young, and D. R. Urban. 1974. "Soil Survey of Henry County, Ohio." Washington, D.C.: U.S. Department of Agriculture.

Hanson, J., I. Fung, A. Lacis et al. 1988. "Global Climate Changes as Forecast by Goddard Institute for Space Studies Three-Dimensional Model." *Journal of Geophysical Research* D93: 9341–64.

Kaiser, H. M. 1985. *An Analysis of Farm Commodity Programs as Risk Management Strategies for Minnesota Corn and Soybean Producers.* Ph.D. dissertation, University of Minnesota.

Kaiser, H. M., and J. Apland. 1989. "DSSP: A Model of Production and Marketing Decisions on a Midwestern Crop Farm." *North Central Journal of Agricultural Economics* 11: 157–69.

Katz, R. W. 1985. "Probabilistic Models," in A. H. Murphy and R. W. Katz, eds., *Probability, Statistics, and Decision Making in the Atmospheric Sciences.* Pp. 261–88. Boulder, Colo.: Westview Press.

Murray, J. J. 1985. "Soil survey of Redwood County, Minnesota." Washington, D.C.: U.S. Department of Agriculture.

Rae, A. N. 1971. "Stochastic Programming, Utility and Sequential Decision Problems in Farm Management." *American Journal of Agricultural Economics* 53: 448–60.

Richardson, C. W. 1981. "Stochastic Simulation of Daily Precipitation, Temperature, and Solar Radiation." *Water Resources Research* 17: 182–90.

Richardson, C. W., and D. A. Wright. 1984. "WGEN: A Model for Generating Daily Weather Variables." Washington, D.C.: Agricultural Research Service, U.S. Department of Agriculture. (ARS Publication 8.)

Rind, D., R. Goldberg, and R. Ruedy. 1989. "Change in Climate Variability in the 21st Century." *Climatic Change* 14: 5–37.

Rosenthal, W. D., R. L. Vanderlip, B. S. Jackson, and G. F. Arkin. 1989. *SORKAM: A Grain Sorghum Crop Model.* Texas Agriculture Experiment Station Computer Software Documentation Series. College Station, Texas: Texas A and M University.

Stockle, C., and G. S. Campbell. 1985. "A Simulation Model for Predicting Effect of Water Stress on Yield: An Example Using Corn." *Advances in Irrigation* 3: 283–311.

van Keulen, H., and J. Wolf, eds. 1986. *Modelling of Agricultural Production: Weather, Soils and Crops.* Wageningen, The Netherlands: Pudoc.

Waggoner, P. E. 1989. "Anticipating the Frequency Distribution of Precipitation if Climate Change Alters its Mean." *Agricultural and Forest Meteorology* 47: 321–37.

Wilkerson, G. G., J. W. Jones, K. J. Boote, K. T. Ingram, and J. W. Mishoe. 1983. "Modeling Soybean Growth for Crop Management." *Transactions of the American Society of Agricultural Engineers* 26: 63–73.

7

A Sensitivity Analysis of the Implications of Climate Change for World Agriculture

Sally M. Kane, John M. Reilly, and James Tobey[1]

Introduction

The phenomenon known as the greenhouse effect is currently the subject of intense national and international political debate. Discovery of ozone "holes" over the polar caps and increased concentrations of CO_2, methane (CH_4), nitrous oxide (N_2O), and chlorofluorocarbons (CFCs) in the atmosphere suggest that human activity can affect the global environment and major earth systems. And, although not conclusively linked to human activities, record hot weather in the 1980s, combined with extreme heat waves, floods, and droughts occurring in 1988, focused increased public attention on the impact that human activity may have on the earth's climate.

Tremendous uncertainty exists in linking human activities with the climate system. Although increased trace-gas concentrations have undeniable effects on the radiative balance of the Earth, the specific effects on temperature, precipitation, and other climatological phenomena are only beginning to be unraveled. Despite this uncertainty, global environmental issues—including climate change—are now a topic of many international meetings of policy makers and scientists, and substantial research efforts are being conducted to understand their economic and social implications. The purpose of this paper is to provide preliminary estimates of the potential effects of climate change on global agriculture. The focus is on identifying

[1] From the U.S. Department of Commerce, National Oceanic and Atmospheric Administration, Office of the Administrator, Washington, D.C. (S.M.K.), and the Economic Research Service, U.S. Department of Agriculture, Washington, D.C. (J.M.R., J.T.). The views expressed are those of the authors and do not necessarily reflect the views of the U.S. government.

the sensitivities of the system, the types of market responses one might expect from the agricultural system, and issues requiring further research.

Sources of Uncertainty

Substantial uncertainty is associated with the climate-change phenomenon and sometimes clouds the real issues involved. By distinguishing among the different types of uncertainties, analytic results can be placed in the appropriate context, and the scope for meaningful statements can be better understood. The following categories of uncertainty can be identified: scientific, physical responses, and economic.

The uncertainty surrounding *climate predictions* can be traced to our scientific knowledge of the climate system and to difficulties encountered in modeling the system. Model results from general circulation models (GCMs) show a few-degree (centigrade) rise in global mean surface temperature for a doubling of greenhouse-gas concentrations in the atmosphere, but knowledge is still inadequate in many areas.

The translation of climate predictions into *physical responses* is another major source of uncertainty. Calculating physical responses, such as crop-growth changes, is made complicated by the lack of knowledge about the combined effects of a carbon-enhanced atmosphere and changes in climate.

Assessing the *economic implications* of physical responses is particularly difficult because of the long time horizon of the problem—in the range of 50–100 years. The response of the agricultural sector is in large part dictated by the development of new production technologies that can efficiently exploit changes in growing conditions. Making credible projections of technological development along with other important factors, such as the structure of the agricultural sector and the pattern of international agricultural-commodity trade over a long time scale, is extremely difficult.

Climate Predictions and Features of Climate Models

The Intergovernmental Panel on Climate Change (IPCC) recently released a scientific assessment of climate change, which was critiqued by hundreds of scientists worldwide (IPCC, 1990a). The report concluded that the gradual accumulation of greenhouse gases in the atmosphere will lead, in general, to an additional warming of the Earth's surface at a predicted rate of ≈ 0.3 °C per decade on the basis of results of large climate-model experiments; this rate of change exceeds that observed over the past 10,000 years (IPCC, 1990a). This assessment is consistent with other climate-change predictions that suggest that over the next 50–100 years, the likely increase in global mean temperature would range between 1.5 °C and 4.5 °C [*see* the overview paper by Albritton (1992) in this volume for additional information].

It is widely held by physical scientists that the warming will be greatest in the polar regions and least in the tropics. Consequently, the temperature difference between the poles and the tropics is expected to decline. There is greater uncertainty with respect to other changes in climatic features such as storm intensity, wind speed, and soil moisture.

Climatologists use GCMs—large, complex computer models—to test the sensitivity of the climate system to increases in greenhouse-gas concentrations in the atmosphere. Several problems arise with the use of GCM predictions for impact studies. Among the most important features of the models that require additional refinement include the timing of changes in climate, the geographical and time scales of predictions, and seasonality. Climate models have been developed to project equilibrium change in climate, a noneconomic measure. Economic studies require information on the timing of the changes in climate, both the timing of the adjustment of the climate system to the doubling of gas concentrations and the timing of the fluctuations in climate during the transition (because changes in global average temperature may not be linear). To refine estimates of timing, greater understanding of the future trends in greenhouse-gas emissions scenarios and of the lags between changes in trace-gas concentrations and climate effects is required.

Currently, GCMs agree strongly in direction for many globally averaged phenomena, the best example of which is surface air temperature (Table 7.1). However, on regional scales, there are significant differences among models. Agricultural production is sensitive to soil moisture level, which is dictated by precipitation. Because precipitation is a localized climate feature, it is not as well simulated by GCMs. Assessment of impacts on agriculture will become more reliable with more refined estimates for smaller geographical units.

Even though specific predictions by GCMs may not be ideal for economic impact analysis, the distribution of climate effects suggested by the models is suitable as a bench mark for economic evaluation of direction of change and net consequences.

Crop-Yield Response to Climate Change

The broad changes in climate projected by GCMs offer some guidance for assessing agricultural effects, but, to evaluate regional-specific effects on crop growth, they must be complemented with more-detailed information on specific daily and seasonal patterns of temperature and precipitation. Mathematical crop-growth models simulate plant growth rates for a

TABLE 7.1 Projected Impact of Climate Change on Crop Yields in the United States, by Crop and Climate Model (% change)

Climate Model[a]	Corn Dry	Corn Irrigated	Soybeans (Dry)	Winter Wheat (Dry)
GISS	−23.7	−24.2	−34.6	−16.0
GFDL	−54.7	−28.5	−59.7	−30.9

[a]GISS, Goddard Institute for Space Studies, National Aeronautics and Space Administration; GFDL, Geophysical Fluid Dynamics Laboratory, National Oceanic and Atmospheric Administration.

Source: U.S. EPA (1990).

particular crop under modified weather conditions by combining information on physical conditions (sunlight, temperature, rainfall, and soil type) with growth processes.

Estimates of crop response to climate change embody considerable scientific uncertainty of two types: the uncertainty associated with GCM climate predictions and the uncertainty associated with the translation of climate changes into yield changes. In general, the current generation of crop-growth models inadequately incorporates several factors, including scope for adaptation strategies and the effect of increased carbon in the atmosphere on plant growth (the CO_2 effect).[2] Crop-growth models also assume that soil nutrients and pests are not limiting. Lastly, where the direct effects are modeled, they may be overstated if climate effects from greenhouse-gas doubling precede doubling of carbon concentrations in the atmosphere. These considerations may prove important in determining the direction and magnitude of climate-induced-yield changes.

Three studies, U.S. EPA (1990), Parry et al. (1988), and Santer (1985), examined regional specific changes in yields induced by changes in climate suggested by GCMs under existing cropping patterns, management practices, and production technologies; typically, carbon-enhanced growth is not considered in the calculations. A selective summary of the results of these studies is shown in Tables 7.1, 7.2, and 7.3. Although each producing area was examined by a different team of experts using different models and

---

[2] For more information on the role of adaptation and carbon fertilization, see Kane et al. (in press).

TABLE 7.2 Climate Change and Impact on Crop Yields

Country, Region	Climate Change (°C [% precipitation])	Crop Yields (% change)					Southern Wheat	Rice
		Hay	Pasture	Rye	Barley	Oats		
Canada								
Saskatchewan	+3.4 °C [+18%]	—	—	—	—	—	−18%	—
Iceland	+3.9 °C [+15%]	+64%	+48%	—	—	—	—	—
Finland								
Helsinki	+4.1 °C [+73%]	—	—	—	+9%	+18%	+10%	—
Oulu	+5.0 °C [+109%]	—	—	—	+14%	+13%	+20%	—
USSR								
Leningrad	+4.2 °C [+52%]	—	—	−13%	—	—	—	—
Cherdyn	+2.7 °C [+50%]	—	—	—	—	—	−3%	—
Saratov	+3.3 °C [+22%]	—	—	—	—	—	+13%	—
Japan								
Hokkaido	+3.5 °C [+5%]	—	—	—	—	—	—	+5%
Tohoku	+2.9 °C [+12%]	—	—	—	—	—	—	+2%
Australia	+1.0 °C [+50%]	—	—	—	—	—	+10–20%	—

Source: Parry et al. (1988).

TABLE 7.3 Projected Impact of Climate change on Crop Yields
in European Community Countries

Country	Average Yield Wheat and Spelt[a] (1975–1979)		BMO[b] Cumulative Yield Changes (%)[c]	GISS Cumulative Yield Changes (%)[c]
	(kg/m^2)	(dt/ha)		
Denmark	5.25	52.5	+18.7	+1.1
Netherlands	5.82	58.2	+1.2	+0.3
Luxembourg	3.09	30.9	+7.8	+6.1
Belgium	4.72	47.2	−9.5	−6.8
France	4.38	43.8	−9.8	−12.3
F.R.G	4.66	46.6	−1.1	−8.6
Italy	2.58	25.8	−0.8	−1.2

[a]Spelt is a cereal intermediate between wheat and rye; dt/ha, dry
tons per hectare.
[b]BMO is a GCM developed at the British Meteorological Office.
[c]Percent of 1975–1979 average yields.

Source: Santer (1985).

methods of analysis, the findings of these studies generally support the
conclusion that middle-latitude yields will fall and northern-latitude yields
will rise with a doubling of CO_2 concentrations.

Table 7.1 shows the negative effects of climate change on yield for the
United States. Soybean yields were most severely reduced, whereas winter
wheat was affected the least (U.S. EPA, 1990). The subarctic European
USSR and Canada were the only regions within the northern latitudes that
displayed negative impacts despite projected increases in precipitation and
temperature, as shown in Table 7.2 (Parry et al., 1988). In general, positive
effects were found in northern areas in Europe and negative ones in
southern areas, as shown in Table 7.3 (Santer, 1985).

The Static-World-Policy Simulation Model

As the concern over climate change has increased, so has the number of
analytical studies of the potential economic effects of climate change on
agriculture. Reports by the IPCC Working Group II (Impacts) (IPCC,
1990b), the U.S. Environmental Protection Agency (U.S. EPA, 1990), and
Parry et al. (1988) describe a variety of climate-change factors that may

impinge on agricultural productivity across regions in the United States and the world. In most cases, such as the case studies presented in Parry et al. (1988), the economic effects of climate change on agricultural producers and consumers are inferred from national yield-change estimates.

Unlike the nonmarket approaches described above, market analysis captures supply responses along with economic effects on the consumer side. For example, Adams et al. (1988, 1990) incorporated climate change into a spatial-equilibrium model to determine its effects on U.S. agricultural supply and demand. Adams et al. (1990) estimated that these consumer effects on welfare dominate producer effects by a factor of 10. Kellogg and Severin (1990) have also modeled agricultural markets—in this case, the Soviet market. They showed that climate change substantially benefits agricultural possibilities, but not enough to overcome basic structural problems that have plagued production since the mid-1980s.

Our analysis takes the Adams et al. work a step further by examining global, rather than only domestic, market effects. In an open world economy, the effect of climate change on U.S. agriculture cannot be considered in isolation from the rest of the world. Changes in agricultural production in other countries, especially from producers, affect global agricultural prices and trade flows that in turn affect consumer and producer welfare in the United States. These second-round effects have not been adequately explored in the literature to date.[3]

To derive the estimates of price and welfare effects of changes in agricultural yield, we use the Static World Policy Simulation Model (SWOPSIM). SWOPSIM is a partial-equilibrium model that describes world agricultural markets through a system of supply-and-demand equations that are specified by matrices of own- and cross-price elasticities (*see* Roningen, 1986).

SWOPSIM was chosen for its ability to calculate the welfare effects of agricultural-production disturbances.[4] (For more information on welfare measures used in SWOPSIM, *see* Kane et al., in press.) SWOPSIM also has the desirable feature of encompassing all regions of the world and 22

---

[3] The only other study of this nature that we are aware of is "Yield Changes of Agricultural Commodities: Climate Change and International Agriculture Project," currently being funded by the U.S. EPA and coordinated by the Goddard Institute for Space Studies in New York and the Atmospheric Impacts Research Group, University of Birmingham, U.K.

[4] In contrast, most empirical models of agriculture ignore traditional welfare and resource-efficiency measures [some widely used agricultural models in this category include FAPSIM (Gadson et al., 1982), WHEATSIM (Holland and Sharples, 1981), FAPRI (Meyers et al., 1986), and POLYSIM (Ray and Richardson, 1978)].

agricultural commodities, including eight crop, four meat-livestock, four dairy-product, two protein-meal, and two oil-product categories. The world model that was created using SWOPSIM separately identifies the United States, Canada, the European Community, Australia, Argentina, Pakistan, Thailand, China, Brazil, the USSR, other European countries (Sweden, Finland, Norway, Austria, and Switzerland), and Japan. All other countries are grouped together. This level of disaggregation covers the major agricultural importing and exporting regions of the world and also highlights those areas projected to be the most strongly affected by climate change.

Sensitivity Analysis

As our discussions above suggest, there are some important limitations of both the crop-response models and the GCMs they rely on. As a result, rather than specifying a particular set of specific yield effects for any region or country, it may be more appropriate to view climate-change yield effects as falling within a range of possibilities. For this reason, it is useful to explore the price sensitivity of world agriculture to a broad range of potential yield assumptions.

Four sets of climate-change scenarios were constructed to test the robustness of world agriculture to hypothetical decreases in agricultural-commodity supplies in some of the most important grain-producing areas in the world (Table 7.4). In each scenario, yields in the United States, Canada, and the European Community are progressively decreased. The four scenarios are described as follows: scenario 1 assumes no corresponding yield changes in the rest of the world; scenario 2 assumes yield increases of 25% in countries in the northern latitudes (the USSR and northern Europe), China, Japan, Australia, Argentina, and Brazil; scenario 3 assumes that a 25%-yield loss occurs in the rest of the world; and scenario 4 assumes that yields in the rest of the world decline by 25%, while yields in the northern latitudes (the USSR and northern Europe), China, Japan, Australia, Argentina, and Brazil increase by 25%.

The basic assumption of decreasing crop yields in the United States, Canada, and the European Community in all scenarios is based on the general scientific consensus that drying in the northern middle latitudes combined with warming will likely lead to negative crop-yield effects. The scenarios were also designed to reflect possible increased yields in northern-latitude countries, China, Japan, Australia, Argentina, and Brazil and uncertainty in crop-yield effects in the rest of the world. Because crop-response models have generally incorporated neither agricultural-management responses to climate change nor the fertilization effect on plant

TABLE 7.4 Climate-Change Sensitivity Analysis: Yield Assumptions of Four Climate Scenarios (% yield change)

Region	1	2	3	4
USSR, northern Europe, China, Japan, Australia, Brazil, Argentina	No change	+25%	No change	+25%
Rest of world	No change	No change	−25%	−25%

growth, assumptions of crop-yield declines in the United States, Canada, and the European Community may be considered conservative.

Aggregate primary-crop price effects of the four scenarios are illustrated in Figure 7.1. In scenario 1, composite primary agricultural prices increase. This is not surprising, because the United States, Canada, and the European Community account for the bulk of cereal grain and oilseed production. It is interesting that world agricultural prices fall by almost 20% in scenario 2 before eventually rising above base-price levels when there are positive yield effects in the northern latitudes, China, Japan, Australia, Argentina, and Brazil. Even with a 30%-yield loss in the United States, Canada, and the European Community, prices decrease 10% below base world levels. Scenario 4 shows that if yield effects are not very negative in the United States, Canada, and the European Community, world agricultural prices still fall when there are yield reductions in the rest of the world because of compensatory yield increases in the northern latitudes, China, Japan, Australia, Argentina, and Brazil. The largest price increases occur under scenario 3.

These scenarios show that reduced production potential in the United States, Canada, and the European Community may be balanced by gains in other areas when the negative yield effects are relatively modest. However, under stronger assumptions of crop-yield declines, world agricultural prices and the world's total food-producing capability are more severely disrupted.

The global and U.S. welfare effects as a percentage of world gross domestic product (GDP) from these scenarios are quite modest (Table 7.5). The welfare effects are larger (but still <3% of GDP) in some individual countries where either agricultural production is a relatively large share of total production, or where food expenditures are a large share of total

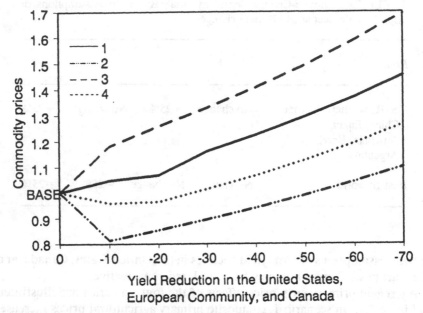

FIGURE 7.1 Sensitivity of world commodity prices (expressed as a value-weighted index for primary agricultural commodities) to yield changes. 1, neutral effect in all other countries; 2, yield increases in the USSR, northern Europe, the People's Republic of China (PRC), Japan, Australia, Argentina, and Brazil (+25%); 3, yield decreases in the rest of the world (−25%); 4, yield decreases in the rest of the world (−25%) and yield increases in the USSR, northern Europe, PRC, Japan, Australia, Argentina, and Brazil (+25%).

expenditures (e.g., China). The net effect on welfare in any country depends on the direction and magnitude of yield changes, the magnitude of changes in consumer and producer surplus following changes in world commodity prices, and the relative strength of the country as a net agricultural importer or net exporter.

Consider the case in which there are two countries, one a large net importer and one a large net exporter, and a climate change that produces yield declines in both countries and an increase in world agricultural prices. These two cases are shown diagrammatically in Figure 7.2. For the large net exporter, the loss in consumer surplus is given by area A. The gain in producer surplus is given by (A + B) − (E + F). Thus, if area B is greater than area (E + F), there is a net gain in consumer-plus-producer surplus. For the large net importer, the loss in consumer surplus is given by (A + B + C), and the gain in producer surplus is given by (A − E). Thus, if

TABLE 7.5 Welfare Effects for the World and the United States as a Percentage of 1986 Gross Domestic Product for Five Cases of Yield Reductions in the United States, Canada, and the European Community (%)

	-10%	*-20%*	*-30%*	*-40%*	*-50%*
World					
Scenario 1	−0.03	−0.06	−0.10	−0.13	−0.17
Scenario 2	0.09	0.06	0.04	0.01	−0.01
Scenario 3	−0.02	−0.24	−0.29	−0.34	−0.40
Scenario 4	−0.07	−0.07	−0.10	−0.14	−0.17
United States					
Scenario 1	0.00	−0.01	−0.03	−0.07	−0.13
Scenario 2	0.02	0.02	0.01	−0.01	−0.05
Scenario 3	−0.01	−0.03	−0.07	−0.12	−0.21
Scenario 4	−0.01	−0.01	−0.65	−0.12	−0.20

(B + C) is greater than E, there is a net loss in consumer-plus-producer surplus.

Australia and Japan are examples of very large net exporters and importers, respectively. Under the assumption of rising world agricultural prices and yield reductions in these countries (following scenario 3) we find, as expected, that welfare increases in the net-exporting country and decreases in the net-importing country (Table 7.6).

In sum, when the effect of climate change on world agricultural prices is taken into account, a country's net welfare outcome depends not only on the relative size of domestic-yield effects, but it also depends heavily on the relative size of the agricultural-producing and -consuming sectors and on the direction and magnitude of world-price effects. Figure 7.3 summarizes the nature of the interdependence among yield changes, world price changes, and economic surplus.

Concluding Comments

Several findings are worth highlighting. First, it is abundantly clear that any analysis of the consequences of climate change on world agriculture must be considered preliminary, because considerable scientific uncertainty about climate predictions and the impact on agriculture of increasing atmospheric concentrations of trace-gas emissions remains. Second, the

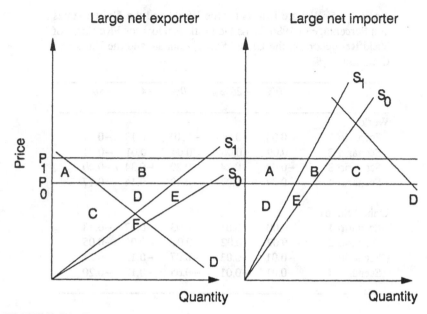

FIGURE 7.2 Effects of climate change on welfare. P, price; S, supply; other letters denote areas.

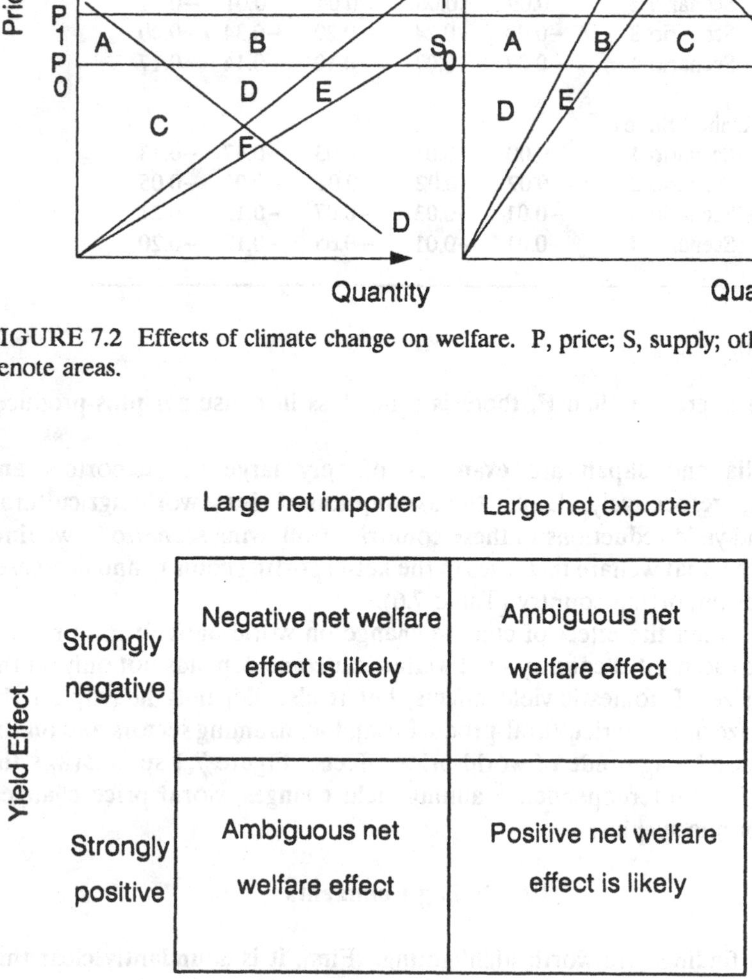

	Large net importer	Large net exporter
Strongly negative	Negative net welfare effect is likely	Ambiguous net welfare effect
Strongly positive	Ambiguous net welfare effect	Positive net welfare effect is likely

FIGURE 7.3 Net welfare effects of climate change, assuming an increase in world agricultural prices.

empirical study reported in the present paper also shows that the evaluation of climatic-change gainers and losers cannot be made on the basis of domestic-yield effects alone. Although some fairly large regional reductions in the production of agricultural commodities are predicted by existing crop-response models, they are not likely to severely disrupt world food supplies because reductions in some areas are offset by increases in other areas. Rather, it also depends on the relative size of the agricultural-producing and -consuming sectors and on the direction and magnitude of world-price effects.

The empirical findings of this report do not incorporate technical and management adaptive responses to changes in climate. We suggest that these responses could be just as important as the actual physical weather changes in determining domestic-crop-yield effects. In addition, we have not attempted to predict economic and population growth rates that could alter the structure of demand-and-supply conditions of the estimating model. For these reasons, the results of SWOPSIM should not be viewed as an accurate representation of the agricultural consequences of climate change on specific economies. Rather, they highlight general directions and the order of magnitude of change, as well as demonstrating some straightforward, but important, general economic principles. Despite the difficulties that arise, impact assessments such as this, when used carefully, can provide useful information to policy makers.

TABLE 7.6 Welfare Effects for Japan and Australia as a Percentage of 1986 Gross Domestic Product for Five Cases of Yield Reductions in the United States, Canada, and the European Community for Scenario 3 (%)

	−10%	−20%	−30%	−40%	−50%
Japan	−0.17	−0.22	−0.28	−0.35	−0.42
Australia	0.33	0.46	0.60	0.78	0.99

References

Adams, R., C. Rosenzweig, R. Peart et al. 1990. "Global Climate Change and U.S. Agriculture." *Nature* 345(17): 219–23.

Adams R, B. McCarl, D. Dudek, and J. Glyer. 1988. "Implications of Global Climate Change for Western Agriculture." *Western Journal of Agricultural Economics* 13(2): 348–56.

Albritton, D. L. 1991. "The Science of Global Change," in J. Reilly and M. Anderson, eds., *Economic Issues in Global Climate Change: Agriculture, Forestry, and Natural Resources.* Pp. 3–23. Boulder, Colo.: Westview Press.

Gadson, K., J. Price, and L. Salathe. 1982. *Food and Agricultural Policy Simulator (FAPSIM): Structural Equations and Variable Definitions.* Washington, D.C.: U.S. Department of Agriculture. (ERS Staff Report AGES820506.)

Holland, F., and J. Sharples. 1981. *WHEATSIM: Model 15 Description and Computer Program Documentation.* West Lafayette, Ind.: Purdue University. (Agriculture Experiment Station Bulletin 319.)

IPCC. 1990a. "Report of the Subgroup on Agriculture, Forestry, and other Human Activities." Report of the Intergovernmental Panel on Climate Change, Working Group III (Response Strategies). Washington, D.C.: Island Press.

IPCC. 1990b. "Report on Agriculture and Forestry." Report of the Intergovernmental Panel on Climate Change, Working Group II (Impacts). Washington, D.C.: Island Press.

Kane, S., J. Tobey, and J. Reilly. In press. "Climate Change: Economic Implications for World Agriculture." Washington, D.C.: Economic Research Service, U.S. Department of Agriculture. (Agricultural Economic Report 647.)

Kellogg, R., and B. Severin. 1990. "Weather and Soviet Agriculture: Implication of an Economic Model." Paper prepared for the American Agricultural Economics Association (AAEA) annual meeting in Vancouver, B.C., Aug. 4–8.

Meyers, W., S. Devadoss, and M. Helmar. 1986. *Baseline Projections, Yield Impacts and Trade Liberalization Impacts for Soybeans, Wheat, and Feed Grains: A FAPRI Trade Model Analysis.* Iowa City: Center for Agricultural and Rural Development, University of Iowa. (Working Paper 86-WP2.)

Parry, M., T. Carter, and N. Konijn, eds. 1988. *The Impact of Climatic Variations on Agriculture, Volume 1: Assessments in Cool Temperate and Cold Regions*, and *Volume 2: Assessments in Semi-Arid Regions.* Boston: Kluwer Academic Publishers.

Ray, D., and J. Richardson. 1978. "Detailed Description of POLYSIM." Stillwater, Okla: Oklahoma State University. (Agriculture Experiment Station Technical Bulletin T-151.)

Roningen, V. 1986. "A Static World Policy Simulation (SWOPSIM) Modeling Framework." Washington, D.C.: U.S. Department of Agriculture. (ERS Staff Report AGE860625).

Santer, B. 1985. "The Use of General Circulation Models in Climate Impact Analyses—A Preliminary Study of the Impacts of a CO_2 Induced Climate Change on West European Agriculture." *Climatic Change* 7: 71–93.

U.S. EPA. 1990. *The Potential Effects of Global Climate Change on the United States, Volume 1: Regional Studies,* Report to Congress. Washington, D.C: U.S. Environmental Protection Agency.

8

Government Farm Programs and Climate Change: A First Look

Jan Lewandrowski and Richard Brazee[1]

Introduction

Economists have long been interested in the market-distorting effects of government farm programs. Farm programs distort agricultural markets by increasing the expected profitability of and/or lowering the risk associated with a subset of the farmer's production possibilities. In recent years, economists have also started to consider the potential costs and benefits relating to agriculture from possible climate change (Dudek, 1989; Adams et al. 1988; Arthur and Abizadeh, 1988).[2] A changing climate could shift the current map of agricultural production by altering regional (as well as international) comparative advantages in the production of commercially important crops and livestock. The extent of any resulting economic impacts will, in large part, depend on how the agricultural sector adapts to the changing environment.

To date, little has been said about how government farm programs might affect the economic impacts of climate change. Farm programs could mitigate or aggravate these impacts by encouraging or discouraging specific farm-sector adaptations to new environmental conditions. For example, one often-discussed future climate scenario (based on current trends in greenhouse-gas emissions) predicts that by the year 2020, the southeastern United States will be 2–3 °C warmer and significantly more prone to

---

[1] From the Resources and Technology Division, U.S. Department of Agriculture, Washington, D.C. The views expressed are those of the authors and do not necessarily reflect the policies or views of the U.S. Department of Agriculture.

[2] A much larger literature exists among noneconomists assessing the effects of climate-change agriculture. The results and methods of 19 of these studies are summarized in Smit et al. (1988). *See also* Wilks (1988).

drought (Hansen et al., 1988).[3] This implies that over the next three decades, farmers in the Southeast will face an increasing probability of losing their crops in any given year if they continue to farm as they do today. In a free market, risk-averse farmers would incorporate this higher likelihood of drought loss into their decision framework, and we would expect to see behavioral changes aimed at guaranteeing some minimal level of income (e.g., more investments in irrigation, the increased use of crop insurance, and the allocation of some land to crops with lower expected returns but with higher tolerances to heat and drought). If, however, farmers believe that government disaster payments will provide this minimal level of income when their crops are lost, they will have little incentive to purchase additional risk protection. By reducing the farmers' incentives to adapt, the disaster-payments program could increase the frequency with which such payments have to be made.

Our purpose here is to initiate a line of research into the relationship between farm programs and the economic impacts of climate change. Because this paper is a first step, our objectives are simply to outline the nature of the relationship and to discuss the issues involved. We start with a discussion of climate change as it relates to U.S. agriculture. The main point is that it is impossible to be confident about existing predictions concerning either climate change or its effects on agriculture. Next, to familiarize readers with the objectives, structure, tools, and costs of today's farm policy, we give a brief overview of government farm programs. Finally, we consider some of the ways in which farm programs could either mitigate or aggravate the social costs of climate change.

Climate Change and U.S. Agriculture

In any given area, the nature of agricultural production is largely defined by the climate. The choices of inputs, outputs, and methods of production all reflect farmers' expectations concerning upcoming temperature, precipitation, growing-season, and soil-moisture patterns. These are also the variables whose means and variances, many believe, are going to be altered as a result of the increasing levels of greenhouse gases in the atmosphere. Agriculture, then, stands to be one of the most affected sectors of the economy should climate change occur.

---

[3] The greenhouse gases are carbon dioxide (CO_2), methane (CH_4), chlorofluorocarbons (CFCs), nitrous oxide (N_2O), and lower-atmospheric ozone (O_3). Of these, CO_2 is by far the most important. The primary sources of CO_2 emissions are fossil-fuel burning in the developed countries and deforestation in the tropics.

Despite its potential for causing major shifts in regional farm practices, current predictions of climate change are expected to be quite inaccurate. As a result, it is impossible to say with any confidence what the exact effects on U.S. agriculture will be. It is useful to review the uncertainties that underlie today's climate-change forecasts in order to keep specific predictions in perspective and to identify sets of predictions that are considered plausible.

Recent attempts to estimate the potential impacts of climate change on U.S. agriculture have generally been based on future-climate scenarios produced by general circulation models (GCMs)[4] (Adams et al., 1988; Dudek, 1989; Rosenzweig, 1985; Wilks, 1988; Smith and Tirpak, 1989). GCMs are elaborate computer models that mathematically simulate global weather and climate conditions over several decades for given values of various climate-forcing parameters (including atmospheric CO_2). Although these models are state-of-the-art for estimating the long-run environmental impacts of greenhouse-gas accumulation in the atmosphere, they are not very reliable.[5] At present, little is known about how ocean currents, regional cloud cover, convection, ground hydrology, and evapotranspiration affect climate patterns; as a result, these processes are greatly simplified in GCM simulations (Hansen, 1989; Hillel and Rosenzweig, 1989). Those who build GCMs state these simplifications up front and caution that as climate processes become better understood, the models will be improved and their predictions will change.

Two caution flags highlight the unreliable nature of GCM predictions. First, the models do not reproduce current climate conditions very well (Hansen et al., 1988; Hansen, 1989; Hillel and Rosenzweig, 1989). Globally, they tend to overpredict mean surface temperatures; regionally, the simulated weather patterns often disagree with those that are observed. Second, for many regions of the world, the different GCMs produce a wide range of climate-change predictions (Hansen, 1989). In the continental United States, for example, the GISS, GDFL, and NCAR models *(see* footnote 5) predict mean temperature increases between 2 and 8 °C, changes in mean precipitation between −1 and 1 mm/day, and decreases in average

---

[4] Some authors refer to GCMs as "global climate models"; the terms mean the same thing.

[5] The most widely used GCM is probably the GISS model, developed at the Goddard Institute for Space Studies. Other GCMs are those developed by the Geophysical Fluid Dynamics Laboratory (the GFDL model), Oregon State University (the OSU model), the National Center for Atmospheric Research (the NCAR model), and the United Kingdom Meteorological Office (the UKMO model).

soil moisture between 0 and -30 mm (Rosenberg, 1989). Seasonally, all three models predict that North America will have moister soil conditions between October and April; between April and September, however, the GDFL model predicts drying everywhere, the GISS model predicts late-summer drying, and the NCAR model predicts moister soil conditions (Rosenberg, 1989).

Their shortcomings aside, it is still useful to consider the predictions of GCMs in developing plausible future climate scenarios. For one thing, there are few alternatives for modeling the long-run effects of today's greenhouse-gas emissions on future climate conditions. Also, the predictions of the different models do become more consistent over larger global areas. In the middle latitudes, for example (which include the contiguous United States), the different GCMs tend to agree that within 30–50 years, the effects of climate change will appear as increases in mean surface temperatures, precipitation levels, and frequencies of droughts and severe storms (Hansen, 1989). Finally, all the GCMs agree that the regional impacts of climate change will differ.

The predictions of warmer temperatures and higher precipitation make sense intuitively. If the levels of heat-trapping gases in the atmosphere increase, then it might, on average, get hotter. If it gets hotter, then evaporation rates will increase and there will be more precipitation. The predictions of more-frequent droughts and severe storms are less clear. Both are consistent with the simulation results of several GCMs (specifically, the GISS, GDFL, and NCAR models; *see* footnote 5), but the empirical support is weak, being limited to the historical correlation between severe droughts and abnormally hot years (Hansen, 1989). Still, the possibility of more-variable weather should be of concern to the farm sector. The timing of heat waves, droughts, downpours, and killing frosts can greatly reduce agricultural yields, even in years with normal weather averages.

The lack of agreement among atmospheric scientists about the likely physical effects of today's emissions of greenhouse gases is only one source of uncertainty concerning climate change for agriculture. Even assuming that the more-common GCM predictions are basically correct, neither the direction nor magnitude of the impact on U.S. agriculture is obvious. Viewed in isolation, several of the forecasted effects of climatic change would have both positive and negative effects on farm production. Higher levels of atmospheric CO_2, for example, would promote photosynthesis and increase plants' water-use efficiency. Many crops, then, could create biomass more quickly, and the need for irrigation water might be reduced. On the other hand, the increased availability of CO_2 could also promote weed growth. Increased precipitation would be a benefit if the extra precipitation is delivered evenly throughout the growing season; this would reduce irrigation requirements and might expand dryland farming opportunities. If, however, the additional rainfall comes in the form of more severe spring storms, it could kill seedlings and/or delay planting dates.

Similarly, the impact of higher temperatures could be positive or negative. In the northern half of the United States, longer growing seasons would increase production possibilities and, in some areas, could allow multiple plantings. In the southern United States, the impact would probably be negative because there would be more days with temperatures above critical plant-tolerance levels. Hotter summers would also increase irrigation requirements and bid up the opportunity cost of irrigation water (i.e., the water's value in its next-best alternative use).

Viewing the predicted impacts of climate change individually, however, is very misleading. Ecosystems are dynamic, and the individual elements of any new climate would interact with each other as well as with a host of other environmental processes. Again, there are many uncertainties. Water, for example, increases the heat tolerance of many crops. Hence, in areas of the South that have adequate water supplies, the negative effects of hotter temperatures could be largely offset. Higher levels of atmospheric CO_2 in combination with more precipitation during the growing season could mean an increase in the frequency of bumper crops but also more problems with weed growth. Higher plant growth rates would also increase the demands for soil nutrients. Commercial farming, then, might become more dependent on chemical herbicides and fertilizers. The use of pesticides would also likely increase because hotter temperatures would enable many agricultural pests (insect and other) to expand their current ranges. The U.S. Environmental Protection Agency (Smith and Tirpak, 1989) has identified the potato leafhopper and the corn earworm as two major pests whose ranges would likely spread northward given a warmer climate; it also predicts a northward expansion of Rift-Valley fever and African swine fever among livestock.

Finally, there are uncertainties concerning the timing and speed of climate change. Some scientists, and many outside the scientific community, believe that the early signs of climate change are observable now. Six of the seven warmest years on record have occurred since 1981, and the drought of 1988 caused the worst agricultural losses in decades. Most published estimates, however, suggest that the impacts of climate change on agriculture will not be evident for 20–50 years. With respect to speed, the usual assumption is that the impacts will show up gradually. A distinct possibility is that the environment has some tolerance level for greenhouse gases, and once this threshold is exceeded, the effects will appear relatively quickly.

Although there is a consensus that the levels of greenhouse gases in the atmosphere are increasing, there is much less agreement regarding the impact of those increasing levels on future climate conditions. Three alternative scenarios serve as benchmarks: (1) no change with respect to present conditions, (2) moderate increases in mean temperature and precipitation, and (3) moderate increases in both the means and variances of temperature and precipitation. If either the second or third scenario occurs, it is still uncertain what the impacts on agriculture would be. What now seems most likely is that the impacts would differ from region to region; in some areas, farming conditions would improve, and in others, they would deteriorate.

We believe, then, that it is best not to focus on one future-climate scenario or one set of hypothetical effects in considering how farm programs might affect the social costs of climate change. Rather, we focus on how the various components of farm programs broadly influence market conditions. This allows us to say something about how different components might aggravate or mitigate the social costs of farm-sector adaptations to assumed changes in the environment. Before discussing specific components, we first give a quick overview of current farm programs.

U.S. Farm Programs—An Overview

For decades, U.S. agriculture and U.S. farm programs have evolved in close association with one another. Although any starting date on this relationship could be debated, a plausible candidate is the period 1933–1938. Culminating with the Agricultural Adjustment Act of 1938, the legislation enacted during these years greatly expanded the government's ability to influence agricultural markets and set the foundation on which all subsequent farm policies have been based.[6] The Agricultural Adjustment Act of 1938 was the first piece of comprehensive price-support legislation. Its features included nonrecourse loans, land allotments, and marketing orders; it also contained the concepts of "basic" crops, price parity, and income parity ("basic" crops are wheat, corn, rice, sugar, peanuts, and tobacco). It had as an objective the provision of an abundant and stable supply of key agricultural commodities.

---

[6] The Agricultural Adjustment Act of 1933 authorized the U.S. Department of Agriculture (USDA) to enter into voluntary agreements with producers to reduce cropland or production, and, with processors, to control prices. It also authorized government expenditures to expand markets and remove surpluses. In 1933, the Farm Credit Administration was established, and the government became active in distributing surplus food. In 1935, the Resettlement Administration (now the Farmers Home Administration), the Rural Electrification Administration, and the Soil Conservation Service were established, and the U.S. president was given the authority to impose import quotas when imports were negatively affecting adjustment programs.

Today, the primary goals of U.S. farm policy continue to be to maintain farm incomes and to provide for a stable supply of agricultural products. Other important goals are promoting U.S. farm exports and reducing the negative impacts of agriculture on the environment. Farm policy is implemented through myriad government programs, most of which emphasize price- and/or income-supporting interventions in the markets for specific commodities. By law, USDA must operate support programs for dairy products, wheat, corn, sorghum, oats, barley, rye, cotton, rice, peanuts, tobacco, soybeans, sugar, wool, mohair, and honey. Table 8.1 details the required (R) and optional (O) policy tools available to the secretary of agriculture for selected commodity programs. The different programs allow for varying degrees of market intervention. Typically, upper and lower limits on the use of the required policy tools are specified, but within these guidelines, the degree of intervention is largely up to the secretary. Also evident in Table 8.1 is the importance placed on dairy products and the "basic" crops.

In addition to supporting commodity prices and farm incomes, USDA also extends farmers lines of credit and insurance at below-market rates, provides numerous marketing services (e.g., inspection and grading, and commodity-promotion programs), and funds a national network of agricultural research and extension. The Department of the Interior (DOI) also provides many farmers in the western United States with heavily subsidized irrigation water and low-cost use of public lands for crops and livestock.

Farm programs are usually justified on the grounds that society receives large nonmarket benefits from a stable supply of high-quality agricultural products. For these benefits, society pays a substantial opportunity cost. In 1989, USDA outlays included ≈$14 billion for price- and income-support programs, ≈$3.4 billion for agricultural credit programs, ≈$1.73 billion for the Conservation Reserve Program, and ≈$1.35 billion for agricultural research and extension [U.S. Office of Management and Budget (OMB), 1990]. Table 8.2 gives a more disaggregated account of the support-program expenditures. The $6.6 billion in direct payments to producers of feed grains, wheat, rice, and dairy products and the $2 billion in commodity purchases highlight the emphasis on commodity-market intervention. Not apparent in Table 8.2 are the costs of the import quota on sugar and the irrigation subsidies for western farmers. In 1989, the world price of refined sugar averaged ≈37.82 cents/kg [17.15 cents/lb. (cpp)]; during the same

TABLE 8.1 Policy Tools of Selected Farm Programs[a]

Dairy: purchases and government-held stocks of processed-milk products (R);milk-marketing orders (R); export-incentive program (R); dairy-indemnity plan (insuring producers against losses due to contamination) (O); herd-reduction program (O); milk-diversion program (O)

Wheat: nonrecourse loans and government-held stocks (R); target prices and deficiency payments (R); national and farm-program land-area limits (R); marketing loans (O); disaster payments (O); land-area-reduction programs (O)

Feed grains (corn, barley, oats, grain sorghum, rye): nonrecourse loans and government-held stocks (R); target prices and deficiency payments (required for corn, not available for rye, optional for oats, barley, and sorghum); national and farm-program land-area limits (R); marketing loans (O); land-area-reduction programs (O); disaster payments (O)

Cotton (upland): nonrecourse loans and government-held stocks (R); target prices and deficiency payments (R); national and farm-program land-area limits (R); marketing loans (R); marketing certificates (R); import quotas (R); land-area-reduction programs (O); disaster payments (O)

Rice: nonrecourse loans and government-held stocks (R); target prices and deficiency payments (R); national and farm-program land-area limits (R); marketing loans (R); land-area-reduction program (O); disaster payments (O)

Soybeans: nonrecourse loans and government-held stocks (R); marketing loans (O); disaster payments (O)

Sugar (beet and cane): nonrecourse loans and government-held stocks (R); import quotas (R); protection of producers (R); disaster payments (O)

[a]R, required; O, optional.

Source: Glaser (1986).

TABLE 8.2 Funding of Selected Agricultural Price/Income Support Activities of
the Commodity Credit Corporation for 1989

Activity	Amount ($ thousands)
Commodity purchases and related inventory acquisitions	2,146,384
Storage, transportation, and other obligations	
not included above	988,831
Producer-storage payments	481,794
Direct-producer payments	
Feed grains	5,034,964
Wheat	626,725
Rice	482,136
Cotton	356,382
Dairy	168,240
Crop-disaster payments	3,385,946
Livestock assistance	532,579

Source: USDA (1990b).

period, U.S. wholesale and retail prices averaged, respectively, 64.78 and
88.27 cents/kg (29.38 and 40.03 cpp) (USDA, 1990a). As a result of the
quota, U.S. sugar prices are typically two to three times the world level. The
costs of the irrigation subsidies are obscured because they are, in large part,
foregone government revenues. DOI generally charges western farmers <4
cents/m^3 [$50/acre foot (af)] for federally supplied water; in many instances,
the prices are as low as 0.16–0.41 cents/m^3 ($2–5/af). Recent studies
suggest that much of this water is priced well below its market value
(Moore, 1991). This is particularly true near urban areas, where water rights
have recently traded at prices between 4.05 and 25.13 cents $\cdot$ m^{-3} $\cdot$ year^{-1}
(between $50 and $310 $\cdot$ af^{-1} $\cdot$ year^{-1}) (Moore, 1991).

Government farm programs play a major role in determining the behavior
of U.S. agricultural markets. By altering the expected returns to and/or risks
associated with specific production decisions, these programs distort the
messages of scarcity, abundance, and risk that free-market prices would
otherwise convey. Today's farm programs entail a large opportunity cost to
society. The magnitude of this cost, however, could change significantly if
the climate patterns that underlie the present map of U.S. (and world)
agriculture are altered. Depending on the accumulation of scientific
evidence, it could become increasingly necessary for policy makers to
account for the effects of climate change when developing and implementing

farm programs. We now consider how farm programs might affect the social costs of adapting to climate change.

Farm-Sector Adaptation to Climate Change

It is generally, although not unanimously, assumed that the farm sector could do much to adapt to all but the most pessimistic climate-change scenarios.[7] Losses that might now be incurred under warmer and/or drier growing conditions could be offset through the use of alternative crops, different cultivars, more efficient irrigation technology, and minimum tillage practices. Unfavorable late summer weather could be avoided by earlier plantings and harvests. Longer growing seasons might allow for multiple plantings. Increased threats from weeds, insects, and plant diseases and the demand for more soil nutrients could be addressed with heavier applications of agrichemicals. Farmers could also adapt to climate change by entry and exit, leaving the industry where conditions become unfavorable to agriculture and entering it where conditions improve.

Conceptually, farm programs could affect the economics of climate change either by encouraging or discouraging the types of adaptations described above. Most of today's farm programs have elements that would do both, so it is useful to focus on the individual tools of farm policy and how they affect market conditions. Table 8.3 groups the tools of farm policy according to whether their main effect is to reduce farmer risk, control market supply, reduce production costs, or protect the environment.

Given the more popular predictions concerning climate change, today's farm programs appear susceptible to large cost increases in several areas. First, several important programs transfer risk from the farm sector to society. Disaster payments, crop insurance, and the dairy-indemnity plan all reduce farmer exposure to production risks. Market risks associated with surplus production (i.e., low prices), imperfect information, and exogenous market shocks are transferred to society via target prices and deficiency payments, nonrecourse loans, government purchases of surplus production,

---

[7] The opposing view is summarized by Ward et al. (1989). These authors point to the generally poor soils in areas where agriculture is expected to expand; the reliance of many optimistic adaptation viewpoints on technical change (i.e., on discoveries that have not been made yet) and on the availability of irrigation water; and the huge outlays that would be required for water-efficient irrigation equipment and agricultural-infrastructure relocation.

TABLE 8.3. Grouping Farm-Policy Tools by Main Effect

Reduce risk to farmers	Control Supply
Price supports	Government stockpiles
Nonrecourse loans	Land-area restrictions
Purchases of surplus	Program land-area limits
production	Land-area-reduction
	programs
Income supports	Production/marketing quotas
Target prices	Import restrictions
Deficiency payments	Export programs
Disaster payments	
Crop insurance	Reduce production costs
Producer protection	Subsidized water sales
(cotton program)	Agricultural research and
Milk-indemnity program	extension
Subsidized credit (other	Marketing loans
than crop insurance)	

Reduce environmental damage
Conservation Reserve Program
Swampbuster provisions
Sodbuster provisions

and (for cotton) producer protection. Table 8.2 indicates how costly this risk transfer is to society. In 1989 USDA outlays for commodity purchases, direct payments to producers, and disaster payments amounted to >$12.7 billion.

If climate change manifests itself as more-variable weather patterns with higher likelihoods of both droughts and severe storms, then agriculture will become more risky. Relative to today, farmers would face an increased probability of incurring large crop losses from unfavorable weather. Additionally, there may be an increase in the frequency of bumper-crop years from favorable weather; this would increase the likelihood of years with low agricultural prices. Current farm programs would allow most of these added risks to be to be passed on to society. Disaster payments and crop insurance would protect farmers against large crop losses. Price and income supports would protect farmers against the negative economic impacts of large surpluses. By shielding farmers from the risks associated with climate change, society would remove the incentive for farmers to engage in other risk-reducing behavior. The cost of footing this bill could be quite large. In 1990, a good year for agriculture, crop disaster payments were estimated to be $6 million (USDA, 1990b). On the other hand, 1988 was a year of severe drought in much of the United States. Disaster payments related to the 1988 drought were ≈$3.4 billion in 1989 (USDA, 1990b).

A second area in which current farm policy is subject to large cost increases resulting from climate change arises from the structure of the commodity programs. These programs, which account for the bulk of USDA outlays, strongly discourage participants from altering their output mix. Switching crops is generally viewed as one of the most obvious and least costly adaptations farmers could make in response to changes in weather patterns. Some evidence suggests that climate change would favor regional shifts in the production of many program crops. Higher levels of atmospheric CO_2 would enable plants in the C3 plant group to increase biomass more than those in the C4 plant group (Hillel and Rosenzweig, 1989). On the other hand, C4 plants would increase their water-use efficiency more than would C3 plants (Hillel and Rosenzweig, 1989). Commercially important C3 crops include wheat, rice, soybeans, legumes, and root crops; important C4 crops include corn, sorghum, and sugarcane. In terms of comparative advantage, then, we might expect climate change to favor shifts to C4 crops where conditions get drier, and to C3 crops where water availability stays constant or increases. Additionally, historical evidence suggests that wheat is particularly susceptible to extreme heat (Rosenberg, 1989). Hence, we might also expect wheat production to decrease in much of the South.

While regional shifts in crop production could help reduce the social costs of climate change, current commodity programs would make these benefits difficult to realize. A farm's allowable land area for most program crops is based on the average area it has allocated to the crop over the previous few years. Hence, it takes time to build program land area in new crops; additionally, farmers are penalized for several subsequent years for any major reduction in one year's program land area. To the extent that farm programs discourage the production of the crops best suited to local conditions, the social costs of climate change will increase.

Another farm-policy tool that could significantly increase the social costs of climate change is the provision of subsidized water to western farmers. By supplying farmers with subsidized irrigation water, the government lowers the private costs of production. This allows marginal producers to enter the industry and competitive producers to earn economic profits. Government-subsidized irrigation water is particularly important because it is one area where today's actions could significantly affect the costs of adapting to climate change. Historically, the government has provided this water under long-term contracts (typically for ≈40 years). Many contracts are now expiring, and western farmers are pushing to have them renewed. At the same time, population growth in the West (particularly in southern California and the urban areas of Denver, Salt Lake City, and Sparks-Reno) has greatly increased nonagricultural demand for water. Future shifts in this demand will become more pronounced if, in addition to population growth, the West becomes hotter and/or drier. As the nonfarm demand for water increases, the opportunity cost of selling federally controlled water to farmers at prices fixed well below current market values will increase. The potential magnitude of these cost increases can be appreciated by considering that in some areas, nonfarm users will now pay 16–24 cents/m^3 ($200–300/af) for water the government sells to agriculture for <4 cents/m^3.

Finally, within the structure of current farm policy, the social costs of climate change could be aggravated by the more frequent use of import restrictions. As discussed earlier, U.S. sugar producers can compete in the domestic market only because the government limits the quantity of imported sugar. The social costs of this protection are largely incurred in the form of higher consumer prices. GCM simulations suggest that similar situations could arise in other agricultural markets. Specifically, GCM simulations predict that the effects of climate change may be less in Argentina, Brazil, and Australia; in Canada, the impacts may be substantial but favorable to agriculture. If production costs in the United States rise relative to these other countries, U.S. farmers may become less competitive in the production of other important commodities. As with sugar today, we may be technically able to meet domestic demand but not able to do so at a lower cost than foreign producers. There will, therefore, be an obvious temptation to protect domestic farmers with import quotas, which will lead to higher consumer costs.

Although many of today's farm-policy tools appear to leave society vulnerable to climate-change-driven cost increases, others could have a mitigating effect. If agricultural production becomes subject to more-frequent booms and busts, then, in the absence of more government intervention, agricultural prices will exhibit more year-to-year volatility. Storage programs could help reduce this instability by smoothing the market supply of agricultural products between poor and abundant harvests. There also seems to be the potential for storage programs to reduce farm-sector risk without large resource transfers from society. Farmers, for example, could be given program-production limits (e.g., a share of national production). Output above this level would be put in storage until a bust year, when it would be returned to farmers. If producers were not paid for surplus production, as they are now, then society would incur only transportation and storage costs. In 1989, USDA spent ≈$1.5 billion storing and transporting surplus production. Given the alternative of highly unstable agricultural markets, a two- to three-fold increase in these outlays might be socially justifiable.

Publicly financed research and extension could also help mitigate the costs of climate change. So far, we have carefully avoided including technological breakthroughs in the list of possible farm-sector adaptations. This is because of the uncertainty of results from experimental research. Still, the recent advances in biotechnology make it reasonable to assume that some breakthroughs are on the horizon that would help agriculture adapt to hotter and/or drier environments. Given the likelihood that the social returns from climate-change research will exceed the private returns, a case can be made for expanding this research with public funds.

We have considered how climate change might affect the social costs of maintaining the current set of farm programs. Two other types of costs—really opposite sides of the same coin—are also relevant to this discussion. First, given the occurrence of climate change, how could current farm programs be modified to achieve their goals at a lower cost? Second, how costly might climate-change-inspired modifications to farm policy be if the anticipated changes do not materialize?

Answering the second question first, the numerous and important uncertainties regarding the impacts of climate change on agriculture argue against undertaking expensive actions geared toward adaptation right now. Almost any farm-sector adaptation one can think of could be accomplished within 5 years; by most estimates, this is well before the impacts of climate change are predicted to become apparent. At the same time, it is useful to consider modifications to today's programs that would facilitate adjustments to climate change but that also would have more immediate justifications. In addition to the already-mentioned expansion of climate-change research, two other adjustments may be worth considering. First, implementing flexibility in the commodity programs would allow participants to choose from a range of crops without directly affecting the participants' level of support. Flexibility has several potential environmental benefits, including protecting ground-water supplies and promoting crop rotations (McCormick and Algozin, 1989). Should climate change become a reality, farmers would have more incentive to shift to crops that are better suited to their new environment.

A second adjustment to farm policy that may be worth considering is reducing the costs to farmers of acquiring more water-efficient irrigation equipment. This equipment is expensive; a center-pivot sprinkler system that can irrigate 1.295 km^2 (320 acres) costs ≈$90,000 (Ward et al., 1989). Farmers who currently have access to adequate water supplies are unlikely to undertake this investment themselves. Where these supplies are publicly subsidized or where water withdrawals now exceed natural replacement, there may be social benefits to reducing irrigation-water use in excess of the private benefits. This would justify assistance to farmers today and would promote adaptation to climate change should it occur.

Conclusions

The magnitude of the costs of climate change may well depend on how fast farmers adapt to new environmental conditions. Farm programs could play a role in determining how quickly this adaptation process occurs. Because farm programs are resource-intensive and because they have proven to be very durable, now may be the time to start considering how these programs might affect the costs of adapting to climate change.

On balance, today's farm programs seem susceptible to large cost increases should climate change occur. With the exception of not renewing long-term water contracts in the West, however, it seems too early to modify farm policy to specifically account for climate change; we do not yet know what types of modifications would be desirable. Some changes to current farm programs that would facilitate climate-change adaptations might warrant consideration for other reasons. These changes include implementing flexibility in the commodity programs, assisting farmers in some water-scarce regions to acquire water-efficient irrigation equipment, and expanding research that will help agriculture adapt to climate change.

References

Adams, R. M., B. A. McCarl, D. J. Dudek, and J. D. Glyer. 1988. "Implications of Global Climate Change for Western Agriculture." *Western Journal of Agricultural Economics* 13(2): 348–56.

Arthur, L. M., and F. Abizadeh. 1988. "Potential Effects of Climate Change on Agriculture in the Prairie Region of Canada." *Western Journal of Agricultural Economics* 13: 216–24.

Dudek, D. J. 1989. "Assessing the Implications of Changes in Carbon Dioxide Concentrations and Climate for Agriculture in the United States," in Congressional Research Service, Library of Congress, eds., *Agriculture, Forestry, and Global Climate Change—A Reader*. Prepared for the Senate Committee on Agriculture, Nutrition, and Forestry. Pp. 205–36. Washington, D.C.: U.S. Government Printing Office.

Glaser, L. K. 1986. "Provisions of the Food Security Act of 1985." Washington, D.C.: National Economic Division, Economic Research Service, U.S. Department of Agriculture. (Agricultural Information Bulletin 498.)

Hansen, J. E., I. Fung, D. Rind, S. Lebedeff, R. Ruedy, and G. Russell. 1988. "Global Climate Changes as Forecast by Goddard Institute for Space Studies Three-Dimensional Model," *Journal of Geophysical Research* 93(D8): 9341–64.

Hansen, J. E. 1989. "Modeling Greenhouse Climate Effects," statement to the United States Senate Committee on Commerce, Science, and Transportation, Subcommittee on Science, Technology, and Space, May 8, provided by NASA Goddard Institute for Space Studies, New York.

Hillel, D., and C. Rosenzweig. 1989. "The Greenhouse Effect and Its Implications Regarding Global Agriculture." Amherst, Mass.: Massachusetts Agricultural Experiment Station. (Research Bulletin 724.)

McCormick, I., and K. A. Algozin. 1989. "Planting Flexibility: Implications for Groundwater Protection." *Journal of Soil and Water Conservation* 44(5): 379–83.

Moore, M. R. 1991. "The Bureau of Reclamation's New Mandate for Irrigation Water Conservation: Purposes and Policy Alternatives." *Water Resources Research* 27(2): 145–55.

Rosenberg, N. J. 1989. "Global Climate Change Holds Problems and Uncertainties for Agriculture," in Congressional Research Service, Library of Congress, eds., *Agriculture, Forestry, and Global Climate Change—A Reader*. Prepared for the Senate Committee on Agriculture, Nutrition, and Forestry. Pp. 180–95. Washington, D.C.: U.S. Government Printing Office.

Rosenzweig, C. 1985. "Potential CO_2-Induced Climatic Change Effects on North American Wheat Producing Regions." *Climatic Change* 7: 367–89.

Smit, B., L. Ludlow, and M. Brklacich. 1988. "Implications of a Global Climatic Warming for Agriculture: A Review and Appraisal." *Journal of Environmental Quality* 17(4): 519–27.

Smith, J. B., and D. Tirpak, eds. 1989. "The Potential Effects of Global Climate Change on the United States, Report to Congress." Washington, D.C.: Environmental Protection Agency. (EPA-230-05-89-050.)

U.S. Department of Agriculture. 1990a. *Sugar and Sweeteners Situation and Outlook Yearbook*, Washington, D.C.: Commodity Economic Division, Economic Research Service, U.S. Department of Agriculture.

U.S. Department of Agriculture. 1990b. *Budget Estimates for the United States Department of Agriculture for the Fiscal Year Ending September 30, 1991.* Washington, D.C.: U.S. Government Printing Office.

U.S. Office of Management and Budget. 1990. *Budget of the United States Government*, FY 1991 issue, Washington, D.C.: U.S. Government Printing Office.

Ward, J. R., R. A. Hardt, and T. E. Kuhule. 1989. "Farming in the Greenhouse: What Global Warming Means for American Agriculture." Washington, D.C.: National Resources Defense Council, 33 pages.

Wilks, D. S. 1988. "Estimating the Consequences of CO_2-Induced Climatic Change on North American Grain Agriculture Using General Circulation Model Information." *Climate Change* 13: 19–42.

9

Modeling Western Irrigated Agriculture and Water Policy: Climate-Change Considerations

Noel R. Gollehon, Michael R. Moore, Marcel Aillery, Mark Kramer, and Glenn Schaible[1]

Introduction

In the arid and semiarid American West, where much of crop agriculture depends on irrigation, consequences of a changing climate will likely be significant. A significant change in climate threatens the physical, economic, and political integrity of western water-supply and use systems.

Irrigated agriculture is the greatest consumptive user of western water. Climate change will affect irrigated agriculture both directly and indirectly. Direct effects will include changing yields and changing irrigation-water requirements and supplies, and reducing or enhancing pressures on irrigation development. Indirect change will include changes in crop prices from climate-induced shifts in production patterns and in nonagricultural water demands.

This paper has three goals. The first is to establish a qualitative linkage between western water policy and climate change. The first section describes the effects of climate change on water availabilities. The second section then characterizes the dimensions of policy-imposed water conservation in the West and the influence of climate change, both as an initiator of policy change and as a consideration in the policy process. The second goal is to

---

[1] From the Water Branch, Resources and Technology Division, Economic Research Service, U.S. Department of Agriculture. The views expressed are the authors' and do not represent the policies or views of the U.S. Department of Agriculture. The authors thank Richard M. Adams, John M. Reilly, and John Hostetler for reviewing previous drafts of this chapter and Olivia Wright for manuscript preparation.

present an aggregate economic model, currently under development, to be used for analysis of irrigation water policy and conservation. The third section outlines the Western Agricultural Water Analysis (WAWA) Model, including its scope, components and structure, data, and analytical capabilities. The third goal, achieved in the section after the WAWA Model description, is to describe in more detail how the direct effects of climate change can be introduced and analyzed using the WAWA modeling framework, emphasizing the role of climate and weather variables in irrigated-crop-production functions. The last section summarizes the paper and discusses implications for future research.

Climate Change and Irrigated Agriculture

Basic ideas of water supply and demand help illustrate the direct influences of climate change on irrigated agriculture. In terms of water supply, climate change—in the form of an increase in mean global air temperature—would increase spring runoff volumes, yet decrease summer runoff volumes (Gleick, 1988; Lettenmaier et al., 1989; Rango and van Katwijk, 1990; Schaake, 1990). The combined effect would be negative in terms of effective agricultural water supply, since the physical infrastructure is designed to capture spring snowmelt to augment low summer flows. A warmer climate thus would shift the supply function of irrigation water, given current reservoir capacities (Figure 9.1).

Global warming would also directly alter irrigation-water demand by changing water requirements, crop yields, and development pressures (Figure 9.1). Greater irrigation-water applications per unit land might be necessary to meet crop transpiration and cooling requirements to maintain a constant yield in a warmer climate. Then, if a change in atmospheric CO_2 concentration alters plant growth and crop yields, plant water needs may shift again. [Rosenzweig (1989) and Peart et al. (1989) have found that the needs for irrigation water increase to maintain yields in warmer climates that have doubled CO_2 concentrations.] A warmer climate would also lengthen the growing season, increasing the water requirements per unit land for perennials. In addition, altering precipitation timing and amount will change the demand for irrigation water to supplement precipitation, thus increasing pressures on irrigation development (Peterson and Keller, 1990). Over time, however, demand shifts may be dampened as more heat- and drought-resistant crops are introduced.

The stylized example in Figure 9.1 shows an increasing value of water and equilibrium water use that could increase or decrease depending on the relative shifts in D^1 and S^1. The figure does not reflect indirect effects. Commodity price changes would increase (or decrease) returns per unit land

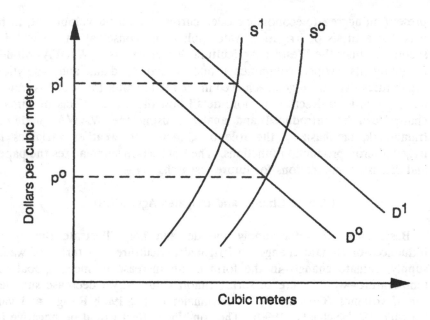

FIGURE 9.1. Supply of and demand for irrigation water. D^0 to D^1, climate-induced shift in demand; S^0 to S^1, climate-induced shift in supply.

and encourage (or discourage) irrigation expansion. Higher temperatures might also increase urban water demand. Changes in the timing and levels of stream flow would intensify pressure for instream flow to protect biological systems.

The effect of global change on the demand for irrigated area and irrigation water was evaluated for California (Dudek, 1989), the western United States (Adams et al., 1988), and the nation (Adams et al., 1990). Common general findings were that incentives to expand (or maintain) irrigated area increase and that crop-water needs along with irrigation demands per unit land increase in response to rising temperature.

A common thread in previous research on the effects of climate change on agriculture is the need to evaluate the effects of climate change on the water-allocation system (Rosenzweig, 1989; Peart et al., 1989; Adams et al., 1990; Brown, 1988; Frederick and Kneese, 1990; Dudek, 1989). Climate change could shatter the basic premise of a water-distribution system based on sufficient water being available to satisfy most water claimants in a "normal" year. By reducing the water quantities available, a changing climate will increase the pressure to reallocate western water resources and to reform the prior appropriation doctrine (Dudek, 1989). [Dudek (1989) examined one reallocation alternative (water markets) in one location

(California). His findings indicate that the "flexibility provided by market incentives ... enables efficient reallocation of resources" with an alternative climate specification. General conclusions for the West, however, must await more evidence.]

Water Supplies for Western Agriculture

Current Water-Policy Environment

In the arid and semiarid western United States, large-scale development of surface-water resources for irrigation and other uses was promoted to spur economic growth and settlement of the West. Today, agriculture dominates water-resource use, irrigating ≈ 153.8 billion m^2 (≈ 38 million acres), which accounts for 80–90% of total ground- and surface-water consumption in the 17 western contiguous United States. Further, the dominant institutions of western water allocation—the prior-appropriation doctrine and the U.S. Bureau of Reclamation (USBR, the federal water-development agency in the West)—are geared toward irrigation-water development and security of agricultural water supplies.

The western states administer surface-water allocation with an annual quantity-based permit system allowing water-right holders' use of a specified water quantity. Most states administer ground water in a similar fashion. However, because annual surface-water flows are subject to stochastic variation, holders who have acquired their water rights recently may receive no water in some years. Thus, the precise quantity received depends on water flow and reservoir-storage levels. Irrigators generally know their annual endowment of both surface and ground water before the crop-production season.

Whereas irrigation development was a defining feature of western water policy between 1900 and 1980, surface-water reallocation replaced development as a defining feature during the past decade (Weatherford, 1982). Physical limits to additional water supply, combined with increasing nonagricultural demands, create increasing economic incentives and political pressures for water reallocation, as indicated by recent changes and innovation in institutions. With continuing growth of other economic sectors, the relative economic prominence of irrigated agriculture is actually declining in the West.

Surface-water reallocation implies irrigation-water conservation. This is a truism, since agriculture currently consumes the bulk of water resources and, in some cases, irrigates crops with low financial returns. A more critical dimension of the reallocation process involves choices among alternative mechanisms of reallocation. Alternatives include voluntary market transfers of water rights, quantity-based regulation that reduces the water volumes of existing rights, price-based (tax-based) regulation to increase water prices and reduce use, and water-conservation subsidies or penalties that reduce use. Voluntary transfers are negotiated between parties, whereas regulatory-imposed conservation increases public policy involvement in the reallocation process. The irrigated agricultural sector is also concerned with production and distributional effects associated with alternative conservation mechanisms.

At the state level, a basis for quantity-based regulation exists in the beneficial-use provision, a central tenet of the prior-appropriation doctrine. The provision requires that water uses not be wasteful, e.g., through excessive application of water to crops (Sax and Abrams, 1986). Some researchers (e.g., Shupe, 1982) have recommended that the provision be used as a coercive policy mechanism to implement irrigation-water conservation. The public-trust doctrine provides a second rationale for states to adopt quantity-based regulation. The doctrine has been applied in California in an attempt to reduce existing appropriative rights in order to acquire water for ecosystem protection (Sax and Abrams, 1986). No state has developed a water-conservation tax or subsidy policy.

Another trait of western water allocation is the paucity of unregulated water markets for guiding allocation decisions. Although local water markets function adequately in a few areas, interstate and regional water markets have not developed (Howe et al., 1986). Trade in water rights generally does not occur across state lines nor across service-area boundaries of irrigation-water districts (especially those districts served by the USBR).

At the federal level, federal reclamation law and policy offers another institutional setting for water policy.[2] USBR provides ≈30.8 billion m^3 (≈25 million acre-feet) of surface water to roughly 40.5 billion m^2 of cropland (10 million acres) in the West (or, roughly one-fourth of the western irrigated area). In the past 2 years, Congress and other federal agencies proposed water-price increases or quantity restrictions on USBR-supplied water (Moore, 1991). In 1988, the Department of the Interior developed a policy statement on water transfers to encourage market activity.

---

[2] The federal government has several needs for surface water, including reserved water rights to enable federal reservations to meet their purposes (e.g., national forests and parks and Indian reservations); water supplies for endangered species; and unpolluted water to meet water-quality standards. These needs heighten interest in analysis of irrigation-water conservation.

Western Water Policy in a Setting of Climate Change

Evaluation of alternative institutional mechanisms of water reallocation generally assumes that initial water endowments are well defined. The prospect of climate change calls into question this basic assumption. Recognition of the potential effects of climate change seems essential for analysis of western water policy.

In fact, policy makers have begun to consider the effects of climate change on other decisions. Currently, various parties are debating contract-renewal terms for water deliveries from USBR's Central Valley Project in California. In light of the possibility of global warming, the U.S. Environmental Protection Agency and the Council on Environmental Quality have recommended analysis of policies that would either decrease irrigation-water-supply volumes or shorten the length of the contract period to less than the standard 40-year reclamation contract (Moore, 1991). This example demonstrates an important point: climate change may influence policy decisions affecting agriculture and water resources without being the central issue in its own right (Ingram et al., 1990).

Research that directly links resource-policy analysis and climatic conditions can be an improvement over independent analyses. The ability to examine water policies over a range of alternative climate scenarios may provide critical information about the stability, or robustness, of alternative water policies. For example, quantity-based regulation and transferable water permits may have similar properties under current climate conditions. With global warming, however, transferable permits may offer irrigators some flexibility in adapting to climate change, relative to rigid, quantity-based regulation. In this case, transferable permits would be the more stable, or robust, policy alternative.

The Western Agricultural Water Analysis Model

The WAWA Model is an economic model for analyzing the potential effects of alternative water policies on irrigated agriculture in the western United States. The model is designed to use consistent data across a wide geographic area to provide both economic efficiency and distributional consequences for the West. This model is still being developed; sensitivity analysis using an initial operational structure is currently being conducted.

In the WAWA Model, the 17 western states can be divided into 61 model subareas (MSAs), the boundaries of which are determined by the interaction of hydrologic and political units (Figure 9.2). Hydrologic regions are based on major watershed divisions to reflect the physical flow of surface water. State boundaries introduce political factors, such as state laws and interstate

FIGURE 9.2. Modeling subareas in the 17 western states.

compacts, that allocate flows of the major river systems. State boundaries are as critical to the movement of water as are hydrologic regions.

The model uses an aggregate modeling approach, which implies homogeneity (e.g., of climate and cultural practices) within the boundaries of a particular MSA (Figure 9.3). The model focuses on irrigated field crops, allowing endogenous determination of crop-output and land- and water-input use in the irrigated production of crops within each MSA.

To facilitate analysis of policies affecting federally developed water supplies, lands served by USBR are treated separately from other irrigated lands. Within applicable MSAs, therefore, two water-supplier categories are included for irrigated land: USBR-supplied and privately (non-USBR) supplied. Nonirrigated production is also included where appropriate.

Structure of the WAWA Model

Objective Function. The WAWA Model is a static, deterministic, partial-equilibrium model that maximizes net returns (quasi-rents) to irrigated and dryland production of agronomic crops. In particular, the model maximizes crop net returns to land, management, existing irrigation infrastructure, and farm overhead. Net returns are calculated as gross revenue per unit land (crop yield times price), minus production costs (including water, preharvest, and harvest costs). Crop alternatives include alfalfa hay, grain corn, cotton, dry beans, hay (other than alfalfa), potatoes, rice, silage, other small grains

Data

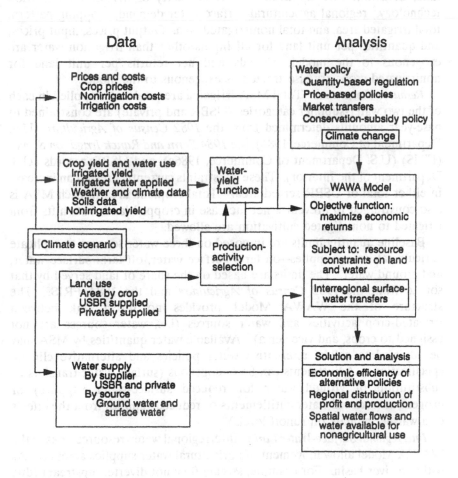

Prices and costs
  Crop prices
  Nonirrigation costs
  Irrigation costs

Crop yield and water use
  Irrigated yield
  Irrigated water applied
  Weather and climate data
  Soils data
  Nonirrigated yield

Climate scenario

Land use
  Area by crop
  USBR supplied
  Privately supplied

Water supply
  By supplier
    USBR and private
  By source
    Ground water and
    surface water

Water-yield functions

Production-activity selection

Analysis

Water policy
  Quantity-based regulation
  Price-based policies
  Market transfers
  Conservation-subsidy policy
  Climate change

WAWA Model

Objective function:
  maximize economic
  returns

Subject to: resource
  constraints on land
  and water

Interregional surface-
  water transfers

Solution and analysis

Economic efficiency of
  alternative policies
Regional distribution of
  profit and production
Spatial water flows and
  water available for
  nonagricultural use

FIGURE 9.3. Schematic of the WAWA Model.

(barley, oats, and rye), sorghum, soybeans, sugar beets, and wheat. The objective function contains a reduced form of econometrically estimated, crop-specific quadratic production functions (a water-yield relationship) for each of these crops.

The WAWA Model allows endogenous determination of yield per unit land, irrigation-water-application rate, efficiency of irrigation-application technology,[3] regional agricultural-surface-water demand, cropping pattern, total irrigated area, and total nonirrigated area. Output prices, input prices, and quantities per unit land for all inputs other than irrigation water are exogenous to the model. Yields and net returns per unit land for nonirrigated crops are also treated as exogenous to the model.

Resource Constraints. Total MSA irrigated area and water applied in each of the two water-supplier categories (USBR and private) are constrained to base-year amounts determined from the *1982 Census of Agriculture* (U.S. Department of Commerce, 1984), the *1984 Farm and Ranch Irrigation Survey* (FRIS) (U.S. Department of Commerce, 1986a), and USBR records (U.S. Department of the Interior). These constraints currently prevent an increase in either total or USBR-served area.[4] Total cropland area in each MSA is also constrained to prevent a net increase in cropped land, but shifts from irrigated to nonirrigated cultivation are allowed.

Baseline quantity limits are imposed on three water sources to replicate agricultural water supplies—on-farm surface water, off-farm surface water, and ground water. The limits are based on the share of land served by that source from the *1982 Census of Agriculture* and the 1984 FRIS. The structure of the WAWA Model provides no direct link between irrigated-crop activities and water sources (i.e., water sources are not assigned to crops, and vice versa). Available water quantities by MSA may be adjusted to reflect alternative water policies and alternative climate specifications. Adjustments can be endogenous (surface-water transfers or substitution of ground water for reduced surface-water supplies) or exogenous (reduced water entitlements or reduced supplies from the effects of a warmer climate on runoff levels).

Interregional Surface-Water Links. Interregional water-resource links in the WAWA Model allow movement of agricultural water supplies across MSAs within a river basin. For example, stream flow not diverted upstream (due to increased environmental flows, water-conservation practices, water-price

---

[3] This aspect of the model is not yet functional, but its incorporation into future versions of the model is planned.

[4] In future analyses, allowance for irrigated-land expansion is anticipated. This will require additional data on the availability of irrigable land and the costs of the irrigation infrastructure.

increases, or water-market developments) is available for downstream use, and vice versa. Interregional links allow the model to respond to regional differences in crop returns by allocating surface-water supplies endogenously across MSAs. Finally, these links are necessary when changing water supplies associated with altered climatic conditions are being considered.

The WAWA Model traces only agricultural water movements to link surface-water supplies across MSAs.[5] This simple specification does not require an accounting of nonmodeled crop-water demands, nonagricultural water demands, or reservoir-storage changes. In this formulation, the baseline regional agricultural-surface-water supply is fixed at the base-year quantity. The actual surface water used in each MSA can be less or greater than that used in the baseline, depending on the policy formulation.[6] Movement of agricultural-surface-water supplies is allowed within river basins, making the MSA agricultural water use endogenous within the basin. In its most simplified form, links could exclude diversion-conveyance losses, transfer losses, and return flows.

Exogenous data required for the simplest specification are estimates of surface-water use for modeled crop production by MSA from the 1984 FRIS, the U.S. Geological Survey (Solley et al., 1988), and USBR. Future research will modify the interregional water-transfer equation system to incorporate coefficients for conveyance and transfer losses and for return flows.

Model Data

Activity Selection and Output. Model production alternatives represent the major field crops, based on the harvested area reported in the *1982 Census of Agriculture.* The combined area in the 13 included crops, plus pasture,

---

[5] Two basic approaches are available for linking surface-water supply. The approach can concentrate solely on agricultural water use for tracing potential water movements, thereby assuming that increases in the quantity of water for other uses will come from agriculture. Alternatively, the approach can concentrate on stream flow, thereby accounting for all uses and sources, including releases from storage facilities and snowpack runoff. A focus on agricultural water supplies offers significantly reduced data cost relative to a stream-flow–balance equation system that requires the timing and quantities of *all* water sources and uses.

[6] The system of equations representing surface-water links imposes an interpretive limitation on some policy results. First, it is not possible to account for the effect on users that do not directly divert, e.g., the agricultural production based on return flows. However, results should not be affected for reallocation policies based on a consumptive-use concept. Second, excluding diversion-conveyance losses will not allow analysis of policy scenarios that involve increasing the efficiency of off-farm water-delivery systems.

accounts for 141.6 billion m² (35 million acres) of irrigated land, or 86% of irrigated area in the western United States relative to the *1982 Census of Agriculture* totals. In addition, modeled crops account for 526.1 billion m² (130 million acres) of harvested nonirrigated land, or 72% of total dryland area (including dryland summer fallow). Crop categories not explicitly defined within the WAWA Model include (1) long-term perennial crops (e.g., orchards and vineyards) for which water use is assumed to be fixed over the life of the established stand and (2) high-valued specialty crops (e.g., vegetables and nursery stock) for which the average value-product of water is high.

Base-year crop-output prices in the WAWA Model reflect expected market prices at harvest, plus a weighted-average farm-program-payment adjustment.

Production Costs. Production-cost data are assembled by crop, separately for dryland and irrigated area. State-level cost data for major farm-program crops and hay crops are primarily from 1984 aggregate-enterprise crop-cost estimates from the U.S. Department of Agriculture's Economic Research Service (USDA-ERS) (Schaible et al., 1989). These cost estimates are based on a West-wide data series to ensure regional consistency within and across crop activities.

Water-use data, irrigation technology, and water-supply costs are based primarily on data from the 1984 FRIS and USBR-project records. On-farm surface-water and ground-water irrigation costs include pumping, purchase, and distribution costs. Purchased-water costs are based on USBR-project records for 1984 and FRIS-reported costs for private water suppliers. Water costs per unit land are calculated, by crop, from water costs and endogenous-water application rates.

Using the WAWA Model to Analyze Water Policies

Several types of policies intended to encourage water conservation can readily be analyzed with the WAWA Model (Figure 9.3). Quantity-based policies constitute a reduction in the water supply available to the agricultural sector. This policy option can be modeled by adjusting downward the exogenous upper bounds on the regional water availability. Federal price-based policies can be modeled by adjusting USBR prices upward. The effect of water markets could be simulated by changing the price for which unused water can be sold, thereby increasing the opportunity cost of water. The water could be "sold" to agricultural users in other MSAs or to users outside the agricultural sector. Subsidization of water-conserving technology could be included through adjustment of technology-adoption costs and use of advanced technologies.

Climatic Adjustments in the WAWA Model

Adjustments in Irrigation-Water Supply

Climatic effects on surface-water supply can be made through adjustments within the interregional surface-water-transfer system. Within this equation system, surface-irrigation-water availability is based on existing agricultural-withdrawal levels. Therefore, changes in stream flow caused by climatic alterations must be translated into changes in surface-water quantities available for agricultural withdrawals. Climatic adjustment to regional agricultural water supplies can be made in a general way by simulating a percentage reduction in MSA baseline-surface-water supplies. For specific climatic changes, hydrologic stream-flow models (e.g., Gleick, 1988; Schaake, 1990; Lettenmaier et al., 1989) may be used to make MSA-level estimates of water-supply changes. (Because most ground water used for irrigation in the West comes from deep aquifers, it will be unnecessary to adjust ground-water-quantity constraints as a result of climate changes).

Adjustments in Irrigation-Water Demand

The most commonly considered adjustment in irrigation-water use from a changing climate is the adjustment in irrigation-water applications to produce a constant crop yield (Rosenzweig, 1989; Peart et al., 1989). This adjustment, however, is only one of several input adjustments that could accompany a changing climate. For example, irrigation levels could remain the same, allowing crop yields to decline. More efficient irrigation technologies could be adopted. Entire cultivation systems may change as weather patterns transform irrigated areas into nonirrigated cultivation,[7] or vice versa. The WAWA Model currently captures these input-substitution possibilities, with the exception of changing technologies, which is planned but not yet operational.

Output substitutions, by changing MSA cropping patterns, offer another response to climate change. The ability to grow less-water-intensive crops currently exists in the multicrop modeling environment. In addition, a warming climate will change the traditional crop growth zones and will necessitate the introduction of additional crops. With realistic assumptions about costs, alternative crops can be introduced at the MSA level to reflect

---

[7] It will be necessary to adjust the exogenous dryland yields in the WAWA Model to more accurately reflect the expected return to nonirrigated cropping alternatives, because the change in yields will affect the margin between irrigated and nonirrigated production in a given region.

climate-induced differences in regional crop choices. By allowing output and input substitution, a more realistic model can be developed of agricultural adaptation to climate change.

Yield-Function Estimation. The water-yield production functions within the WAWA Model enable it to select endogenously an irrigation-water-application rate per unit land and associated yield that maximize economic returns. These functions also serve as a conduit for partially adjusting irrigation-water demands from climate change.[8]

The yield functions are estimated econometrically with data from individual irrigator responses to the 1984 FRIS using western irrigated farms, supplemented with secondary data for soils, weather, and climate. Crop-specific functions are estimated by ordinary least squares for the 13 primary irrigated crops by using the following quadratic form (Moore et al., 1990):

$$q = bw + cw^2 + \sum_{i=1}^{9} d_i x_i + \sum_{i=1}^{9} e_i x_i^2 + \sum_{i=1}^{n} f_i z_i + \sum_{i=1}^{5} g_i v_i$$
$$+ \sum_{i=1}^{5} h_i (v_i w) + \sum_{i=1}^{5} k_i (v_i w^2) + \varepsilon$$

where q is output per unit land; w is centimeters of irrigation water applied per unit land; x_1–x_5 are centimeters of precipitation per unit land for the periods January–February, March–April, May–June, July–August, and September–October, respectively; x_6–x_9 are base 18 °C (65 °F) cooling degree days (CDDs) per unit land (degree-days) for the periods March–April, May–June, July–August, and September–October, respectively; z_i ($i=1,...,n$) are dummy variables for climate, farm structure, and soil quality; v_i is a dummy variable for water-management options; b, c, d_i, e_i, f_i, g_i, h_i, and k_i are coefficients for estimation; and ε is an error term. The functions are estimated without an intercept, constraining functions for irrigation activities to pass through the origin, thereby preventing irrigation activities that apply no irrigation water.

---

[8] The functions used here are estimates based on cross-sectional data. As such, estimates of crop production and irrigation-water use in a "warmer" climate reflect actual technology and physical conditions where that crop is currently grown. This procedure fails to consider the effects of enhanced CO_2 on crop yields, only because no data exist on cross-sectional CO_2 concentrations for use in this procedure. Studies incorporating enhanced CO_2 concentrations use plant-growth simulation models (Peart et al., 1989; Rozenzweig, 1989; Rosenberg et al., 1990). The growth-simulation approach has disadvantages in spatial coverage (generally based on experiment-station plots) and pragmatic adjustments (technology and physical conditions are held constant).

Weather variables, representing the precipitation and CDD conditions that producers faced in 1984, are modeled as nonpurchased inputs rather than as dummy variables. The precipitation variables measure water available for plant growth in addition to irrigation water, and the CDD variables measure energy available for plant growth. The variables are specified as county-level means computed from station-level data in bimonthly periods throughout the pregrowing and growing seasons (U.S. Department of Commerce, 1986c).

Climate variables in the z_i set are based on 30-year average rainfall and CDD. They serve as proxies for unobserved producer decisions affected by climate but made before the 1984 observed weather. Climate variables are specified as dummy variables relative to their average to avoid multicollinearity problems with weather variables. Four climatic dummy variables for rainfall are specified for the ranges 0–30.48, 30.49–45.72, 60.97–76.2, and 76.21+ cm (0–12.00, 12.01–18.00, 24.01–30.00, and 30+ inches) of precipitation measured relative to the long-term average range of 45.73–60.96 cm (18.01–24.00 inches). Four climatic dummy variables for CDD are specified for the ranges 0–300, 301–800, 1301–1800, and 1800+ CDDs measured relative to the long-term average range of 801–1300 CDDs (U.S. Department of Commerce, 1986b).

Other weather-event dummy variables (also in the z_i set) include the number of days that rain exceeds 2.5 cm (1 inch) and the number of days that temperature exceeds 32 °C (90 °F). In addition, four general farm characteristics and three soil-quality variables also enter the yield-function estimation through the use of dummy variables, z_j.

Input substitution for irrigation water is central to a water-policy model. Farm-level variables for water-management practices represent the observed effects of management capability substituting for irrigation water. [*See* Moore et al. (1990) for more details.]

Yield-Function Reduced Forms.   The WAWA Model utilizes all nonirrigation application information by combining the effects of climate, weather, and farm and irrigation management practices into a composite intercept term. This enables the model to incorporate local conditions, while keeping an operational, tractable, crop-specific relationship between applied irrigation water and yield. The functional form used within the WAWA Model is

$$y_{ck} = \alpha_{ck} + \beta_{ck}w_{ck} + \gamma_{ck}w_{ck}^2$$

where c is crop, k is MSA, y is crop yield per unit land, α is a composite term that includes the effects of variables other than irrigation water, β is the effect of irrigation water on yield, w is irrigation water applied per unit land, and γ is the effect of irrigation-water-squared on yield.

To transform estimated functions into reduced-form equations for the WAWA Model, MSA-level averages for each variable are substituted into the functions and multiplied by their respective estimated coefficients.[9] The MSA data for this process are computed from area-weighted county-level data. When MSA-level averages are not available, state-level averages are substituted.

Reduced Forms Adjusted for Climate Change. The current reduced-form equations are estimated by using 30-year-average climatic conditions, effectively a climate scenario reflecting the past. Modifying the equations to incorporate an alternative scenario involves (1) selecting a climate-change scenario, (2) translating the climate-change scenario into an absolute or percentage change in precipitation and CDDs for each MSA, and (3) recomputing the reduced-form water-yield functions using the scenario data.

To demonstrate this process, an example is provided relating climate-induced changes in irrigation-water demands for three crops: alfalfa hay in west-central Colorado, barley for grain in southern Idaho, and sorghum for grain in southwestern Kansas. Reduced-form functions are computed for three climate scenarios: (1) current climatic conditions, (2) a warmer climate, and (3) a warmer and drier climate.

To reflect a warmer climate, climate variables expressed as dummy variables were adjusted by recalculating the proportion of the MSA area in each dummy range, after an assumed 20% increase in CDD. To reflect a warmer and drier climate, the same adjustment was made in CDD, with a similar adjustment reflecting an assumed 10% reduction in precipitation. An alternative climate also involves weather data that represent actual growing-season conditions. Weather data (including values for four CDDs and five precipitation variables) were computed by using average CDD values plus 20% for a warmer climate and current average precipitation values less 10% for a drier climate. The change-in-weather data associated with a 10% decrease in precipitation were 3.81, 4.83, and 6.35 cm (1.5, 1.9, and 2.5 inches) annually for the barley, alfalfa, and grain sorghum areas, respectively. With a 20% increase in CDDs, the changes were 89, 90, and 305 CDDs annually for the barley, alfalfa, and grain sorghum areas, respectively. The CDD changes do not directly translate into temperature changes. Assuming all 62 days in July and August have an average temperature of 18 °C (65 °F) (not a certainty at high elevations), the

---

[9] For example, the alfalfa hay equation estimates a coefficient on March–April precipitation ($\hat{d}_{2,\text{alfalfa}}$) of 0.255. The mean data value for MSA-61 in southern California is 3.3 cm (1.3 inches), and the resulting contribution to α is 26 g/m^2 (0.118 tons/acre).

temperature change computes to be <1 °C in the barley and alfalfa areas and <2 °C in the grain-sorghum area. (The percent-adjustment method used here has the potential to increase July and August temperatures in the desert southwest by 15 °C.)

Some interesting observations emerge from the revised water-yield functions in Figure 9.4, although generalizations should not be drawn on the basis of only three examples. Irrigated grain sorghum is the only crop presented that exhibits a "classic" response: a warmer climate requires more irrigation water to achieve a given yield, and a warmer and drier climate requires even more water to achieve that same yield.

With a warmer climate, both irrigated alfalfa hay and irrigated barley yields increase a small amount for a specified irrigation-water level. Increased yield for a given water level is actually plausible, if not likely, for some crops and locations. Increased energy may benefit crops like alfalfa that currently produce high yields in hot climates, indicating successful utilization of energy with adequate water. Barley is typically grown in relatively cool climatic areas that may not receive enough energy to allow the growth of higher-yielding varieties. For example, the yield of a barley variety grown in California may be higher than one in Idaho, other variables held constant. A warmer climate may allow cultivation of varieties with potential for higher yields. These examples illustrate the significance of output adjustments per unit land from potential climate changes.

Summary

Scientific debate over the magnitude and implications of climate change will continue for the foreseeable future, if not longer. Considering the uncertainty surrounding future climatic conditions, current policy decisions should not be delayed until a conclusive scientific answer is found to the uncertainties associated with climate change. Current and proposed policies need to be assessed in the context of alternative climatic scenarios and their relationship to the human and natural environments.

This paper describes a framework in which alternative climatic scenarios can be considered in evaluating alternative water policies in the American West. The paper defines western water policy as choices among alternative institutional mechanisms of western-water reallocation that will prompt irrigation-water conservation. Climate change offers the prospect of altering the basic parameters of the water-reallocation setting: water supply (including initial water endowments), agricultural water demand, and nonagricultural water demand. Recognition of this prospect is essential for a full analysis of western water policy.

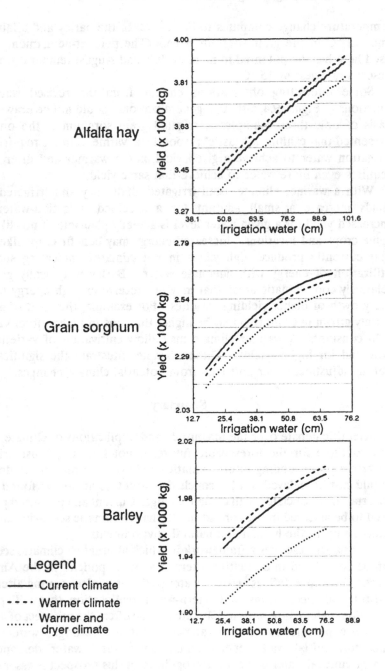

FIGURE 9.4. Examples of climate-induced adjustments in water-yield production functions.

The WAWA Model serves as a tool for evaluating agriculture's response to alternative western water policies given specified climatic scenarios. The model provides opportunities for climatic adjustment in both irrigation-water demand and supply. Alternative climate scenarios may be incorporated explicitly into the model structure by simulating the relationship between a climate's crop yield and water use. Growing different crops may also be an important option in response to water policies and climate change. The WAWA Model determines the irrigation-water demand by choosing the irrigation application level, the crop yield, and the crop mix.

In terms of water supply, the WAWA Model currently links agricultural surface-water supplies within a simple interregional water-transfer system. Because climate change appears likely to reduce irrigation-water supplies, an interregional water-supply link allows the model to move remaining supplies endogenously, if that is economically efficient.

A primary focus of the WAWA Model when it was originally conceived was to analyze agricultural impacts of alternative water-conservation mechanisms in the current era of water reallocation in the western United States. However, effects of alternative climate scenarios can be assessed with only modest modifications to that model. The ability to consider climate scenarios in conjunction with Western water-policy analysis should enhance the model's ability to foster sound economic research.

References

Adams, R. A., C. Rosenzweig, R. M. Peart et al. 1990. "Global Climate Change and US Agriculture." *Nature* 345(6272): 219–24.

Adams, R. A., B. A. McCarl, D. J. Dudek, and J. D. Glyer. 1988. "Implications of Global Climate Change for Western Agriculture." *Western Journal of Agricultural Economics* 13(2): 348–56.

Brown, B. G. 1988. "Climate Variability and the Colorado River Compact: Implications for Responding to Climate Change," in M. J. Glontz, ed., *Societal Responses to Regional Climate Change: Forecasting by Analogy*. Pp. 279–305. Boulder, Colo.: Westview.

Dudek, D. J. 1989. "Climate Change Impacts upon Agriculture and Resources: A Case Study of California," in J. B. Smith and D. A. Tirpak, eds., *The Potential Effects of Global Climate Change on the United States: Appendix C, Agriculture*. Pp. 5-1-5-38. Washington, D.C.: Office of Policy, Planning and Evaluation, U.S. Environmental Protection Agency.

Frederick, K. D., and A. V. Kneese. 1990. "Reallocation by Markets and Prices," in P. E. Waggoner, ed., *Climate Change and U.S. Water Resources*. Pp. 395–419. New York: John Wiley & Sons.

Gleick, P. H. 1988. "Climate Change and California: Past, Present, and Future Vulnerabilities," in M. J. Glontz, ed., *Societal Responses to Regional Climate Change: Forecasting by Analogy*. Pp. 307–28. Boulder, Colo.: Westview Press.

Howe, C. W., D. R. Schurmeier, and W. D. Shaw. 1986. "Innovative Approaches to Water Allocation: the Potential for Water Markets." *Water Resources Research* 22(4): 439–45.

Ingram, H. M., H. J. Cortner, and M. K. Landy. 1990. "The Political Agenda," in P. E. Waggoner, ed., *Climate Change and U.S. Water Resources*. Pp. 421–43. New York: John Wiley & Sons.

Lettenmaier, D. P., T. Y. Gan, and D. R. Dawdy. 1989. "Interpretation of Hydrologic Effeclts of Climate Change in the Sacramento-San Joaquin River Basin, California," in J. B. Smith and D. A. Tirpak, eds., *The Potential Effects of Global Climate Change on the United States: Appendix A, Water Resources*. Pp. 1-1-1-52. Washington, D.C.: Office of Policy, Planning and Evaluation, U.S. Environmental Protection Agency.

Moore, M. R. 1991. "The Bureau of Reclamation's New Mandate for Irrigation Water Conservation: Purposes and Policy Alternatives." *Water Resources Research* 27(2): 145–55.

Moore, M. R., N. R. Gollehon, and D. H. Negri. 1990. "Alternative Forms for Production Functions of Irrigated Crops." Washington, D.C.: Water Branch, Economic Research Service, U.S. Department of Agriculture (manuscript).

Peart, R. M., J. W. Jones, R. B. Curry, K. Boote, and L. H. Allen, Jr. 1989. "Impact of Climate Change on Crop Yield in the Southeastern USA: A Simulation Study," in J. B. Smith and D. A. Tirpak, eds., *The Potential Effects of Global Climate Change on the United States: Appendix C, Agriculture*. Pp. 2-1-2-54. Washington, D.C.: Office of Policy, Planning and Evaluation, U.S. Environmental Protection Agency.

Peterson, D. F., and A. A. Keller. 1990. "Irrigation," in P. E. Waggoner, ed., *Climate Change and U.S. Water Resources*. Pp. 269–306. New York: John Wiley & Sons.

Rango, A., and V. van Katwijk. 1990. "Water Supply Implications of Climate Change in Western North American Basins," in J. E. Fitzgibbon, ed., *Proceedings of the Symposium on International and Transboundary Water Resources Issues*. Pp. 577–86. Bethesda, Md.: American Water Resources Association.

Rosenberg, N. J., B. A. Kimball, P. Martin, and C. F. Cooper. 1990. "From Climate and CO_2 Enrichment to Evapotranspiration," in P. E. Waggoner, ed., *Climate Change and U.S. Water Resources*. Pp. 151–75. New York: John Wiley & Sons.

Rosenzweig, C. 1989. "Potential Effects of Climate Change on Agricultural Production in the Great Plains: A Simulation Study," in J. B. Smith and D. A. Tirpak, eds., *The Potential Effects of Global Climate Change on the United States: Appendix C, Agriculture*. Pp. 3-1-3-43. Washington, D.C.: Office of Policy, Planning and Evaluation, U.S. Environmental Protection Agency.

Sax, J. L., and R. A. Abrams. 1986. *Legal Control of Water Resources*. St. Paul: West Publishing Co.

Schaake, J. C. 1990. "From Climate to Flow," in P. E. Waggoner, ed., *Climate Change and U.S. Water Resources*. Pp. 177–206. New York: John Wiley & Sons.

Schaible, G. D., M. P. Aillery, and P. Canning. 1989. "A User's Manual for the Irrigation Production Data System (IPDS)." Washington, D.C.: Water Branch, Economic Research Service, U.S. Department of Agriculture. (Staff Report AGES 89-10.)

Shupe, S. 1982. "Waste in Western Water Law: A Blueprint for Change." *Oregon Law Review* 61: 483–95.

Solley, W. B., C. F. Merk, and R. R. Pierce. 1988. "Estimated Use of Water in the United States in 1985." Washington, D.C.: U.S. Government Printing Office. (U.S. Department of the Interior, Geological Survey Circular 1004.)

Weatherford, G., ed. 1982. *Water and Agriculture in the Western U.S.: Conservation, Reallocation, and Markets.* Boulder, Colo.: Westview Press.

U.S. Department of Commerce, Bureau of the Census. 1984. *1982 Census of Agriculture.* Washington, D.C.: U.S. Government Printing Office.

U.S. Department of Commerce, Bureau of the Census. 1986a. "1984 Farm and Ranch Irrigation Survey." Washington, D.C.: U.S. Government Printing Office. (Report AG84-SR-1.)

U.S. Department of Commerce, National Climatic Data Center. 1986b. "Climatography of the U.S. No. 20, 1951–1980 Period of Record." Asheville, N.C.: National Climatic Data Center.

U.S. Department of Commerce, National Climatic Data Center. 1986c. "Summary of the Month Cooperative, TD-3220." Asheville, N.C.: National Climatic Data Center.

U.S. Department of the Interior, Bureau of Reclamation. *1982 Summary Statistics: Water, Land and Related Data.* Denver: Division of O&M Technical Services, Economics and Statistics Branch, Bureau of Reclamation.

10

Methodology for Assessing Regional Economic Impacts of and Responses to Climate Change: The MINK Study

*William E. Easterling, Pierre R. Crosson,
Norman J. Rosenberg, Mary S. McKenney,
and Kenneth D. Frederick*[1]

Introduction

Rapid and sweeping changes in climate from a steadily strengthening greenhouse effect are possible in the next two to five decades (Bolin et al., 1986; IPCC, 1990; Schneider and Rosenberg, 1989). Scientific uncertainties cloud our understanding of the exact course that climate will take in the future. However, scientists generally agree that the rate and magnitude of climate change from greenhouse warming could be unprecedented in human experience. Surely efforts are justified to identify the kinds of effects that might occur should society decide to allow climate to change unabated or if society cannot abate the change.

---

[1] From Resources for the Future, Washington, D.C. W.E.E. is now with the Department of Agricultural Meteorology, University of Nebraska at Lincoln. This paper is based on a manuscript by these authors submitted to the Center for Growth Studies of the Houston Area Research Center as an entry to the 1990 George and Cynthia Mitchell Prize Competition for Sustainable Development. This research is part of an effort sponsored by the U.S. Department of Energy, Office of Health and Environmental Research. The analytical work conducted at Resources for the Future involved several people who do not appear as authors: Michael D. Bowes, Roger A. Sedjo, Joel Darmstadter, Kathleen Lemon, and Laura A. Katz. Space precludes mentioning the many other people who have helped with this work.

Scientific uncertainties remain in our understanding of the climatic changes that may follow from greenhouse warming. Nevertheless, large and rapid changes in regional climates are conceivable. The impacts of such changes as the 8 °C increase in mean summer temperature in the central United States accompanied by a 1 mm/day decrease in mean precipitation (Manabe and Wetherald, 1987) would be severe. This prediction is more radical than others that have been made (Schlesinger and Mitchell, 1985). Nonetheless, as long as the direction of change is credible, efforts are warranted to identify the kinds of impacts to expect if society chooses to allow climate to change or cannot stop it from changing and what might be done to adjust to those impacts.

Additionally, laboratory and growth-chamber studies show that when CO_2, one of the primary greenhouse gases, increases in concentration in the atmosphere, plants respond with increased rates of photosynthesis and reduced rates of evapotranspiration and, thus, with improved water-use efficiencies, i.e., yield per unit of water consumed (Rosenberg, 1982; Kimball, 1983, 1986; Cure and Acock, 1986). However, the strength of the CO_2 effect under field conditions is still uncertain.

The research we report here does not deal with the issue of scientific uncertainty. Our aim instead is to improve the research method base for assessing the regional impacts of climate change and of responses to it. In the following pages, we explain the need for improving methodology and describe the analytical framework we have developed to accomplish this. We illustrate our approach by examining the impacts of climate change, CO_2 enrichment, and adaptation on the productivity of agriculture and the reliability of water resources in a particular region now and in the future. We focus on a region instead of a single resource sector (e.g., agriculture) because many interdependent resource sectors in a region are likely to be affected by climate change. Finally, we analyze the implications of these impacts and responses for the current and future economies of the region chosen for study.

Methodological Limitations of Previous Studies

Four methodological limitations typify most studies of the regional impacts of climate change. First, the climate of tomorrow is imposed abruptly on the world of today. Second, the natural temporal and spatial variability of climate characteristic of large regions is ignored. Third, the full range of available technologies, management techniques, and policy tools that can lessen (or capitalize on) the impacts of climate change are not fully considered. Fourth, links between resource sectors affected by climate change are not explicitly considered.

The Analytical Framework

To overcome the limitations described above, we developed an analytical framework that aims to provide base-line information on current functioning of regional-scale economies and how they might develop in the future in the absence of climate change; analyze how different kinds of climatic change may alter base-line resource productivity (e.g., crop production, forest output, runoff to rivers, and water storage); study the ways in which the primary enterprises affected (e.g., farms, timber companies, and water resource districts) may respond to these first-order effects; and study how the responses of primary enterprises may in turn affect the regional economy as a whole.

The full study deals with all of the important resource sectors identified above and with all of the orders of effects and links. However, because of space limitations, in this paper we demonstrate our methodology by focusing on the agricultural and water-resources sectors and their links with the rest of the economy (Rosenberg et al., 1991). The analysis consists of four specific tasks and a set of subtasks listed in Table 10.1.

Task A is to develop a base-line description of a region showing the current structure of the economy of the region, with special attention to agriculture, water resources, forestry, and energy.

Task B is to examine how the region would be influenced qualitatively and quantitatively were a particular climate change to occur with today's resource base, technology, economy, and institutional structure. Task B thus shares with other studies of climate-change effects a limitation we noted above: imposing the change on the existing situation. In our study, however, task B is only a first step toward our ultimate goal of analyzing the impacts of change on a regional economy as it may exist several decades in the future. This analysis is undertaken in tasks C and D.

Task B consists of three subtasks. In subtask B_1, we impose a climate change on the selected region and assess the impacts on several sectors and on the economy as a whole, assuming no adaptive responses. In essence, subtask B_1 provides a worst-case condition.

In subtask B_2, we study the direct and indirect regional impacts assuming that climate change occurs with an increase in the atmospheric concentration of CO_2 from the current 350 μmol/mol to 450 μmol/mol. Again, we assume no adaptation.

In subtask B_3, we allow for adaptations to the analog climate change without ($B_{3.1}$) and with ($B_{3.2}$) the 100-μmol/mol CO_2-concentration increase. The only adaptations permitted, however, are those that are available today and are simple and inexpensive. Subtask B_3 includes an analysis of the economy-wide impacts of climate change on several economic sectors.

TABLE 10.1 The MINK Study: Analytical Tasks

Task A   Current base-line description of the region

Task B   Imposition of a climate change on the current base line

   B_1 Climate change only, no adaptation

   B_2 Climate change plus 100-μmol/mol (100-ppm) increase in atmospheric CO_2 concentration

   $B_{3.1}$ B_1 with currently available adaptation techniques

   $B_{3.2}$ B_2 with currently available adaptation techniques

Task C   Base-line description of the region in the future

Task D   Imposition of a climate change on the future base line

   D_1 Climate change only, no adaptation

   D_2 Climate change plus 100-μmol/mol increase in atmospheric CO_2 concentration

   $D_{3.1}$ D_1 with future adaptation technologies

   $D_{3.2}$ D_2 with future adaptation technologies

In task C, we develop a new base line of the economic, technical, and institutional structure of the selected region as it might be in 2030, absent climate change.

In task D, we assume the regional climate changes gradually, taking the analog form by 2030. We then study the difference this pattern might make to the performance of the economy developed in task C by developing several subtasks.

In subtasks D_1 and D_2, we impose the analog climate change without and with the 100-μmol/mol CO_2-concentration increase, again assuming no additional adaptive responses to task C conditions. In subtask D_3, we allow for adaptive responses with technologies and management practices that could develop by 2030 under conditions of a gradually changing climate. Subtask $D_{3.1}$ studies these responses without and $D_{3.2}$ studies them with the

100-μmol/mol CO_2 increase. Finally, we examine the economy-wide effects of the initial climate-induced change in economic activity.

Our aim has been to develop a sound methodology that can be applied to the assessment of the regional impacts of climate change and of the possible responses to such change. The framework of this methodology must be theoretically sound. Its subcomponents need not be fixed, however. Each database, model, or analytical tool used should be open to improvement or to replacement as better tools become available.

The Region of Study

The methodological design described above was applied, in the first case, to the region composed of four states in the central United States—Missouri, Iowa, Nebraska, and Kansas (hereafter, the MINK region). MINK is economically and topographically homogeneous, with no deserts or maritime areas within or adjacent to it. Moreover, compared with the rest of the nation, the MINK economy is specialized in the natural resource–based sectors most likely to be affected by climate change. Finally, some general circulation models (GCMs) predict dire changes in climate for the region (e.g., Manabe and Wetherald, 1987).

Current Climate and Scenario of Climate Change

The MINK region, far removed from moderating influences of large bodies of water, is typically continental and characterized by large seasonal swings in temperature and precipitation. Winters are cold and dry, and summers are hot, with moisture and precipitation declining from east to west.

Precipitation in the MINK region is controlled by two physiographic features: the Rocky Mountains, which remove moisture from maritime Pacific air from the west, and the Gulf of Mexico, which provides moisture-laden maritime-tropical air to the region. Distance from the Gulf and the Rocky Mountain rainshade explain the sharp decline in precipitation from east to west. Precipitation is lowest in winter and greatest in summer, except in Missouri, where the peak occurs in spring (Table 10.2).

Scenarios of climate change can be developed by simulation, such as with GCMs, paleoclimatic reconstructions, or through use of the historic climatic records (Lamb, 1987). We chose to use the last approach and have drawn from the historic record a segment, the 1930s. We believe this choice is appropriate because the 1930s in MINK were hotter and dryer than today, characteristics that approximate those that GCMs predict will occur in the MINK region sometime in the first half of the coming century, when the

TABLE 10.2 Seasonal Mean Monthly and Mean Annual Temperature and Mean Seasonal and Annual Precipitation Totals and Their Standard Deviations in the MINK States, 1951–1980

	Winter (D,J,F) Temp. (°C)	Precip. (mm)	Spring (M,A,M) Temp. (°C)	Precip. (mm)	Summer (J,J,A) Temp. (°C)	Precip. (mm)	Fall (S,O,N) Temp. (°C)	Precip. (mm)	Annual Temp. (°C)	Precip. (mm)
Missouri										
Mean	0.3	152	12.7	299	24.4	291	13.9	246	12.8	989
SD	1.8	48	1.0	74	0.9	71	1.2	77	0.7	179
Iowa										
Mean	-5.7	77	8.9	236	22.3	318	10.6	184	9.0	815
SD	1.8	26	1.2	60	0.7	55	1.2	72	0.7	133
Nebraska										
Mean	-3.6	43	8.9	179	22.6	238	10.5	105	9.6	566
SD	1.7	15	1.0	49	0.9	51	1.1	42	0.7	100
Kansas										
Mean	0.1	57	12.1	202	25.2	264	13.6	161	12.8	684
SD	1.6	26	1.1	59	1.0	79	1.2	56	0.7	157

total radiative equivalent of a doubled preindustrial CO_2 concentration may occur.

In all of the analyses that follow, the climate of 1931–1940 is compared with that of the period currently defined by World Meteorological Organization (WMO) convention as normal, i.e., 1951–1980.[2] Differences between these two periods are shown by state and by season in Table 10.3. Temperatures were higher across the region in the 1930s, particularly in summer. Winters were unusually warm in Iowa and Missouri, as were autumns in Nebraska and Kansas. On average, mean monthly precipitation in the 1930s was lowest (compared with the control period) in spring and summer. Winters on average were a bit wetter. Although precipitation was lower in the 1930s, the interannual variability in that decade was also less. Interannual variability in temperature was greater, however.

It is incorrect to declare, simply, that the climate of the 1930s was hotter and drier throughout the MINK region than during the current 30-year normal. Climatology assures us that that could not have been the case everywhere in the region all of the time. Experience with drought since 1931 has been different in each of the four states. The western states descended rapidly into severe drought in the 1930s and did not emerge until the beginning of the 1940s. Although droughts did occur in the eastern states in the 1930s, they were less protracted and severe than in Nebraska and Kansas. In fact, drought was more severe in Missouri and Iowa in the 1950s than in the 1930s.

The 1940s were generally benign in the MINK region. The 1940s are excluded from consideration as part of the climatic normal by convention, while the drought years of the 1950s are included. Thus, the contrast between the climate of the 1930s and that of the climatological normal is less than it would have been had all years since 1940 been included.

Task A: Agriculture and Water Resources in MINK—Current Base Line

Agriculture in the MINK Region

We illustrate our approach by highlighting the agricultural sector and its links to the rest of the economy and provide information on water resources to the extent that they influence the future of the agricultural sector. First, however, we describe agriculture and water resources in the MINK region as they are today. Additionally, in the crop-modeling work reported below,

---

[2] By definition, the last three decades for which there are complete records. Climatic means for the 1980s have not yet been computed. After means are available, the climatic the normal will be 1961–1990.

TABLE 10.3 Differences Between 1931–1940 and 1951–1980 in Seasonal Mean Monthly and Mean Seasonal and Annual Precipitation Totals and Their Standard Deviations in the MINK States

	Winter (DJF) Temp. (°C)	Precip. (mm)	Spring (MAM) Temp. (°C)	Precip. (mm)	Summer (JJA) Temp. (°C)	Precip. (mm)	Fall (SON) Temp. (°C)	Precip. (mm)	Annual Temp. (°C)	Precip. (mm)
Missouri										
Mean	1.0	16	0.0	−23	1.1	−21	0.6	−1	0.7	−28
SD	0.3	−3	−0.0	18	0.3	7	0.2	12	0.2	−57
Iowa										
Mean	1.1	6	0.3	−53	1.2	−28	0.6	16	0.8	−60
SD	0.9	−6	−0.2	−4	0.1	11	0.2	20	0.4	−28
Nebraska										
Mean	0.6	4	0.7	−23	1.6	−54	1.0	−21	1.0	−93
SD	0.8	−2	0.1	7	−0.1	−13	−0.1	−13	0.2	−30
Kansas										
Mean	0.9	0	0.6	−19	1.3	−59	1.0	−24	0.9	−102
SD	0.4	−12	−0.1	−6	0.1	25	0.0	−1	0.1	−78

we emphasize those crops grown under both dryland and irrigated conditions: corn, sorghum, and wheat. For analysis of the overall impact of climate change on the regional economy, we include information on soybeans as well.

In the base period 1984–1987,[3] MINK accounted for 34% of the value of the nation's production of corn for grain, 30% of its soybeans and winter wheat (the only wheat grown in the region), and 50% of its grain sorghum. Average yields of these four crops were higher and production costs per unit volume (e.g., per bushel) were less in MINK than in the rest of the country. Nebraska had 64% and Kansas 30% of the irrigated land of the region. Corn in Nebraska accounted for 53% of all the irrigated land in MINK. MINK also accounted for 28% of national production of cattle and calves and 42% of its hogs. Nebraska produced 35% of MINK's cattle and calves, and Iowa 63% of the region's hogs.

Value added by agriculturally related manufacturing industries was 77% of value added on the region's farms. Meat packing alone accounted for 31% of value added in agriculturally related manufacturing during the base-line 1984–1987 period.

Analysis of links between agriculture and the rest of the regional economy showed that in the base-line period, a change of $1.00 in on-farm production generated an additional change of $0.65 in production in the rest of the economy. The regional impact of activities tied to agriculture, such as meat packing, was greater, e.g., a $1.00 change in meat-packing output generated an additional change of $1.57 in the rest of the economy.

The MINK economy is much more specialized in agriculture than is the rest of the nation: the ratio of farm income to total income in MINK is 3.4 times the ratio of farm income to total income in the nation as a whole. Measured by the share of farm income in total income, Nebraska is the most agricultural and least industrial of the four states, and Missouri is the least specialized in agriculture. Manufacturing is the most important single sector in Iowa, Kansas, and Missouri, followed closely by services. Finance, insurance, and real estate are most important in Nebraska, followed by services and manufacturing.

Although MINK clearly is specialized in agriculture compared with the rest of the country, that sector's share of regional income (3.7%) suggests that MINK is, nonetheless, not primarily an agricultural region. On the contrary, it could more accurately be called a manufacturing-services region, because the shares of manufacturing and services in the economy are 4 times

---

[3] This is the "base period" against which a future scenario of the MINK economy is developed. Statistics available when the MINK study began carried only through 1987.

as great as those of agriculture. These interpretations, however, seriously understate the importance of agriculture to the MINK economy and the region's vulnerability to climate change for two reasons: a significant amount of manufacturing in MINK is directed to processing farm outputs (meat packing) and manufacturing farm inputs (fertilizers, machinery), and agriculture's percentage contribution to regional exports—the key element in the economic base of the region—is several times as high as its contribution to regional income.

Water Resources of the MINK Region

The four MINK states are not an independent hydrologic region. Because they include parts of four major water-resource basins (the Missouri, the Arkansas-White-Red, and the Upper and Lower Mississippi), the water available to MINK depends not only on events in MINK but also on circumstances in states up river. Moreover, water use in MINK may be restricted by obligation to downstream states. Instream flow needs for navigation, fish, and wildlife habitat are large compared with withdrawal uses and mean stream flows.

Irrigation accounts for about half of the water withdrawal and 89% of water consumption within MINK. Nebraska and Kansas account for 97% of all irrigation water use within MINK. Groundwater provided 80% of all irrigation water and 48% of all water withdrawals in 1985. Intensive pumping and low recharge rates have resulted in the mining of water from the Ogallala formation underlying western Kansas and much of southwestern Nebraska. Conflicts over alternative uses of available water supplies have increased in recent years. Institutions for allocating water supplies within the region were stressed by drought during the 1980s and by a growing demand for more water for recreation and fish and wildlife habitat. Any climate change that diminished runoff to rivers and reservoirs, reduced recharge of groundwater where recharge is consequential, or increased demand for water for irrigation or for competing uses could have profound effects on agriculture.

Modeling Effects of the Analog Climate

The Representative Farms

One important feature of our modeling approach has been to consider the impacts of climate change on representative farms, which are treated as economic entities, rather than on individual crops. The regional impacts of climate change are an aggregate of the representative farms. A representative farm is a cohesive, functional enterprise that typifies most of the farms in its particular region. Forty-eight such farms in 11 major land resource areas (MLRAs) throughout the MINK region were designed in consultation with agricultural experts in each of the MINK states. Although alfalfa, soybeans, and wheatgrass are included in the analysis, in this paper we present results for only three crops—corn, sorghum, and wheat—which are grown in the region under both rain-fed and irrigated conditions.

Crops: The EPIC Model

A mechanistic model of crop growth is needed to allow us to predict the response of current crops to a climate change in the MINK region (task B). We use it also for predicting the productivity of crops in ≈2030 (task C) and how crop productivity would be affected by climate change in 2030 (task D). We chose to work with a family of simulation models known as EPIC (Erosion Productivity Impact Calculator). The EPIC models were developed at the Texas Agricultural Experiment Station in Temple by the Agricultural Research, Soil Conservation and Economic Research Service of the U.S. Department of Agriculture and are described in Williams et al. (1984).

EPIC accommodates numerous variables for weather, crop phenology and physiology, soil physical and chemical condition, farming practices including tillage, irrigation, and pest control, and farm economics. The model operates on a daily time step, so the daily weather is required as an input.

A major reason for our choice of EPIC in the MINK analysis is its flexibility, which permits simulation of alternative crop maturities, rotations, tillage practices, planting dates, irrigation, and fertilization strategies that might be used by farmers attempting to adapt to climate change. Preparation of EPIC for the MINK study required modification to account for the direct effects of increasing atmospheric CO_2 concentration, characterization of the representative farms, and validation of the model's results. CO_2 enrichment of the atmosphere increases photosynthetic rate, especially in C3 species (the small grains, legumes, and most trees and grasses) and reduces evapotranspiration (ET) in both C3 and C4 plants (e.g., corn, sorghum, and millet) (Rosenberg et al., 1990). To allow us to evaluate the effect not only of forecasted climate change (or its analog) but also of the direct effects of CO_2 enrichment, we had to modify EPIC to account for direct effects of changing CO_2 concentration on photosynthesis and evapotranspiration. These modifications are described in Stockle et al. (in press).

Validation of EPIC

We tested EPIC for realism in three ways. First, the means of a 60-year EPIC run of annual yields assuming 1984–1987 technology and using the weather records for 1951–1980 were compared with actual yields for the period 1984–1987 (extracted from USDA, 1984–1987). Second, the simulated means were compared with expert judgments of what the yields should be on the representative farms with current technologies. Third, the means, standard deviations, and range of yields and ET generated by EPIC were compared with results of pertinent experiments identified in an extensive review of the agronomic literature.

The results of these tests are discussed in detail in Rosenberg et al. (in press). We concluded that EPIC simulations of yields and ET are sufficiently realistic to permit use of the model to compare crop production in the base-line period with production that might occur under the analog climate.

Water Supply and Demand

Irrigation is the principal withdrawal use of stream flow in the Missouri basin today and the principal consumptive use of water in the states of Nebraska and Kansas. In this paper, we use EPIC to estimate changes in irrigation requirements caused by the analog climate. The economic demand for water, however, would not necessarily equal requirements.

Ideally, to estimate the effects of the analog climate on stream flow, we would compare natural flows in the 1930s with those during the control period. However, flows in the basins today are affected by diversions, reservoirs, and kinds of consumption that did not exist in the 1930s. For example, most of the 12,000 reservoirs in the MINK region and many more in the upstream states were built after the 1930s. Evaporation from the reservoirs is certainly greater than if the land were still vegetated or bare; and the impact of increased temperature and lower humidity on evaporation is amplified because of the presence of the reservoirs.

Consequently, only a very small number of the thousands of gauging stations in the region appear to have been unaffected since the 1930s by human activities. Those stations can be used as a proxy for estimating natural stream flows for the regions of interest.

We estimated gross evaporation in the reservoirs under control and analog conditions by two well-known methods—Harbeck (1962) and Penman (1948). Precipitation for each period was then subtracted from gross evaporation to yield net evaporation. Harbeck's method is highly empirical but considers reservoir surface area, which would likely be lower under the analog climate. Penman's method has a firmer theoretical base but does not consider reservoir surface area. Results of the analyses of water supply in the MINK region are presented below.

Task B: Impacts of a Climate Change on the MINK Region

Crop Yields

In task B_1, we impose the analog climate without a 100-μmol/mol increase of CO_2, and in task B_2, we impose it with the increase. In task B_3, a set of adaptations judged by expert opinion to be economically feasible was applied to the grain crops, again with and without the CO_2 effect. These adaptations include earlier planting, except for wheat, use of longer-season varieties of annual crops, and furrow diking for dryland row crops. Crop substitutions were also permitted in the B_3 runs, including shifts of irrigated corn land in Nebraska and Kansas to dryland production of wheat and sorghum. In all runs including irrigation, water is applied in accordance with EPIC's estimates of requirements determined by climatic conditions. In essence, the crop is irrigated when soil-available water in the root zone has been depleted by 10%.

The EPIC simulations of harvestable yields were averaged by crop across all representative farms in the MINK region. All inputs other than weather were identical (e.g., cultivars, tillage). In the B_1 case (analog climate only), yields of dryland corn and sorghum are reduced by 25% and 22%, respectively. This yield loss is due primarily to a shortened growing season caused by the more rapid accumulation of heat units in the warmer analog climate. In the B_2 case the CO_2 increase diminishes yield loss by reducing evapotranspiration and stimulating photosynthesis. The mean yield of dryland wheat is unaffected by the analog climate, primarily because that crop's growing season lengthens as the winter warms and dormancy breaks earlier. Additionally, it appears that under control conditions the number of nitrogen-stress days was greater than under analog conditions, depressing control yields unrealistically and thereby minimizing the apparent response of wheat to the analog climate. With the CO_2 increase, simulated wheat yields are higher than under the control climate.

Loss of yield in corn and sorghum due to the analog climate is considerably smaller under irrigation than on dryland (-8% and -11%, respectively) and is almost eliminated by the CO_2 increase. The yield of irrigated wheat increases by 8% under the analog climate and by 11% with the addition of CO_2.

The use of currently available adjustment measures (e.g., earlier planting and longer-season cultivars) is most effective for sorghum, both dryland and irrigated. Yield loss is eliminated in case $B_{3.1}$, and yields are increased by $\approx$15% above control results in case $B_{3.2}$ under irrigation, where CO_2 and adjustment interact favorably. Adjustments improve yields of dryland corn slightly in $B_{3.1}$ and eliminate losses in irrigated corn. Carbon dioxide increase plus adjustments in $B_{3.2}$ raise corn yields further, but yields remain

$\approx$12% below control yields on dryland, although $\approx$5% above control yields under irrigation. Earlier planting does not apply to winter wheat, and the longer-season cultivars encounter temperatures too high for beneficial growth. Furrow diking is also not appropriate for wheat. Therefore, on both dryland and under irrigation, with or without the CO_2 increase, wheat yields are not improved by the simple adjustments made, which may say more about the selection of adjustments than about the ability of wheat to be adapted to the climate change. When adjustments are ineffective or even deleterious, we assume unrealistically that farmers would not adjust to the climate change.

Water Demand and Supply

Average changes in irrigation requirement across all irrigated farms are shown in Table 10.4. With no adjustment, the analog climate increases irrigation requirement by 11% in wheat to 24% in corn. The demand is moderated by CO_2 for the corn and sorghum and is eliminated in wheat. Adjustments increase irrigation requirements because of the longer growing season that results. CO_2 moderates this effect by reducing evapotranspiration.

On the basis of a scaling-up of the individual EPIC farm simulations in Table 10.4 for all irrigated farms—according to the area they represent and assuming all demand could be met by current irrigation systems and water supplies—water withdrawals under the analog climate (B_1) would increase by 39% in Nebraska and 14% in Kansas. Specifically, irrigation requirements increase 39% for corn in Nebraska and 12% for corn, 11% for wheat, and

TABLE 10.4 EPIC-Simulated Percentage Changes in Base-Line (Task A) Irrigation Requirement under the Analog Climate

	No Adjustments		With Adjustments	
	No CO_2 Effect (D_1)	CO_2 Effect (D_2)	No CO_2 Effect ($D_{3.1}$)	CO_2 Effect ($D_{3.2}$)
Corn	+24	+15	+43	+32
Sorghum	+20	+9	+39	+26
Wheat	+11	−1	+4	−3

20% for sorghum in Kansas. The effect of the CO_2 increase (B_2) is to lower irrigation demand 4–12% on the irrigated farms and to lower total consumptive use for irrigation by $\approx$6%.

On the supply side, our analysis indicates that during 1931–1940, natural stream flows were 72% of their 1951–1980 averages for both the Missouri and Upper Mississippi basins and 93% for the Arkansas basin. Increased evaporation from the surface of current reservoirs would decrease availability of stream flow only an additional 1%. On the other hand, the CO_2 increase could also reduce evaporation from the watersheds contributing to the stream flow. However, we did not assess this possibility.

It is unlikely that the increases in irrigation implied by the EPIC simulations under the analog climate will be realized. Competition for water in Nebraska and Kansas suggests that irrigators will not be able to acquire additional surface water rights. It is also unlikely that farmers will pump more water to produce the lower yields implied by the EPIC simulations.

Instead, irrigators are likely to adapt to scarce and more costly water by investing in more efficient irrigation techniques, switching to more drought-tolerant cultivars (if available) or to different crops, or by abandoning irrigation. The analog climate change would likely accelerate the decline of irrigation in those portions of Kansas and Nebraska where farmers are already adversely affected by rising energy costs, increasing pumping depths, and declining well yields. In the B_3 scenario, we assume that some currently irrigated land in corn in these states is shifted to dryland production of wheat and sorghum.

Other impacts of the analog climate on water resources might affect agriculture. Competing demands of recreation, hydropower, and fish and wildlife habitat would likely reduce water supplies for navigation on the Missouri River with a consequent rise in the costs of shipping agricultural inputs and products.

Task C: Future Base Line for the MINK Region

Crop Yields in 2030

To use EPIC to estimate the impact of the analog climate in the D scenario, we must have projections of crop yields in the absence of climate change, that is, in the 2030 base-line scenario. We constructed yield projections by incorporating in EPIC new technologies that could be economically available to farmers by 2030 (U.S. Congress, Office of Technology Assessment, 1986; Ruttan, 1989; personal communication with Alan Jones, one of the developers of EPIC, 1990).

For example, although opinions vary on the likelihood that significant increases in photosynthetic efficiency will occur over the next 40 years, improvements are at least theoretically possible. The desert plant *Camissonia claviformis*, for example, fixes absorbed noon sunlight with an efficiency of 8.5%, ≈80% of that theoretically possible (Mooney et al., 1976). Most crops currently fix far less than 1%. In EPIC, we modeled increased photosynthetic efficiency by increasing light-use efficiency by 10%.

Increases in harvest index (ratio of harvestable organs to total biomass) have been responsible for much of the yield increase seen in the past few decades. Although there are likely to be physiological constraints to improvement in this characteristic, a continuation of past trends seems likely; we impose an increase in harvest index of 10%. Much research now is directed to developing improved techniques for pest control; we programmed a 15% improvement in pest control. Earlier leaf development would provide greater photosynthetic capacity in the spring, when water stress is generally less limiting; we simulate a shift in the timing of leaf development such that maximum leaf area is achieved 5% earlier in the growing season. Technologies that reduce field losses during harvest increase the harvested yield; we simulated an improvement of 10% in harvest efficiency.

The yields from EPIC incorporating these technologies are shown in Table 10.5. In this paper, we use the EPIC yield projections for 2030 instead of yield trend projections, because this assumption permits us to estimate the yield impacts of the analog climate under simulated conditions of 2030. The five technological improvements would increase yields of dryland and irrigated corn and sorghum 55–74% compared with the current base line. In soybeans, dryland yields would increase 60% and total yields, 64%, reflecting a small increase in irrigated soybean production in the eastern part of MINK. In wheat, the increases would be 58% for dryland and 93% for irrigation. The greater response for wheat may be due to the constrained supply of nitrogen in the current base-line (task A) simulations of that crop. In all future (task C) runs, nitrogen is made to be nonlimiting, because by 2030, biological N fixation might be possible in nonleguminous as well as leguminous plants and N application and conservation technology will be much better than today.

Water Demand and Supply

Although the EPIC simulations show that water requirements for irrigation in 2030 in the absence of climate change would be little different from the current base line, water supplies for irrigation could be considerably reduced. Specifically, changes in upstream water use and new reservoir development may reduce the quantities of surface water available to MINK in 2030. The greatest uncertainties and largest potential impacts

TABLE 10.5 Base-Line Crop Yields in 1984–1987 and in 2030 as Projected by Trends and by EPIC

	1984–1987 Base Line (kg/m²)	2030 Base Line		Percent Change 1984–1987 to 2030	
		Trend (kg/m²)	EPIC (kg/m²)	Trend (%)	EPIC (%)
Corn	780	1260	1100	+62	+41
Irrigated	990		1720		+74
Dryland	640		990		+55
Wheat	310	370	500	+19	+61
Irrigated	450		870		+93
Dryland	310		490		+58
Sorghum	360	770	610	+114	+69
Irrigated	600		1000		+67
Dryland	330		570		+72
Soybeans	250	290	410	+16	+64
Irrigated	—[a]		610		—
Dryland	250		400		+60

[a]Irrigation soybean production was insignificant.

Source: Trend values are extrapolations of yield trends established in 1972–1988 by using yield data for each crop and each state in the USDA's Agricultural Statistics for various years in the 1970s and 1980s. The 1984–1987 and 2030 base-line estimates are from EPIC. The estimates for 2030 include the yield-increasing effects of new technologies that could be in general use by 2030, as described in the text.

result from the Pick-Sloan Missouri basin program, authorized as part of the Flood Control Act of 1944 (Campbell, 1984). The program has already transformed the region with more than 4 billion m² now flooded by reservoirs that provide downstream flood protection and more reliable water supplies for many instream and offstream uses. The remaining finished or planned projects under Pick-Sloan would increase irrigated area substantially within the upper basin. However, budgetary and environmental concerns have stalled most major new irrigation projects in recent years, and their eventual outcome is in doubt. If irrigation projects are not funded, surface-water supplies to the MINK states will be severely curtailed.

During the mid-1970s, net groundwater depletions from the 15 water-resource subregions located at least in part within the MINK states, particularly from the Ogallala aquifer underlying the Nebraska and Kansas High Plains, averaged ≈18 million m^3/day. Depletion continued into the 1980s at rates that, unless checked, will force major reductions in groundwater use in some parts of MINK well before 2030.

Unlike in Nebraska and Kansas, current water use poses little or no strain on groundwater supplies in Missouri and Iowa. Groundwater use in these states could be maintained or increased in most areas with little impact on the quantity in storage.

The increasing scarcity and cost of groundwater in western Nebraska and Kansas imply substantial reductions in groundwater withdrawal for irrigation in those parts of MINK by 2030. Surface water for irrigation may also decline slightly in coming decades. These waters are already scarce, and new withdrawals for agriculture will have to overcome formidable economic and environmental barriers. Water supplies will not impose such constraints on irrigation in the eastern portions of Nebraska and Kansas and in Iowa and Missouri. If economic conditions are favorable, these areas might experience a substantial percentage increase in irrigation.

Task D: Impacts of the Analog Climate on MINK in 2030

Crop Yields

Table 10.6 shows the percentage impact of the analog climate on 2030 base-line yields with and without the CO_2 and on-farm adjustment effects. The results are similar in amount and pattern to those found when we imposed the analog climate on the current base line in the B_1, B_2, and B_3 cases: the impact is most severe when no allowance is made for the CO_2 effect and on-farm adjustments and becomes progressively less severe when these conditions are relaxed. However, the change from the D_1 scenario to the D_3 scenario with the CO_2 and on-farm adjustments is more marked than the comparable change from the B_1 to the B_3 scenarios because the D_3 scenarios incorporate two technological innovations not reflected in the B_3 scenarios: increased irrigation efficiency and increased resistance to drought (modeled in EPIC as increased stomatal resistance). We introduced these innovations in the D scenarios because, if the climate in MINK moved in the direction of the analog, farmers, as well as scientists in both public and private research institutions, would have increased incentive to find ways to offset the increasingly hotter and drier conditions of the analog climate. More research might be directed to overcoming the technical and economic obstacles to widespread adoption of such methods as trickle irrigation, for

TABLE 10.6 EPIC-Simulated Percentage Changes in 2030 Base-Line (Task C) Crop Yields Under the Analog Climate (Task D)

| | No Adjustments | | With Adjustments | |
| | No CO_2 Effect | CO_2 Effect | No CO_2 Effect | CO_2 Effect |
	(D_1)	(D_2)	$(D_{3.1})$	$(D_{3.2})$
Corn				
Dryland	−28	−20	−16	−6
Irrigated	−12	−7	+2	+8
Sorghum				
Dryland	−21	−11	+11	+25
Irrigated	−10	—[a]	+7	+18
Wheat				
Dryland	−12	+3	+1	+17
Irrigated	−1	+9	+6	+18
Soybeans				
Dryland	−25	−15	−11	+2
Irrigated	—	—	+1	+12

[a]Less than 1%.

Source: Calculated from EPIC.

example, and to expanding knowledge of how to make plants drought resistant.

We believe that the study of climate-change impacts should explicitly include research of and technology development for the likely social responses to the impacts.

Water Demand and Supply

As shown in Table 10.7, the imposition of the analog climate on the agriculture of 2030 increases irrigation requirements. Without CO_2 (D_1) wheat requires 11% more water and corn and sorghum 22% and 20% more, respectively. CO_2 enrichment (D_2) reduces that requirement, nearly eliminating the increase in wheat. The increased need for irrigation is also reduced by adjustments, especially in combination with increased CO_2. Improved irrigation efficiency is partly responsible, although the major

TABLE 10.7 EPIC-Simulated Percentage Changes in Future (Task C) Irrigation Requirement Under the Analog Climate (Task D)

| | No Adjustments | | With Adjustments | |
	No CO_2 Effect	CO_2 Effect	No CO_2 Effect	CO_2 Effect
	(D_1)	(D_2)	$(D_{3.1})$	$(D_{3.2})$
Corn	+22	+12	+14	+5
Sorghum	+20	+6	+16	+5
Wheat	+11	+2	−9	−23

effect, especially in wheat whose irrigation requirement is reduced, is attributable to the increased stomatal resistance, which reduces ET.

Even if the most optimistic of these projections $(D_{3.2})$ were realized, water-supply problems would remain. As noted above, the analog climate implies significantly less water for the MINK states and the river basins of which they are a part. The impacts are particularly great in the Missouri River basin, where flows at the outflow point are only 69% of the long-term average. Flows in 2030 would be the same, barring changes in net exports and increased evaporation from manmade reservoirs.

The amount of water stored in the Ogallala aquifer underlying the Kansas and Nebraska High Plains in 2030 is not likely to differ significantly as a result of climate. Under both the control and analog climate cases, increasing water costs are likely to curb groundwater use to amounts approaching sustainable supplies by 2030. Although the quantity of water stored in the aquifer in 2030 will probably not depend on climate, the path taken to reach that quantity and the sustainable amount of pumping will depend on climate.

The hotter and drier conditions of the analog climate would tend to improve the economics of irrigated agriculture compared with dryland agriculture. EPIC simulations of representative farms in Iowa and Missouri suggest that irrigated farming would be more profitable than dryland farming in these states. Net revenues would increase by 42–49% in Iowa and 28–31% in Missouri. These projections consider only base-line (task A) production costs; allowing for expected increases in energy costs reduces but does not eliminate the advantage of irrigation. Pumping depths in Iowa and Missouri are likely to be much less than the 70–90 m typical of the High Plains.

Although the hotter and drier analog climate might improve the economics of irrigation agriculture, the increased scarcity of water would constrain its expansion. Under the control climate, water is available for a major expansion of irrigation in the two eastern MINK states without any sizable depletion of groundwater stocks. Under the analog climate, however, water use exceeds mean assessed stream flow in all but 2 of the 15 water-resource subregions. Although groundwater stocks are large in Iowa and Missouri, recharge rates would be lower under the analog climate. Moreover, pumping from alluvial aquifers would tend to draw down surface flows. The net effect of the increased profitability of irrigation and the increased water scarcity on Iowa and Missouri is unknown. But it seems unlikely that irrigation would grow much faster than it did from 1965 to 1985 (3.7%/year).

Economic Impacts of the Analog Climate

The B Scenarios

Effects on Crop Production. The changes in crop yields under the analog climate in the B scenarios would tend to change the costs of producing the crops. This change would affect the amount of crop production in two ways: Production would be cut back because the previous level would no longer be profitable under the higher costs, and some farmers would shift entirely out of a no-longer-profitable crop to a profitable one.

In our B_1 and B_2 scenarios, only the first factor affects production. We lacked the resources to estimate the cost curves for each crop, so we assumed that costs would increase such that the decline in production would be proportional to the decline in yields. Our B_3 scenario reflects both the direct yield effect on production as well as the conversion of some land in irrigated corn production to dryland production of wheat and sorghum. Consequently, in the B_3 scenario, corn production declines proportionately more than the decline in corn yields, and production of wheat and sorghum declines proportionately less than the decline in their yields.

Table 10.8 shows the changes in the value of production of corn, wheat, sorghum, and soybeans under the B_1 scenario and the B_3 scenario with and without the CO_2 effect. The B_1 scenario represents a worst-case situation because it excludes the CO_2 effect and adjustments that farmers would make to the analog climate. The B_3 scenario with the CO_2 effect represents a best-case situation because it includes on-farm adjustments as well as the CO_2 effect. We include the B_3 scenario without CO_2 because comparison of it with the B_1 scenario indicates the effect of on-farm adjustments in easing the impacts of the analog climate.

TABLE 10.8 Changes in the Value of Crop Production Under the Analog Climate

	B_1 Scenario[a]		B_3 Scenarios[b] Without CO_2 ($B_{3.1}$)		With CO_2 ($B_{3.2}$)	
	Value (millions 1982 $)	Percent Change in 1984–1987 Production (%)	Value (millions 1982 $)	Percent Change in 1984–1987 Production (%)	Value (millions 1982 $)	Percent Change in 1984–1987 Production (%)
Corn	−1644	−21.3	−1729	−22.4	−1,236	−16.0
Wheat	−14	−0.8	+361	+19.7	+139	+7.6
Sorghum	−215	−17.1	−35	−2.8	+178	+14.1
Soybeans	−789	−23.0	−542	−15.8	−48	−1.4
Total	−2662	−18.7	−1945	−13.7	−967	−6.8

[a]Assumes no CO_2 effect and no on-farm adjustments. It represents, therefore, a worst-case impact.
[b]Assumes on-farm adjustments. This scenario, with the CO_2 effect, represents a best-case impact.
[c]Calculated from average 1984–1987 production valued in 1982 prices.

Table 10.8 indicates that the combination of on-farm adjustments and the CO_2 effect reduces the decline in the value of production of the four crops from $2662 million to $967 million. The CO_2 effect accounts for 58% of the difference and on-farm adjustments, for 42%.

Effects on the Regional Economy. Because crop production is an integral part of the MINK economy, changes in crop output will affect production in other sectors. These effects can be represented quantitatively by "multipliers," numbers that show the effect on regional production of both the initial change in crop output and the change in the rest of the economy induced by the initial change.

We calculated such multipliers from data provided in IMPLAN, a county-level input-output model of the entire United States developed by the U.S. Forest Service (1989). The IMPLAN model for MINK was built by aggregating the county-level data for the region as a whole. We present an account of IMPLAN and the aggregation process in our report (Rosenberg et al., 1991). Here we deal only with the results the model gives in estimating the multiplier effects of the decline in corn and soybean production that would occur under the analog climate. We focus on corn and sorghum production because the IMPLAN model indicates a strong multiplier relationship between these feed grains and animal production, which in turn has a strong multiplier relationship with the rest of the economy.

In Table 10.9, we present two sets of estimates of the multiplier effects of the decline in corn and sorghum production imposed by the analog climate. In one set, we assume that the increase in costs of feed-grain production caused by the climate-induced losses of yield results in a reduction in export demand for the crops equal to the simulated decline in crop production. Animal producers in MINK are assumed to substitute lower-priced grains from outside MINK, so animal production in the region is not affected by the decline in grain output.

The other set of multiplier estimates assumes that the full amount of the decline in grain production is borne by animal producers in MINK, with exports of the crops unaffected.

Neither of the two underlying assumptions is fully realistic. Both export demand and internal MINK demand for grains would be reduced by the increased cost of producing these crops under the analog climate. However, given the resource limitations of this study, we had no way of estimating this more realistic outcome. The estimates obtained with the two underlying assumptions should be regarded as polar cases setting the limits within which the actual multiplier effects would fall.

Table 10.9 shows two estimates for each of the two underlying assumptions. One set assumes no CO_2 effect or adjustments by producers

TABLE 10.9 Effects of Climate-Induced Reductions in Crop Production on Total Production in MINK[a]

	Assuming the Feed-Grain Production Decline Falls on Exports				Assuming the Feed-Grain Production Decline Falls on Animal Producers in MINK			
	No CO$_2$ Effect, No Adjustments (B$_1$)		CO$_2$ Effect and Adjustments (B$_{3.2}$)		No CO$_2$ Effect, No Adjustments (B$_1$)		CO$_2$ Effect and Adjustments (B$_{3.2}$)	
	($ millions)	(% total production)	($ millions)	(% total production)	($ millions)	(% total production)	($ millions)	(% total production)
	−4,123	1.3	−1,403	0.5	29,881	9.7	17,622	5.7

[a]Production is expressed in 1982 prices.

Source: The declines in production of corn and sorghum shown in Table 10.8 and regional production multipliers derived from the IMPLAN model (USDA Forest Service, 1989).

(B_1 case); the other assumes both the CO_2 effect and adjustments ($B_{3,2}$ case). Under worst-case conditions—no CO_2 effect, no on-farm adjustments, and the total burden of the decline in feed-grain production falling on animal producers in MINK—the climate-induced decline in feed-grain production would reduce total production of all sorts in the region by $29.9 billion, or 9.7%. Under the best-case conditions—CO_2 effect with on-farm adaptations and the total burden of the feed-grain production decline falling on exports—total regional production would decline by $1.4 billion, or 0.5%.

The main reason for the difference between the two polar cases is that when the feed-grain production decline is absorbed internally, it reduces animal production, which in turn reduces output of the meat-packing industry. This industry is not only the largest single manufacturing activity in MINK, but it also has the largest multiplier effect on the rest of the economy because it draws heavily on locally produced animal and other inputs.

We believe that the effect of the analog climate on the MINK economy would be closer to the lower than to the higher polar case for three reasons. First, farmers would surely adjust to the changed climate, as in the B_3 scenario. Second, some benefits of the CO_2-concentration increase in the atmosphere, which we have estimated conservatively, would be likely. Third, the assumption that the burden of reduced grain production would fall on exports is more realistic than that it would fall wholly on MINK animal producers. The higher grain-production costs under the analog climate would weaken the competitive position of MINK grain outside the region, leading to a decline in its exports. MINK animal producers at the same time could import cheaper feed grain from outside the region. Thus, the impact on animal production in the region, and thus on the meat-packing industry, would be less than in the worst-case outcome. Over the long term, however, some animal production might shift to other grain-producing regions less affected by climate change. This shift could induce a decline in meat packing in MINK because that activity tends to locate near animal production.

The D Scenarios

Size of the Regional Economy in 2030. To represent the MINK economy in 2030 without climate change, we used population and economic projections for the MINK states developed by the Bureau of Economic Analysis (BEA), U.S. Department of Commerce (U.S. Department of Commerce, 1990). We note here only two aspects of these projections: They show the MINK economy in 2030, measured by real personal income, to be 75% larger than it was in the 1984–1987 base-line period, and farm income's share of total regional income, only 3.1% in 1984–1987, declines to 2.8% in 2030. (However, for reasons presented below, we believe the BEA projections may substantially underestimate the growth of MINK agriculture from now until 2030.)

The Agricultural Economy. Wheat, feed grains, and soybeans dominate world crop production, and they dominate world agricultural trade in crops even more. MINK agriculture, therefore, is integrally tied to world agriculture, and its future cannot be adequately evaluated except in the context of the future of world agriculture.

We performed this evaluation in several steps described in detail in Rosenberg et al. (in press). Our analysis of prospective global supply-and-demand conditions for grains and soybeans and of the U.S. and MINK's competitive position in world markets for these crops suggests several plausible scenarios. In the one presented here, world trade in grains and soybeans grows in proportion to world demand for them (driven primarily by population and per capita income growth in the developing countries), the U.S. shares of world trade remain at 1980s levels, and MINK's shares of U.S. production remain as they were in the 1984–1987 base-line period.

Under these conditions, MINK's production of grains and soybeans would be as indicated in Table 10.10. In this scenario, U.S. demand for grains and soybeans grows with population— ≈25% from 1984–1987 to 2030 (United Nations, 1989). On the plausible assumption that MINK's production is divided between domestic and export demand in the same proportion as in the nation, most of the increase in the region's production of wheat and soybeans would be for export abroad, as would well over half the increase in sorghum production, and the projected 40% increase in corn production would be split almost evenly between domestic and export demand.

In this scenario, production of the four crops grows 67% over the base-line amount, not much less than the BEA projection for regional income (75%). Thus, we believe that the BEA projections may overstate the decline in agriculture's share of regional income.

Effects on Crop Production. As in the B scenarios, the impacts of the analog climate on 2030 base-line yields depicted in the D scenarios would change the base-line costs of crop production, with consequent effects on the amount of production. Also, the changes in production would reflect both the yield effects and the effects of shifts of some land out of irrigated corn production into dryland production of wheat and sorghum.

Table 10.11 shows the changes in the value of crop production from the 2030 base line. The table corresponds to Table 10.8 in showing both the worst-case and best-case situations, as well as a scenario that permits calculation of the difference made by the on-farm adjustments. Comparison of the worst-case scenarios in Tables 10.8 and 10.11 shows that without the

TABLE 10.10   Production of Grains and Soybeans in MINK in 1984–1987 and Projected for 2030, Without Climate Change

	1984–1987		2030		
	(millions kg/m²)	*(millions 1982 $)*	*(millions kg/m²)*	*(millions 1982 $)*	*Percent Increase 1980–1987 to 2030 (%)*
Corn	7050	7,714	9900	10,832	40
Wheat	1430	1,832	2900	3,715	103
Sorghum	1260	1,258	2100	2,097	67
Soybeans	1640	3,431	3400	7,113	107
Total		14,235		23,757	67

Source: 1984–1987 from U.S. Department of Agriculture (1988); 2030 projections of weight per unit area are described in Crosson et al. (1991).

CO_2 effect or on-farm adjustments (scenarios B_1 and D_1), the analog climate would decrease production of the four crops by 22% from the 2030 base line (Table 10.11) and by 19% from the current base line (Table 10.8). However, because total crop production would be substantially greater in 2030 (Table 10.10), the decline in production—$5285 million in 1982 prices—would be almost double the decline from the current base line—$2662 million (Table 10.8).

Comparison of the best-case scenarios shows that production actually would increase from the 2030 base line (Table 10.11) while it declined by ≈7% from the current base line (Table 10.8). The increase in production from the 2030 base line results from the favorable effects of the increased CO_2 concentration and from the assumed success of researchers and farmers in the region in devising technologies to counter the effects of the analog climate. Although we think the emergence and adoption of these technologies would be plausible, they would contribute to greater crop production in MINK only if other competing regions lacked access to them. The record shows that new agricultural technologies diffuse rapidly around the United States (and the world). Should the technologies that boost MINK output in our best-case D scenario actually become available by 2030, they more likely would result in lower grain and soybean prices than increased production on the scale shown in the scenario.

Effects on the Regional Economy. The production multipliers showing the impact of changes in agricultural production on the rest of the regional economy reflect technical and economic conditions determining MINK's

TABLE 10.11. Changes in the Value of 2030 Base-Line Crop Production Under the Analog Climate

	D_1 Scenarios		D_3 Scenarios			
			No CO_2		CO_2	
	No CO_2	No	With Adjustments		With Adjustments	
	(millions	Adjustments	(millions		(millions	
	1982 $)	(%)	1982 $)	(%)	1982 $)	(%)
Corn	−2621	−24	−1711	−16	−682	−6
Sorghum	−419	−20	+310	+15	+539	+26
Wheat	−424	−11	+126	+3	+639	+17
Soybeans	−1821	−26	−754	−11	+149	+2
Total	−5285	−22	−2029	−9	+645	+3

Source: Percent-yield changes from EPIC times the value of production shown in Table 10.10.

economic structure. To develop projections of the multipliers would require a detailed analysis of these underlying conditions, which is beyond the scope of this study. Something useful can nonetheless be said about the regional economic impacts of the D scenarios. The analysis of the B scenarios showed that if all of the decline in crop production were reflected in reduced exports, the effect would be to reduce total regional production by roughly 0.5–1.5% (Table 10.9). This result and the assumption of growth of the MINK economy to 2030 (75% according to the BEA projections) suggest that even if the crop-production multipliers were to double from the current to the 2030 base line, the decline in crop production in the worst-case D scenario would have a small percentage impact on the regional economy, if all the decline occurred in crop exports.

However, if the decline in crop production resulted in an equal reduction of feed-grain supplies to animal producers in MINK, and if the crop-production multipliers were to increase, then the analysis of the B scenarios suggests that the worst-case D scenario would imply a decline in total regional production of >10%—no longer a trivial amount. This worst-case result for the economy as a whole probably is less likely than the best-case result, for the same reason as in the B scenario: The decline in crop production is more likely to show up as reduced exports than as reduced feed-grain supplies to local animal producers.

The worst-case result in the D scenarios also assumes that animal production in MINK remains closely related to feed-grain production and that the animal–meat–packing relation remains important. The high multiplier effect of the decline in crop production in the worst-case result for the regional economy assumes continuation of the strong locational relationship in MINK between feed-grain production, animals, and meat packing.

Of course, if crop production in MINK should increase, as in the D_3 scenario with the CO_2 effect and on-farm adjustments, then the impact of the analog climate on the regional economy would be positive, at least for agriculture.

Summary and Conclusions

A methodology has been developed to allow assessment of the regional economic impacts of climate change and to assess the efficacy of responses to climate change. In the study reported above, we imposed an analog of the world climate change anticipated to follow greenhouse warming onto the four-state MINK region as it is today and as we project it may be 40 years from now when the climate change may actually be upon us. Under both present and future circumstances, the impacts of the climate-change analog on crop production and water demand and supply are calculated with and without adjustments by farmers and with and without an increase of 100 μmol/mol in atmospheric CO_2 concentration (above the current 350 μmol/mol).

Crop-yield response to the climate change–CO_2 adjustment cases was simulated with the EPIC model, especially adapted to deal with increased CO_2 and increased aridity. Changes in surface-water supplies were calculated by relating the flow in the 1930s at several gauging stations unaffected by human activities to the flow during the control period (1951–1980) at the same stations.

Were the climate change to occur now, yields of grain crops, except for wheat, would be lowered significantly. The reduction would be tempered by CO_2, and simple adjustments would lessen the yield reduction further still. Under the most optimistic of these scenarios, however, the productivity of the region's agriculture would be significantly diminished.

The number of days of water stress for crops would increase under the analog climate, indicating an increased requirement for irrigation to avoid losses of yield. However, surface-water flows would be considerably reduced, especially in the Missouri River basin, so that the increasing irrigation demand would probably not be met with greater withdrawals of surface water, for which there are already many competing demands. In much of the western MINK states, irrigation is based on depletable ground-water sources. A climate change would deplete these sources still faster.

Crop productivity and irrigation requirements in the absence of climate change were projected to 2030 by EPIC simulations. We adjusted EPIC to consider a set of foreseeable technologies. Yields of the grain crops considered will be 70–90% greater in 40 years, according to EPIC.

Irrigation requirements in the absence of climate change will not be much different in 2030 from the current situation, but by that time additional constraints on water supply are likely.

When the analog climate is imposed on the MINK region of 2030 with no adjustments by farmers and no CO_2 effect, yields are reduced in percentages not very different from those under the same conditions in the present situation. However, total production remains higher in 2030 than at present. The CO_2 effect and simple agronomic adjustments moderate the analog climate effects as before, but two additional adjustments that might be driven by an awareness of oncoming climate change, e.g., drought resistance and irrigation-efficiency improvements, tend to counteract the climate-induced losses more effectively. Indeed, after allowance for all these improvements and the CO_2 effect, crop production in MINK might be even a bit higher with the analog climate than without it.

The increased demand for irrigation water under climate change would be still more difficult to meet in 2030 because surface flows will be lower, and the aquifers underlying much of the region will by then be depleted or regulated for use at no more than sustainable rates.

The climate-induced declines in crop production would reduce total MINK production because of the links between crop production and the rest of the economy. If the decline in crop production resulted mostly in a reduction of crop exports, the decline in total regional production would be ≈1–2%. However, if the crop-production decline resulted mostly in reduced feed-grain supplies to local animal producers and, thus, of supplies of animals to the meat-packing industry, total regional production could fall 10% or more. The actual decline in total production likely would be closer to 1–2% than to 10%. Should crop production rise, as it does in one scenario, total regional production would be higher under the analog climate than without it.

We caution the reader not to think of the foregoing as predictions of what will be happening in the MINK region over the coming 40 years, but rather as illustrations of what might happen, given the region's agricultural and water-resource vulnerabilities to climate change and its economic structure.

Our methodology provides a framework for analysis that overcomes important limitations of previous studies. We recognize that our data sources, models, and analytical tools are imperfect, and that better ones need to be developed. There is reason for confidence, however, that such improvements are possible and that their development will keep pace with or lead advances in climate modeling of the kinds needed to provide reliable predictions of the regional distributions of climate change. As confidence grows in GCM predictions, their results can readily be used in the MINK study framework.

References

Bolin, B., B. R. Doos, J. Jager, and R. A. Warrick, eds. 1986. *SCOPE 29: The Greenhouse Effect: Climate Change and Ecosystems.* New York: John Wiley and Sons.

Campbell, D. C. 1984. "The Pick-Sloan Program: A Case of Bureaucratic Economic Power." *Journal of Economics Issues* 18: 449–56.

Crosson, P. R., L. A. Katz, and J. W. Wingard. 1991. *Report IIA—Agricultural Production and Resource Use in the MINK Region Without and With Climate Change.* Washington, D.C.: U.S. Department of Energy, Office of Energy Research.

Cure, F. D., and B. Acock. 1986. "Crop Responses to Carbon Dioxide Doubling: A Literature Survey." *Agriculture and Forest Meteorology* 38: 127–45.

Harbeck, G. E. Jr. 1962. "A Practical Field Technique for Measuring Reservoir Evaporation Utilizing Mass-Transfer Theory." Washington, D.C.: U.S. Government Printing Office. (Geological Survey Professional Paper 272-E.)

Intergovernmental Panel on Climate Change (IPCC). 1990. *Working Group I: Scientific Assessment of Climate Change.* Geneva: WMO/UNEP.

Kimball, B. A. 1983. "Carbon Dioxide and Agricultural Yield: An Assemblage and Analysis of 430 Prior Observations." *Agronomy Journal* 75: 779–88.

Kimball, B. A. 1986. "Influence of Elevated CO_2 on Crop Yield," in H.Z. Enoch and B.A. Kimball, eds., *Carbon Dioxide Enrichment of Greenhouse Crops.* Pp. 105–15. Boca Raton, Fla.: CRC Press.

Lamb, P. J. 1987. "On the Development of Regional Climatic Scenarios for Policy-Oriented Climatic-Impact Assessments." *Bulletin of the American Meteorological Society* 68: 1116–23.

Manabe, S., and R. T. Wetherald. 1986. "Reduction in Summer Soil Wetness Induced by an Increase in Atmospheric Carbon Dioxide." *Science* 232: 626–28.

Mooney, H. A., J. Ehleringer, and J. A. Berry. 1976. "High Photosynthetic Capacity of a Winter Annual in Death Valley." *Science* 194: 322–24.

Palmer, W. L. 1965. "Meteorological Drought." Washington, D.C.: U.S. Government Printing Office. (U.S. Department of Commerce, Weather Bureau, Research Paper 45.)

Penman, H. L. 1948. "Natural Evaporation From Open Water, Bare Soil and Grass." *Proceedings of the Royal Society of London, Series A* 193: 120–45.

Rosenberg, N. J., B. A. Kimball, P. H. Martin, and C. F. Cooper. 1990. "From Climate and CO_2 Enrichment to Evapotranspiration," in P.E. Waggoner, ed., *Climate Change and U.S. Water Resources.* Pp. 151–75. New York: John Wiley & Sons.

Rosenberg, N. J. 1982. "The Increasing CO_2 Concentration in the Atmosphere and its Implication on Agricultural Productivity. II. Effects through CO_2-Induced Climate Change." *Climatic Change* 4: 239–54.

Rosenberg, N. J., P. R. Crosson, W. E. Easterling et al. 1991. "Processes for Identifying Regional Influences of and Responses to Increasing Atmospheric CO_2 and Climate Change: The MINK Project." Report prepared for the U.S. Department of Energy, Office of Health and Environmental Research. Washington, D.C.: U.S. Department of Energy.

Ruttan, V. M. 1989. "Biological and Technical Constraints on Crop and Animal Productivity: Report on a Dialogue July 10–11, 1989." Staff Paper p89-45. St. Paul, Minn.: Department of Agricultural and Applied Economics, University of Minnesota.

Schlesinger, M. E., and J. F. B. Mitchell. 1985. "Model Projections of the Equilibrium Climatic Response to Increased Carbon Dioxide," in M. C. MacCracken and F. M. Lutner, eds., *Projecting the Climatic Effects of Increasing Carbon Dioxide.* Washington, D.C.: U.S. Department of Energy, Carbon Dioxide Research Division. (DOE/WR-0237).

Schneider, S. H., and N. J. Rosenberg. 1989. "The Greenhouse Effect: Its Causes, Possible Impacts and Associated Uncertainties," in N. J. Rosenberg, W. E. Easterling III, P. R. Crosson, and J. Darmstadter, eds., *Greenhouse Warming: Abatement and Adaptation.* Pp. 7–34. Washington, D.C.: Resources for the Future.

Stockle, C. O., J. R. Williams, N. J. Rosenberg, and A. C. Jones. In press. "Estimating the Effect of Carbon Dioxide-Induced Climate Change on Growth and Yield of Crops. I. Modification to the EPIC Model For Climate Change Analysis." *Agricultural Systems.*

Thornthwaite, C. W. 1948. "An Approach Toward a Rational Classification of Climate." *Geographical Review* 38: 55–94.

United Nations. 1989. *World Population Prospects 1988.* New York: Department of International Economic and Social Affairs.

U.S. Congress, Office of Technology Assessment. 1986. *Technology, Public Policy and the Changing Structure of American Agriculture.* Washington, D.C.: U.S. Government Printing Office. (OTA-F-285.)

U.S. Department of Agriculture. 1981. *Land Resource Regions and Major Land Resource Areas of the United States.* Washington, D.C.: U.S. Government Printing Office.

U.S. Department of Agriculture. 1984–1987. *County Level Estimate Tapes.* ASB 111. Washington, D.C.: National Agricultural Statistical Service.

U.S. Department of Agriculture. 1987 and 1988. *Agricultural Statistics 1987 and 1988.* Washington, D.C.: U.S. Government Printing Office.

U.S. Department of Commerce, Bureau of Economic Analysis. 1990. *Regional Projections to 2040. Vol. 1: States.* Washington, D.C.: U.S. Government Printing Office.

U.S. Department of Agriculture, Forest Service. 1989. *Micro IMPLAN Release 89-03 Help File.* Fort Collins, Colo.: U.S. Department of Agriculture, Forest Service, Land Management Planning.

Williams, J. R., C. A. Jones, and P. T. Dyke. 1984. "A Modeling Approach to Determining the Relationship Between Erosion and Soil Productivity." *Transactions of the American Society of Agricultural Engineers* 27: 129–44.

11

Imbedding Dynamic Responses with Imperfect Information into Static Portraits of the Regional Impact of Climate Change

Gary W. Yohe[1]

Introduction

It is becoming increasingly clear, at least on a theoretical level, that modelers of the potential impacts of climate change must impose that change on the world as it will be configured sometime in the future rather than confine their attention to considerations of what would happen to the world as it looks now. Initial base lines that focus on current circumstances are certainly worthwhile points of departure in any study, of course, but the truth is that social, economic, and political systems will evolve as the future unfolds; careful analysis of that evolution across a globe experiencing changes in its climate must be undertaken, as well. In the vernacular of the analysts' workroom, while it may be interesting to try to see what would happen to "dumb farmers," who continue to do things as they always have regardless of what happens, it is critically important to evaluate the need for any sort of policy response to climate change in a world of "smart farmers,"

[1] From the Department of Economics, Wesleyan University, Middletown, Conn. Supported by U.S. Department of Energy contract DE-AC06-76RL01830 and Connecticut Sea Grant. Helpful comments were offered during the early stages of this work by Al Liebetrau and Michael Scott of Pacific Northwest Laboratory in Richland, Wash.; Norm Rosenberg, William Easterling, and Mary McKenney of Resources for the Future, Washington, D.C.; and Thomas Malone of Sigma Xi and North Carolina State University in Raleigh.

who will have observed the ramifications of climate change and responded in their own best interest.

Taking this point from the theoretical to the practical can, however, be problematical. Models of climate change are growing increasingly complex—so complex, in fact, that it is frequently beyond the scope of the modelers to run even "interesting" time-dependent scenarios of what might happen; and capturing the full flavor of the uncertainty with which we view the future with full-blown probabilistic analyses appears to be even more impossible. There are simply too many random variables and too much structure in the typical model to accommodate even abbreviated Monte Carlo techniques. The best that can be expected in many cases is a short series of portraits of social and economic structures drawn for specific times in the future with the actors having reacted in some, usually equilibrium-based, way to some altered but static new climate.

It must be recognized even in these constrained analyses, however, that future experiences will be time-dependent. Future responses to those experiences will be dynamic adjustments informed not only by what is known at the present time, but also by what will have been learned as the future has unfolded. Reactions at any point in time will not necessarily be perfect reflections of how to best respond to the new and/or emerging climate of that time, though. They will, instead, be imperfect reactions based on a learning process that will have taken the most recent manifestations of that climate as imprecise evidence that the climate *might* have changed. The longer the new climate has been in force, the less imperfect will be the perceptions of the change, but perfect recognition of the new climate and perfect reaction to its ramifications cannot be expected. Just as it is incorrect to analyze the impact of change in a world of "dumb farmers," it is inappropriate to fill a model of the future with "clairvoyant farmers," who are too smart.

The issue to be confronted here, then, is how to create a methodology with which to capture the imprecision of adaptation within a static and time-specific portrait of the future without running a complete set of probabilistically weighted scenarios over the immediately preceding period. Were it possible to draw such a portrait, incorporating this imprecision into the actions of its subjects, a more accurate reflection would appear of what the social and economic evolutionary process will have achieved. It would therefore be an improved foundation on which to base subsequent exploration of the need for policies designed to accomplish (1) some additional adaptive response or (2) the frequently hailed dramatic and preemptory averting response.

This paper does not attempt to describe a general methodology. Instead, it begins the development of such a methodology by suggesting a means by which imperfect information might be translated into incomplete and imprecise reaction for a single, very specific decision—the choice that farmers face to switch crops, or at least their mix of crops, in response to growing evidence that the climate appears to be changing. To that end, the first section presents an outline of a utility-based decision model and demonstrates the type of individual reaction functions that the model supports in an uncertain world. The second section follows with an arbitrary, numerical illustration designed specifically to explore the structure of these reaction schedules. The theoretical underpinnings for a crop-switching decision are thereby established.

The third section of the paper turns practical, applying the same structure used in the preceding sections to data produced by the agriculture team involved in a thorough, methodologically focused analysis of the impact of potential climate change on the resources and economies in a four-state region located in the center of the United States.[2] A specific farm, defined by location variables and confronted with a specific crop-switching decision in light of growing evidence that the "dust-bowl" climate of the 1930s has returned in the wake of greenhouse-induced climate change, is examined. Farm reaction given 5 and 25 years of experience with the new climate is postulated, and methods for aggregating individual farm responses into regional pictures of an agricultural sector in flux are proposed.[3]

Concluding remarks propose some general insights that can be supported from the lessons of the agriculture modeling. There is reason to believe that these insights could turn out to be quite robust; agriculture is, after all, high on virtually every list of economic sectors likely to sustain significant impacts from greenhouse-induced climatic change.[4] Moreover, the general notion

---

[2] The MINK Study is a regional impact assessment that concentrated on a four-state region (Missouri, Iowa, Nebraska, and Kansas), which was sponsored by the U.S. Department of Energy. It began in 1988 and was completed in late 1990. It is the product of a collaborative effort of Resources for the Future, Oak Ridge National Laboratory, Pacific Northwest Laboratory, and Sigma Xi, and much of its attention was focused on issues of methodology. The analytical structure employed evolved into a system of many components (notably, agriculture, energy, forests, and water) with many interactions and many interdependencies. Each interaction was defined fundamentally in terms of regional macroeconomic relationships, which were, in turn, derived directly from the microeconomic foundations of the sectors involved.

[3] The MINK Study is designed to produce portraits of the impacts of climate change in the years 2010 and 2030. It will be assumed, therefore, that dust-bowl conditions appear in the region beginning in the year 2005.

[4] Nordhaus (1990a, 1990b), for example, used the 1989 U.S. Environmental Protection Agency (EPA) report to Congress to review the national income accounts in search of sectors that might be affected by greenhouse-induced climatic change. Only agriculture, forestry, and fisheries fell into the severely impacted category.

that motivates the approach proposed here is a simple one—if the entire model is too complex to incorporate uncertainty and imperfectly informed decision making throughout, then construct an overall impact portrait of a socioeconomic system in flux by (1) introducing uncertainty into the simpler submodules of the larger model, (2) investigating response mechanisms in these simpler contexts, and (3) summing these responses across submodules in an economically consistent way.

Crop Decisions in a Utility Framework

Let a farmer's utility be given by

$$u(y) = \{y^{\beta+1}/(\beta+1)\} \tag{1}$$

where y reflects the profit derived from farm activity, and the parameter β is the usual Arrow-Pratt measure of relative risk aversion.[5] Farm profits are assumed to be derived from a mix of two crops (crops 1 and 2, denoted x_1 and x_2) according to

$$y(a,x_1,x_2) = ax_1 + (1 - a)x_2 \tag{2}$$

where x_i represents the profitability per 10^4 m^2 (or, per hectare) of crop i, and the parameter a represents the proportion of farming activity, expressed in the area cultivated, devoted to crop 1. The two crops can, in fact, represent specific crops or specific rotations. The only critical point here is that they capture alternative uses for a given farm or farmland. Scale does not matter, either, given the structure of equation 1. The farm can, therefore, be assumed to cover only 10,000 m^2 (1 ha) for convenience; but the parameter a can be interpreted more generally as the proportion of farmland within a given homogenous region devoted to crop 1.[6]

Moderately affected sectors included construction, water transportation, real estate, energy, and recreation.

[5] The Arrow-Pratt measure of relative risk aversion is a curvature notion, defined precisely by $R(y)=yu''/u'$; for the schedule given in equation 1, then, $R(y)$ is constant and equal to β. It is proportional to the insurance that an individual would purchase to avoid a random lottery on income, so larger values for β (in magnitude, because $\beta<0$) correspond intuitively to larger aversion to taking risks. *See* Arrow (1970).

[6] More than two crops can be accommodated by simply extending equation 2 through some arbitrary x_n. The joint probability distributions for the $\{x_1,...x_n\}$ could be more complex, of course, but the procedure outlined here would still be perfectly applicable.

Now add uncertainty to the model by representing the joint distribution of profitability of crops 1 and 2 from year to year under the existing climate by $f_o(x_1,x_2)$, and the analogous distribution under a potential new and different climate by $f_n(x_1,x_2)$. These economic-yield distributions can be assumed to be reasonably well-known. They are, perhaps, the result of simulation exercises based on crop-yield models that currently exist or as they will exist after being refined by climate-related research undertaken in the future.[7] It is their respective applicability that is in doubt in the face of possible climate change.

Since farmers, in making their decisions, are likely to look to experts and agricultural stations to provide information about the relative likelihood that the climate has changed, it is appropriate to let π_t be an index of the farmer's subjective perception at time t that the climate has indeed changed so that $f_n(x_1,x_2)$ and not $f_o(x_1,x_2)$ applies. Note that this structure implies that farmers understand the implications of possible climate change on their yields but are unsure of whether or not it has occurred. Experts have provided the $f_j(x_1,x_2)$ distributions to their own satisfaction, in other words, but are in disagreement over the value assumed by π_t. It takes 30 years to define a climate change, and experts have been left to try to digest the content of series of annual weather patterns in the meantime.

The farmer's decision at any point in time can now be characterized as one of selecting a^*, which maximizes the expected utility derived despite uncertain profitability

$$EU[y(a)] = (1-\pi)\int\int u[y(a,x_1,x_2)]f_o(x_1,x_2)dx_1dx_2$$

$$+ \pi\int\int u[y(a,x_1,x_2)]f_n(x_1,x_2)dx_1dx_2$$

subject to the constraints defined in equations 1 and 2. Algebraic manipulation of the appropriate first-order condition reveals that

$$\frac{(1-\pi)}{\pi} = -\frac{\int\int [a^*x_1+(1-a^*)x_2]^\beta[x_1-x_2]f_n(x_1,x_2)dx_1dx_2}{\int\int [a^*x_1+(1-a^*)x_2]^\beta[x_1-x_2]f_o(x_1,x_2)dx_1dx_2} \qquad (3)$$

$$\equiv k\{a^*,\beta\}$$

---

[7] EPIC (for Erosion Productivity Impact Calculator) is a biomass-yield model developed at Texas A&M University and adapted in consultation with Resources for the Future to accommodate climate change with CO_2 fertilization. Evolution of such models in the future would be based on actual experience and some reflection of an associated learning process. Current-likelihood weights can be used to incorporate what might be learned along alternative scenarios into a weighted set of $f_t(x_1,x_2)$ distributions. *See* Yohe (in press a and b).

The interior structure of equation 3 implicitly defines a reaction function $a^*(\pi,\beta)$ by requiring that

$$\frac{(1-\pi)}{\pi} = k \{ a^*(\pi,\beta),\beta \} \qquad (4)$$

For each and every degree of risk aversion, therefore, equations 3 and 4 define a correspondence between the perceived likelihood of climate change and the desired allocation of farming effort between crops 1 and 2.

An Illustrative Example

The meaning and significance of the class of reaction functions characterized in equations 3 and 4 are perhaps best understood when cast in terms of a specific but qualitatively realistic illustration. Climate change is, for example, expected to bring hotter and drier weather to the midwestern portion of the United States.[8] As a result, crop yields and associated profits from farming could easily fall, even as they become more variable from one season to the next. Constructing any illustrative example meant to display possible effects on and responses from the agricultural sector of that region should therefore reflect both trends (i.e., falling and becoming more variable).

Suppose, to that end, that profits per 10^4 m^2 from x_1 and x_2 were distributed from year to year according to

$$x_1 \sim N(10.0;2.0) \qquad (5a)$$

and

$$x_2 \sim N(9.0;3.0) \qquad (5b)$$

prior to any change in climate and, in the aftermath of such a change, would be distributed according to

$$x_1 \sim N(8.0;2.0) \qquad (5c)$$

and

$$x_2 \sim N(7.5;2.5) \qquad (5d)$$

---

[8] *See* Rosenberg and Schneider (1988) for some introduction to the uncertainty with which such a statement can now be made.

These distributions show that mean profitability for both crops would fall and become more variable if the climate was to change.[9] Because the x_1-x_2 ratio of mean yields and t statistics would both fall in the wake of climate change, moreover, it is clear that crop 1 would be more severely effected both in terms of lower yields and increased variability.

The change in means reflected in equations 5a through 5d would certainly make crop 2 relatively more attractive in the new state of nature (i.e., the new climate), but lower relative expected yields alone would not be enough to inspire a change in farming activity. Crop 1 would still dominate, even given the new climate, if the ranking was judged solely on the basis of average profitability (note that 8.0 > 7.5 just as 10.0 > 9.0). No adjustment in the mix of crop should be forthcoming, therefore, unless risk aversion placed some weight on the expected utility cost of increased relative variability.

Figure 11.1 displays this tradeoff by portraying the results of applying equation 4 to the distributional structure of equation 5. The reaction schedules reflected there for various Arrow-Pratt values of relative risk aversion are the product of simulation exercises that assumed perfect correlation across x_1 and x_2 *within any year* for both states of nature. They show the full range of correspondence between the subjective probability that the climate has changed and the proportion of farming effort and land devoted to crop 1.

Consider one of the more extreme cases shown in Figure 11.1. Note that a farmer with low relative risk aversion (RRA = −0.8) would devote 100% of his or her effort to x_1 until it became nearly certain that the climate had changed; even with perfect certainty that the new state of nature had arrived, in fact, the proportion of effort devoted to x_1 would only fall to 60%. This case clearly illustrates the general notion that low risk aversion allows considerable weight to be given to the persistent higher average profitability of crop 1. On the other side of the coin, a farmer with a strong aversion to risk (RRA = −8.0) would specialize in crop 1 under the current climate but would respond quickly and dramatically to even the hint of climate change; effort devoted to x_1 would plunge to 20% with only a 10% perceived chance that the new climate had arrived and would disappear completely when experts were 30% certain.

The intermediate cases reflected in Figure 11.1 are, of course, far less definitive. Given any probability of climate change, Figure 11.1 shows that almost any level of response could be expected depending on the degree of risk aversion of the farmer in question; the specifics of this illustration were, in fact, chosen so that this full range of reactions would emerge. Were it

---

[9] Variability here is measured by the t statistic.

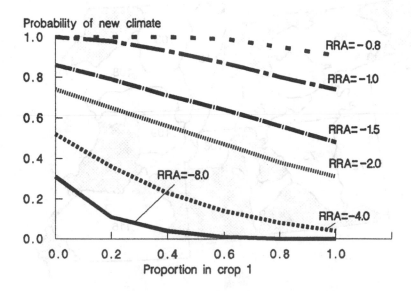

FIGURE 11.1 Results of applying equation 4 to the distributions recorded in equations 5a through 5d.

possible to restrict the potential values assigned to the RRA parameter, perhaps by prescribing initial conditions based on the relative importance of crops 1 and 2 under current conditions, then a more limited range of anticipated responses might be achieved; but specifying risk aversion is only part of the problem. Regardless of the limits placed on the risk parameter, applied work based on the modeling in the previous section must also be supported by methods designed to reflect subjective views of the likelihood of significantly altered climate. The next section addresses both issues—the need to specify ranges for RRA and the subjective probability that climate has changed—in the context of an application to one type of farm captured in the Missouri, Iowa, Nebraska, and Kansas (MINK) Study (Rosenberg and Crosson, 1990).

Application within the MINK Study

The agricultural portion of the MINK Study incorporated extraordinarily detailed microanalyses of typical farms from 11 major land resource areas (MLRAs) scattered across the four-state region. The map portrayed in Figure 11.2 displays their coverage. Together, these 11 subregions can be used to span all of the important crop rotations, soil types, distinct weather patterns, and irrigation practices found in the MINK states. One farm in MLRA 109, for example, is included as a representative of farms located in

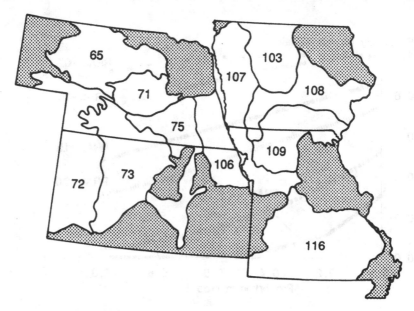

FIGURE 11.2 Major land resource areas (MLRAs) selected for study in the MINK region (i.e., Missouri, Iowa, Nebraska, and Kansas).

southern Iowa and northern Missouri that currently grow soybeans and corn in biannual rotation without extensive irrigation. Experts consulted by the MINK Study group have suggested that sorghum be substituted for corn in the rotation if greenhouse-induced climate change was to return to the dust-bowl weather patterns of the 1930s.[10] This substitution decision will be examined here as an illustration of how the utility structure of the section on crop decisions in a utility framework might be applied within a holistic regional impact assessment.

The Erosion Productivity Impact Calculator (EPIC) agriculture-yield model was first run for the target farm in MLRA 109 (henceforth, farm 109) by using 30 years of actual daily weather experience recorded there from 1951 through 1980. These runs suggest, in the context of current price expectations and commonly accepted cost structures, that the annual

---

[10] Absent any climate, much less weather, information from global-circulation models for regions as small as MINK, the MINK Study has used actual weather from the dust-bowl years as a climate analog representation of the potential impact of greenhouse-induced warming. It was therefore possible to run the agriculture-yield simulations given actual weather experiences and to contrast the results with yields that are supported by the current climate and associated weather. It was also possible to provide diversity in weather within the region that is nonetheless consistent with a clear change in climate.

profitability per 10^4 m^2 of growing corn in the soybean-corn rotation given current climatic conditions is distributed by

$$c_o \sim N(112.9;137.8) \tag{6a}$$

The corresponding distribution of the annual profitability of growing sorghum under current conditions is, meanwhile,

$$s_o \sim N(27.2;73.5). \tag{6b}$$

Repeating the EPIC simulation exercise for three successive decades of dust-bowl climate defined by actual weather patterns observed in MLRA 109 from 1930 through 1939 suggests comparable analog profitability distributions for corn and sorghum given, respectively, by

$$c_n \sim N(-20.2;120.7) \tag{6c}$$

and

$$s_n \sim N(24.3;81.6). \tag{6d}$$

Corn obviously dominates under the current climate for all but the most risk-averse farmer; indeed, sorghum is *not* now an important crop in MLRA 109. Were the change in climate perfectly recognized, however, the switch to sorghum would obviously pay dividends; the experts are right. In a world of uncertainty and imperfect recognition, though, the questions posed here remain. How sure must farmers be that the climate has changed before they decide to switch crops? And how sensitive is that decision to their relative aversion to risk?

The answers to these questions, specific to farm 109 at least, can be gleaned from Figure 11.3. The reaction schedules drawn there were derived by applying the content of equations 2 and 3 to the distributions recorded in equations 6a through 6d.[11] They are the product of simulations that again assumed perfect correlation of profitability for corn and sorghum

---

[11] Variable profitability in each case was measured against a positive base income without loss of generality so that the utility structure of equation 1 could be applied even with the negative returns that emerge in equations 6a through 6d. The notion of relative risk aversion is unaffected.

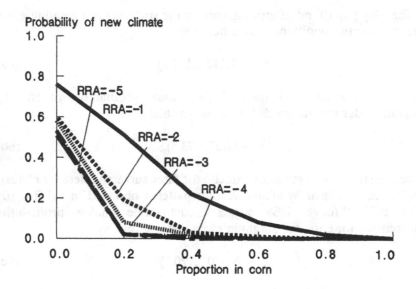

FIGURE 11.3 Results of applying equations 2 and 3 to the distributions recorded in equations 6a through 6d.

within any single year in both states of nature.[12] Notice that complete specialization in corn dominates in all five cases when there is no perceived chance that the climate has changed. For all but the smallest aversion to risk, however, even a small perceived chance that a new climate has arrived precipitates a significant response. A 10% chance, more specifically, inspires at least a 60% shift to sorghum for the other four schedules, and 80% adjustments for three of the RRA values depicted. Perhaps more importantly, though, complete specialization in sorghum is anticipated in four cases when it is perceived to be 60% likely that the dust bowl has returned.

Turning now to the issue of the subjective likelihood that climate has changed, the issue is specifying a reasonable range for the π_t parameter. Suppose that agriculture stations base their climate announcements on the basis of 30-year moving averages of annual rainfall.[13] A portrait of the year

---

[12] The actual correlation coefficient of profitability in both states of nature lies above 0.98, so assuming perfect correlation does little violence to the reality of the EPIC simulation results.

[13] It is widely recognized that it takes 30 years to define climate, so 30-year moving averages are at least reasonable tools for assessing changes in climate.

2010 under dust-bowl weather might reasonably assume that the citizens of MLRA 109 will have experienced 5 years of associated weather and thus rainfall. Drawing 5 years of weather at random from the climate of the 1930s and letting the previous 25 years be drawn similarly from the current climate, the 30-year moving average would assign a minuscule probability value to the null hypothesis that average annual precipitation at farm 109 would match the 1930s' average. If experts then quoted this probability value as the likelihood that the change had occurred so that $\pi_{2005} \approx 0.0$, then Figure 11.3 shows that no shift to sorghum in the rotation should be expected to have occurred. Aggregation across MLRA 109 assuming the "dumb-farmer" base-line scenario would then be appropriate.

If a portrait were drawn for the year 2020 with 15 years of dust-bowl weather having been experienced, however, the story would change. Drawing 15 years of weather from both the 1930s and the current climate, the 30-year moving average would assign a probability value of 0.34 to the null hypothesis. If experts then announced a 34% probability that the climate had changed, then Figure 11.3 shows a significant shift to sorghum. Aggregation across the MLRA assuming that between 75% and 90% of the farming activity was devoted to sorghum and the rest remained in corn could be advanced as a more accurate representation of the subregion in flux—responding to the imperfect but growing realization that the new climate had arrived. It would, in other words, paint a better portrait of the region than either the "dumb-farmer" snapshot, with everyone still growing corn, or the "clairvoyant-farmer" vision, with everyone switched to sorghum.

Clairvoyance would, however, be appropriate 10 years later. Applying the same procedure to the year 2030 with 25 years of experience with dust-bowl conditions would again assign a minuscule probability value, this time to the alternative null hypothesis that the climate had not changed. A value of 0.0% might then be assigned to π_{2030}, and Figure 11.3 shows that a total shift to a soybeans-sorghum rotation should have been accomplished throughout MLRA 109.

Notice, finally, that application of the decision model to all three years produces different but clear pictures of how much response might be expected without severely limiting the range of risk aversion. This clarity is, of course, the fortunate result of bunching among the reaction schedules reflected in Figure 11.3—bunching that was, in turn, generated by the dramatic changes in the profitability distributions associated with moving to the dust-bowl experience. Less dramatic change would produce more ambiguous results, but would also be less important.

Translating the reactions just noted for the years 2010, 2020, and 2030 into expected profitability statistics can, however, suggest that the implications of their incorporation into the aggregate economic picture of the MINK region can prove to be extremely significant. Taking the distributions recorded in equations 6a through 6d as given and accepting the modeler's 20-20 vision that climate has indeed changed, notice that no switch to sorghum with 5 years' experience by 2010 can be expected to reduce average profits for farms represented by farm 109 by over $133/10^4$ m^2 (from $113 to −$20). Summing across all such farms in MLRA 109 would then show considerable economic damage—damage that would filter down to other sectors of the MINK region through macroeconomic feedbacks. Assuming that an 80% switch to sorghum by 2020 would raise average profits from −$20/10^4$ m^2 to $15/10^4$ m^2 (the 20–80% weighted sum of −$20 from corn and $24 from sorghum) and reduce the overall economic impact by 26%. Minimum losses from the new climate, equalling $89/10^4$ m^2 given complete specialization in sorghum, would represent a 33% reduction in overall cost that would be accomplished by 2030. Clearly, better information that could speed the adjustment process would pay large dividends not only for the individual farmer, but also for the MINK region as a whole.

Concluding Remarks

The motivation behind the simple, utility-based decision analysis described here is purely practical. The current debate over how and when to respond to the threat of greenhouse-induced climate change must be informed by modeling exercises that cast the range of potential impacts of that change against the world as it will evolve as the future unfolds. It is not enough to study the possible effects on today's world because individuals, institutions, and even markets will adapt, at least to some degree, to change as it occurs. There is, as a result, a fundamental need for analysts who are evaluating the relative efficacy of exogenous policy responses to the threat of climate change to incorporate into the aggregate economic picture the "moving" pictures of how societies and economies react endogenously, if imperfectly, to the potential ramifications of climate change.

Models designed to support this type of policy analysis have, unfortunately, grown so large and complex that they cannot easily accommodate this demand for incorporating endogenous, intertemporal adaptation. They are typically too large to allow the iterative generation of the necessary time-series trajectories of important state variables; they tend to focus, instead, on static portraits of large systems, which can be produced for specific benchmark years scattered at regular intervals into the future. In the process of drawing these portraits, therefore, modelers are left with the task of deciding on an almost ad hoc basis what sort of adaptation might have taken place by each of those benchmark years.

The present paper accepts this modeling constraint, but suggests that it might actually apply only across an entire system taken as a whole. Systems approaches to climate-change impacts usually rely on the integrated sum of many submodules to produce their aggregate results, and these submodules are usually smaller, easier to manipulate, and come closer to portraying the behavior of the actors who will, in fact, be doing the adapting.[14] Even if it were impossible to conduct probabilistic analyses of adaptation decisions across the entire system, might it not therefore be the case that specific decisions, made over time despite imperfect recognition of the extent to which climate has changed and fundamental uncertainty about its ultimate impact, could be investigated? Consistent integration of these imprecise decisions over the larger system could then add some appropriate sense of flux and transition to the static portrait of the aggregate system.

The theoretical utility structure presented here suggests that the answer to this query can be affirmative. Its application to farmers' decisions to switch crops in the face of the return of severe dust-bowl weather reaffirms that answer for the MINK impact-assessment methodology. It should be noted, however, that careful attention was paid in the development of the MINK Study to the process of aggregating the micro impacts of climate change into catalogs of macroeconomic consequences. This emphasis on bottom-up design certainly makes it easier to accommodate the entire range of sparse, partial, or complete reactions that can emerge from micro decision analyses under uncertainty. Difficult as it is to wade through the details of a bottom-up construction, the resulting ability to look at sectors in flux adds a whole new distributional dimension to impacts analysis. Glimpses of who responds too early and who responds too late can show who loses a little and who loses a lot and can provide some insight into the equity and efficiency consequences of new information.

References

Arrow, K. 1970. *Essays in the Theory of Risk Bearing*. Chicago: Markham.

Liebetrau, A., and M. Scott. 1990 "Strategies for Modeling the Uncertain Impacts of Climate Change." *Journal of Policy Modeling* 13(2): 185–204.

Nordhaus, W. 1990a. "Uncertainty about Future Climate Change." New Haven, Conn.: Yale University. (Yale discussion paper.)

Nordhaus, W. 1990b. "To Slow or Not to Slow." New Haven, Conn.: Yale University. (Yale discussion paper.)

Rosenberg, N., and P. Crosson. 1990. "A Methodology for Assessing Regional Economic Impacts and Responses to Climate Change—The Mink Project: An Overview." Washington, D.C.: Resources for the Future. (RFF discussion paper.)

---

[14] *See* Liebetrau and Scott (1991) for a discussion of how to exploit this modular structure in conducting uncertainty analysis.

Rosenberg, N., and S. Schneider. 1989. "The Greenhouse Effect: Its Causes, Possible Impacts and Associated Uncertainties," in N. J. Rosenberg, W. P. Easterling, P. R. Crosson, and D. Darmstadter, eds., *Greenhouse Warming: Abatement and Adaptation.* Pp. 7–34. Washington, D.C.: Resources for the Future.

Yohe, G. In press a. "The Cost of Not Holding Back the Sea." *Journal of Ocean and Shoreline Management.*

Yohe, G. In press b. "Toward a General Criterion for Selecting Interesting Scenarios with Which to Analyze Adaptive and Averting Response to Potential Climate Change." *Climate Research.*

U.S. Environmental Protection Agency. 1989. *Policy Options for Stabilizing Global Climate.* Washington, D.C.: U.S. EPA.

12

Biological Emissions and North-South Politics

Thomas Drennen and Duane Chapman[1]

Introduction

The Montreal Protocol (UNEP, 1987) was the first substantive international agreement to reduce future emissions of a potent family of greenhouse gases, the chlorofluorocarbons (CFCs).[2] Initiatives are currently under way to forge agreements on other greenhouse gases. Negotiating strategies range from drafting agreements on single gases, such as CO_2, to forging comprehensive agreements that establish composite allowable emission levels for several or all known greenhouse gases.

Historically, attention focused on CO_2 as the primary greenhouse gas. More recently, concern shifted to the other gases, such as methane (CH_4), nitrous oxides (N_2O), CFCs, and tropospheric ozone.[3] One reason for this increased interest is their comparative growth rates: while CO_2 concentrations increased by 4.6% from 1975 to 1985, concentrations of CH_4 increased by 11.0% and concentrations of several of the CFCs more than doubled (Ramanathan, 1988). Many of these gases are more effective than CO_2 on a per-molecule basis at trapping infrared radiation. As a result, Ramanathan (1988) reported, the non-CO_2 gases contributed $\approx 50\%$ to the warming effect for the period 1975–1985.

---

[1] From the Department of Agricultural Economics, Cornell University, Ithaca, N.Y.

[2] The Montreal Protocol's primary purpose is to eliminate chemicals that break down stratospheric ozone, resulting in increased ultraviolet radiation reaching the earth's surface.

[3] In the lower atmosphere, the troposphere, ozone acts as a greenhouse gas, trapping infrared radiation. In the upper atmosphere, the stratosphere, ozone screens out harmful ultraviolet radiation.

Partially as a result of the widespread recognition of these other gases, there has been increased interest in comprehensive agreements. Theoretically, the way such an agreement might work would be to establish an index to weight the global-warming potential of each greenhouse gas, similar to the ozone-depleting-potential index contained in the Montreal Protocol.[4] One possible weighting scheme, suggested by the U.S. Department of State (1990), would assign each molecule of CO_2 a rating of 1, each unit of CH_4 a rating of 25, and each unit of CFC-12 a rating of 15,000. A reduction goal would then be established that would give each country broad latitude in how best to meet the target given its particular needs and cultural values. Consider one view of how this approach might work:

"Some nations might be able to reduce CO_2 emissions below their limit, such as through substitution of non-fossil fuels, but be unable to reduce CH_4 output (e.g., a nation importing oil and dependent on rice crops, but endowed with untapped solar power opportunities). Those nations would meet their net limits by reducing $2CO_2$ more rapidly than CH_4; requiring them to limit each gas by the same amount would prove much more costly (perhaps in terms of lower economic growth, higher taxes, or reduced rice production) and would leave additional affordable CO_2 reductions unexploited. Other nations might find themselves in the opposite situation, able to afford to limit CH_4 more than CO_2 (e.g., a nation dependent on coal reserves) but able to modify the diet of its ruminant animal husbandry." [5]

Through a discussion of the sources of CH_4 and, in particular, the emissions from bovine animals, this paper demonstrates potential problems with implementing the State Department proposal. Four central questions arise. The first concerns the difference between the instantaneous radiative effect used by Ramanathan (1988) and the total long-term effect. A molecule of CH_4 has an instantaneous effect 25 times greater than a molecule of CO_2, but it also has a much shorter atmospheric lifetime, decaying to CO_2 in 10–14 years. Does ignoring this fact overemphasize the importance of CH_4 as a greenhouse gas?

The second question concerns the importance of the origin of the different gases. Is CH_4 released from a cow really the same as CH_4 released from the mining and transmission of natural gas? In the latter case, new carbon is being added to the atmosphere, whereas CH_4 from bovine animals includes carbon that was once in the atmosphere.

Third, what is it likely to cost to reduce emissions of CO_2 compared with CH_4? Comparatively little is known about the costs of reducing CH_4

---

[4] Montreal Protocol, Annex A: Controlled Substances (UNEP, 1987).
[5] U.S. Department of State (1990), pages 15–16.

emissions from bovine animals. Recent estimates are presented that raise the question of whether CH_4-emission reductions would make economic sense.

Finally, there is a question that touches on North-South politics. An international agreement that focuses on reductions in CO_2 emissions would put the largest burden of responsibility on industrialized countries, which to date have been responsible for a large percentage of the increased atmospheric CO_2. However, by including other gases such as CH_4, the emissions of CH_4 from the animal population and rice paddies of developing countries become much more important in terms of their contribution to greenhouse warming.[6] Is this what the United States and other industrialized countries are really pursuing by pushing for a comprehensive agreement?

Sources of Methane

Background

Reaching agreement on meaningful reduction strategies for any greenhouse gas requires a thorough understanding of the sources and sinks for that gas. Consider the sources of CH_4 (Table 12.1). The largest source is natural wetlands and bogs, where CH_4 is continuously formed through anaerobic decomposition of organic matter. Other sources include rice paddies; enteric fermentation (the intestinal fermentation that occurs in animals such as cows); biomass burning; coal mining; the drilling, venting, and transmission of natural gas; and termites. Few, if any, of these sources seem susceptible now to accurate data estimates of emissions, effective regulation, or monitoring of plans for emissions reductions. However, the State Department (1990) targets both rice production and ruminant animals as possible CH_4-reduction sources in its proposal.

Bovine animals in general are poorly understood as a source of CH_4 emissions. Although estimates of the magnitude of this source exist, it is not a precise number and it is certainly not uniform among bovine animals, but depends on such factors as temperature and feed quality and quantity.[7] One

---

[6] It is, of course, true that an agreement regulating CO_2 alone would affect the future growth rates of energy use in developing countries. However, an agreement on CH_4 would have to affect current agriculture practices in these same countries.

[7] A recent article in the *New York Times* (O'Neill, 1990) typifies the increased focus on CH_4 from bovine animals. The article cites "bovine flatulence" as a significant source of CH_4, accounting for "...up to 400 liters of methane [per animal] per day."

TABLE 12.1 Sources of Methane: Annual Emissions of Methane into the Atmosphere

Source	Quantity (Tg)[a]	Proportion of Total (%)
Natural wetlands (includes bogs, swamps, tundras)	115	21.3
Rice paddies	110	20.3
Enteric fermentation (ruminant animals)	80	14.8
Biomass burning (includes fuel wood, agricultural burning, forest fires)	55	10.2
Gas drilling, venting, transmission	45	8.3
Termites	40	7.4
Landfills	40	7.4
Coal mining	35	6.5
Oceans	10	2.1
Fresh waters	5	1.1
Total	535	99.4

[a]Teragrams, or 10^{12} g.

Source: Cicerone and Oremland (1988).

has to ask how an agreement to limit this source would be monitored. The next section clarifies the process of CH_4 production in ruminants and attempts to reconcile estimates by various authors in terms of quantities of CH_4 produced.

Ruminant Production of Methane

The process begins with the ingestion of plant material. Rather than relying on enzymes to break down the plant material, the stomach, referred to as the rumen, relies on microorganisms that ferment the material. The fermentation produces volatile fatty acids, CH_4, and CO_2 (Wolin, 1979). The gases are removed by belching, with a gas composition of $\approx 27\%$ CH_4, 65% CO_2, and traces of other gases (Miller, 1991). The basic reactions are

$$HCO_2OH \rightarrow CO_2 + H_2$$

$$4H_2 + CO_2 \rightarrow CH_4 + 2H_2O$$

There are two widely quoted sources on CH_4 quantities produced by ruminants. The two sources are discussed and compared below.

The first, and probably more widely quoted, source is Crutzen et al. (1986).[8] Crutzen et al. use an energetic approach to calculating worldwide CH_4 emissions. They first examine feeding practices in three representative countries, the United States, Germany (representative of Europe), and India (representative of developing countries). They then take available estimates of energy losses due to CH_4 releases as a function of feed quality and quantity and estimate average emission rates. Table 12.2 summarizes their calculations for the United States. For example, milk cows, which comprise 10% of U.S. cattle, consume an average of 10,150 feed units/day. A feed unit is defined as equivalent to the amount of energy contained in 0.45 kg (1 lb.) of corn. The gross energy intake is equivalent to 230 MJ.[9] Of this amount, $\approx 5.5\%$, or 12.65 MJ, of energy is lost by the belching of CH_4. Assuming that 1 kg of CH_4 is equivalent to 55.65 MJ, this implies an annual emission of 83 kg/animal. For the other two types of bovine animals, feed and range cattle, Crutzen et al. estimate annual CH_4 releases of 65 and 54 kg, respectively. These estimates imply a weighted average of 58 kg of CH_4 per animal per year. Note that this number does not include consideration of the CH_4 content of animal feces.

---

[8] This is the source cited by Wuebbles and Edmonds (1988) and Abrahamson (1989).

[9] 1 megajoule (MJ) = 948 Btu.

TABLE 12.2 Estimated Methane Emissions by U.S. Cattle

Type of Cow	Feed Units	Daily Energy Intake (MJ)	Methane Yields (%)	Proportion of Population (%)
Milk cows	10150	230	5.5	12.5
Feed cattle	6650	150	6.5	12.5
Range cattle	4800	110	7.5	77.5

For developing countries, Crutzen et al. (1986) adopt an average feed consumption of 60.3 MJ, which is much lower than the value for even the range cattle in the United States, and a CH_4 loss due to the low quality of feed of 9%. From these numbers, they estimate an annual CH_4 production rate of 35 kg per animal in the developing world.

Using the Food and Agriculture Organization (FAO) value for the world cattle population of 1.2 billion cattle, 53% of which are in developing countries and 47% in the developed world including Brazil and Argentina, Crutzen et al. conclude that the global CH_4 release to the atmosphere from cattle totals 54 Tg annually (59.4 million tons), 10% of all annual emissions of CH_4.

The other widely cited source of estimates of ruminant CH_4 emissions is Cicerone and Oremland (1988). However, their source is Meyer Wolin at the New York State Department of Health in Albany. Wolin (1979) estimated that the amount of CH_4 produced per day in a 500-kg cow averages ≈ 200 L/day. Liters of gas are easily converted to kilograms by this equation:

$$pV = nRT$$

where p is pressure, V is volume, n is number of moles of gas, R is the universal gas constant per mole, and T is degrees centigrade. Solving for n implies that 200 L of CH_4 contains 7.8 mol of CH_4 (Terry Miller, personal communication, 1990). Because 1 mol of CH_4 contains 16 g of CH_4, 200 L reduces to 0.125 kg/day, or 45.6 kg/year.

For a world total, multiplying 45.6 kg per animal per year times 1.2 billion bovine animals yields 55.9 Tg/year, essentially equal to the Crutzen et al. estimate of 54 Tg/year. Hence it seems that there is fair agreement between these two sources.

An estimate of up to 400 L/day was referenced in the *New York Times* (O'Neill, 1990). This number is probably the upper limit of what could be released during a 24–hour period. Milk cows in the United States come closest to that value; using Crutzen et al.'s estimate of 83 kg/year, we calculate a release of $\approx$360 L/day. Recall, however, that this is but 10% of the U.S. herd size.

The Importance of Ruminant Methane in the Global-Methane Cycle

The next task is to quantify the effect of CH_4 emissions of this magnitude on climate change. Several recent articles suggest that the combined effect of several of the trace gases, CFCs, N_2O, and CH_4, could rival the effect of the greenhouse gas most often mentioned, CO_2 (Rodhe, 1990; U.S. EPA, 1990). This claim stems from the Ramanathan numbers mentioned above. The articles suggest that CH_4's role is $\approx$18% of the total (Figure 12.1).

Ramanathan's numbers are misleading for two reasons. First, they ignore the differences in atmospheric residence times of the gases and, second, they ignore the source of the gases and whether any cycling of gases occurs. These reasons are considered in turn.

Consideration of Atmospheric Residence Times

In a recent article in *Nature*, Lashof and Ahuja (1990) note that other weighting schemes ignore the difference in atmospheric residence times for the different gases. They note that CH_4, with a residence time of 14.4 years (vs. some 230 years for CO_2) is eventually oxidized to CO_2 and H_2O. Rather than use the instantaneous radiative-forcing index[10] of 25 suggested by others, Lashof and Ahuja suggest an index that weights CH_4 at 10 times CO_2 on a weight basis. They conclude that if one uses their proposed index, "carbon dioxide emissions alone account for 80% of the contribution to global warming of current greenhouse gas emissions"[11] (Figure 12.2). Their analysis suggests that the primary emphasis for greenhouse-gas reductions should really remain on CO_2. This conclusion is even more important in light of the recent amendments to the Montreal Protocol that call for a

---

[10] Instantaneous radiative forcing is a measure of the gas's ability to absorb energy (WMO, 1990).

[11] Lashof and Ahuja's estimate of 80% is for "the total contribution of CO_2, including net CO_2 produced from emissions originating as CO and CH_4." *See* Lashof and Ahuja (1990), page 531.

phaseout of most CFCs by the year 2000. If one assumes that this phaseout will occur, the total contribution attributable to CO_2 approaches 90%.

Consider the following calculation, which uses the proposed Lashof and Ahuja (1990) criteria to illustrate two greenhouse-gas-reduction goals. The first is reducing CH_4 production by reducing cattle populations; the second is reducing CO_2 emissions by increasing lighting efficiency. One could phrase the question as: How does a bovine animal compare with a light bulb in terms of its effect on global warming? The answer is that one bovine animal has the same warming effect as a 75–W light bulb operating continuously for 1 year (Figure 12.3).[12] This suggests that a policy of replacing 75–W incandescent light bulbs in industrialized countries with new 18–W compact fluorescent bulbs would go much further toward reducing future climate-change impact than would trying to regulate bovine emissions in developing countries.

Consideration of the Carbon Cycle

The importance of the bovine-animal contribution to climate change can also be overemphasized by ignoring the source of the carbon content of CH_4. Methane released from ruminant animals is not the same as CH_4 released from sources such as leakage from natural-gas pipelines or from coal-mining operations. Methane from these fossil-fuel sources is adding carbon to the atmosphere that was removed tens of thousands of years ago, whereas animals are simply recycling carbon. The following example clearly illustrates the importance of considering both the atmospheric residence times and the source of carbon.

This example looks at the carbon cycle for a steady-state (i.e., mature) 500-kg beef cow (Table 12.3). The cow in this example consumes 9 kg silage/day (dry weight) with an approximate carbon content of 40%. Inputs of carbon amount to ≈ 3600 g. In steady state, the total input and output of carbon fluxes must balance (column 1). Through normal respiration, 2095 g of carbon immediately returns to the atmosphere. Of the remaining quantities, ≈ 173 g is returned in the form of CO_2 and 94 g in the form of

---

[12] The calculations are straightforward. Assume that one U.S. cow emits 58 kg of CH_4/year. This is equivalent to 3625 mol of CH_4. Applying the Lashof index of 3.7, the emissions per cow have the same impact as 13,413 mol of CO_2. Next, note that the conversion of 1 kg of coal to electricity results in 7.56 MJ (2.1 kWh) of electricity and 41.66 mol of CO_2. Therefore, 12,413 mol of CO_2 is the end product of producing 2431 MJ (≈ 676 kWh) of electricity, approximately the electricity consumed by one 75–W light bulb operated for 1 year.

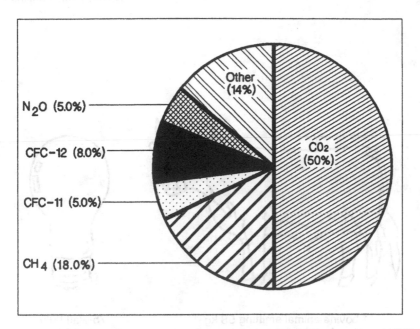

FIGURE 12.1 Greenhouse-gas contributions according to Ramanathan (1988).

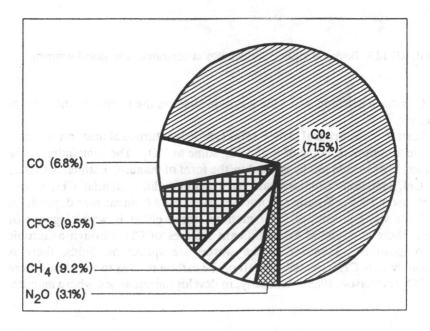

FIGURE 12.2 Greenhouse-gas contributions according to Lashof and Ahuja (1990).

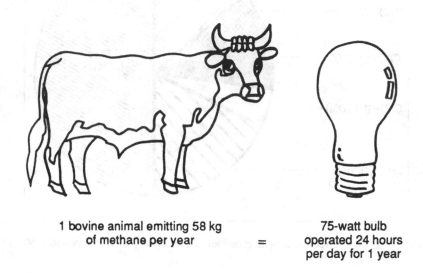

1 bovine animal emitting 58 kg   75-watt bulb
of methane per year   =   operated 24 hours
                                      per day for 1 year

FIGURE 12.3 Bovine animals vs. light bulbs as contributors to global warming.

CH_4 through belching, and 1238 g is deposited on the ground in the form of manure.[13]

In sum, of the original carbon intake, 66% is returned almost immediately to the atmosphere, some of it as CH_4, some as CO_2. The remainder of the carbon is deposited on the ground in the form of manure. Estimates to date of CH_4 emissions from animals have ignored this potential CH_4 source (Patterson, 1989). Whether or not CH_4 is released from manure depends on how the waste is handled. For manure stored either in waste lagoons or piles, there is potential for significant releases of CH_4 through anaerobic decomposition. However, if the wastes are spread on fields, there is probably little CH_4 released, and all of the carbon returns to the atmosphere as CO_2 (Patterson, 1989). Similarly, in developing countries, where manure

---

[13] This assumes a carbon content in manure of 34.4% (Tunney, 1980).

TABLE 12.3 Daily Carbon and Greenhouse-Gas Cycles, Assuming a 500-kg Bovine Animal in Steady State

	Total Carbon (g/day)	CO_2 (g/day)	CH_4 (g/day)	CO_2 Equivalents (g/day)
Input: ≈9 kg silage/day				
(dry weight)	3,600	13,200	—	13,200
Output				
CO_2–belching	173	634	—	634
CH_4–belching	94	—	125	1,250
Manure (1,238 g)				
Carbon released as CO_2	1,238	4,539	—	4,539
Carbon released as CH_4	—	—	—	—
Carbon into soil	—	—	—	—
CO_2–respiration	2,095	7,682	—	7,682
Carbon in urine	N[a]	—	—	—
Total	3,600	12,855	125	14,105

[a]Negligible.

is collected and dried for use as a fuel, carbon is recycled as CO_2 without CH_4 production from manure.

In the example (Table 12.3), proper waste handling is assumed so that there is no CH_4 released from this source. However, entries have been included in Table 12.3 for use in alternative scenarios where one assumes that CH_4 is produced and/or that there is a net addition of carbon to the soil. Here, the contribution of these two items is assumed to equal zero.

Consider the overall effect of this carbon cycle in terms of greenhouse-gas effect. The second and third columns in Table 12.3 indicate the quantities of CO_2 and CH_4 cycled. The last column shows the CO_2 (or, greenhouse-gas) equivalence of the various components of the cycle, using the weighting factors of Lashof and Ahuja (1990). The results are enlightening: while

14,105 greenhouse-equivalent units are released to the atmosphere, 13,200 units are removed from the atmosphere, for a net increase of just 6.9%.

Compare this result with the result obtained by ignoring the issue of CH_4's atmospheric residence time and the cycling of carbon. To emphasize the magnitude of the overemphasis, it is useful to compare CO_2 emissions from automobiles with CH_4 emissions from the steady-state bovine animal. The average automobile emits 280 g of CO_2 for each kilometer driven (1 lb./mile).[14] For a CO_2 weighting factor of 1, this implies a CO_2 equivalence of 280. Traditional calculations of the effect of CH_4 emissions from bovine animals would take the CH_4 emissions from belching and weight this estimate by the instantaneous radiative-forcing effect for a CO_2-equivalence rating of 7250.[15] The result is that a bovine animal, per day, has the equivalent greenhouse effect of driving 26 km. Now consider the result reached in this paper. On the basis of the input-output analysis, the net effect of the cow is 905 (i.e., the difference between total inputs and total outputs) CO_2 equivalents, comparable to driving a vehicle just 3 km/day.

A similar principle applies to every biological source of CH_4, including rice production, termites, and wild animals. If only the emission of CH_4 is considered, and the ecological cycle of atmospheric CO_2 removal is ignored, then the apparent contribution of biological sources to the greenhouse effect will be seriously overstated.

This is an example of recycled carbon. The net effect of each unit of CH_4 from bovine animals is definitely less than that of a unit of CH_4 emitted through fossil-fuel combustion or leakage. In the latter case, we are adding the combined effect of $\approx$10–14 years of CH_4, followed by the effect of $\approx$200 years of CO_2, whereas the former case involves only the increased infrared-trapping effect of the 14 years of CH_4.

Cost Estimates of Various Greenhouse-Gas-Reduction Goals

Table 12.4 presents cost estimates for four different strategies for reducing greenhouse-gas emissions. Three strategies, increased lighting efficiency,[16]

---

[14] This assumes 5.35 lbs. C/U.S. gallon and an average fuel efficiency of 20 mpg.

[15] This assumes a radiative forcing of 58 on a weight basis.

[16] This estimate is based on replacing continuously operated 75-W incandescent light bulbs with 18-W compact fluorescents. It assumes an average electricity cost of $0.0178/MJ ($0.064/kWh), a light-bulb cost of $0.75, and a compact-fluorescent-bulb cost of $15.99.

fuel switching,[17] and tree plantations,[18] target CO_2 emissions. The fourth is an estimate by Adams et al. (1992)[19] for reducing CH_4 emissions by altering the diet of ruminant animals.

The estimate by Adams et al. of $351 per 10^3 kg CO_2 equivalent (in the form of CH_4) is quite high compared with the other alternatives presented and, in general, with those found in the literature for CO_2-reduction strategies. Although this estimate is the result of preliminary work, if further work confirms the magnitude of this reduction strategy, it will be further evidence of the difficulty of pursuing any CH_4-reduction strategies that target bovine animals.

TABLE 12.4 Cost Estimates of Various Greenhouse-Gas-Reduction Goals

Strategy	Cost ($/CO_2 equivalent Mg^a)
Increase lighting efficiency	−56.00
Switch fuels from coal to natural gas	22.00
Forestation	54.00
Modify cows' diet	352.00

^{a}1000 kg, or 1 metric ton.

[17] This number represents the difference in fuel costs for fossil-steam plants operating with natural gas rather than coal. It assumes coal costs of $1.36/GJ and natural gas costs of $2.70/GJ.

[18] This assumes a growth ratio of 1.35 kg· m^{-2}· year^{-1} (6 tons· acre^{-1}· year^{-1}); cost estimates include site preparation, weed control, planting, land rental, fertilizer, harvesting, and removal of trees from the site. It also assumes the use of short-rotation-intensive culture (SRIC), which uses fast-growing trees on managed plantations. *See* Chapman and Drennen (1990).

[19] Adams et al. estimated that to reduce emissions of CH_4 by altering ruminant diets would cost between $2250 and $4900 per 10^3 kg of CH_4. This was converted to greenhouse-gas equivalents by applying the Lashof and Ahuja index and taking an average.

The North-South Political Question

The implications of pursuing CO_2 reductions alone vs. pursuing a comprehensive approach also raise important questions that touch on North-South political issues. For example, which countries should bear the largest burden of responsibility in curbing global warming? Presumably, in negotiating a comprehensive approach, countries would have to settle the question of an appropriate bench-mark level of emissions for the different gases. Regarding CFCs, one can imagine disagreement arising over starting levels or credit for past reductions as achieved under the Montreal Protocol. The United States, the largest single consumer of CFCs,[20] would likely be insistent on gaining recognition and credit for already-achieved reductions in CFC levels. Consider the following numerical example of such a claim by the United States.

U.S. consumption in 1986 of CFC-12 alone was $\approx$140 million kg.[21] Using the index based on instantaneous radiative forcing,[22] this implies a value of 805 billion CO_2-equivalent units. If one assumes a phaseout of CFC consumption by the year 2000, as agreed to, the United States would most likely insist on a credit of 805 billion units/year toward its reduction of greenhouse gases. Compare this estimate with the CO_2-equivalent units emitted by U.S. coal consumption. The United States consumed 19 billion GJ of coal in 1990 (U.S. DOE, 1991), emitting $\approx$1.9 trillion kg of CO_2, or 1900 billion CO_2-equivalent units, just 2.3 times the radiative-forcing effect of current CFC-12 consumption itself.[23] Hence, the United States is likely to claim that by agreeing to the CFC phaseout, it has done its share of reducing the risk of future climate change. Meanwhile, countries with low CFC consumption would not benefit from such a credit. Indeed, it would be these countries, such as India, that would have to make sizable changes in their CH_4 emissions to capture a similar credit.

Whether intentional or not, the effect of pursuing the comprehensive approach might be a failure to reach any accord. Would India or China, who see the industrialized countries as the prime culprits, agree to

---

[20] The United States accounted for 29% of total worldwide consumption in 1986. *See* Shea (1988).

[21] Shea (1988) reports U.S. per capita use rates of 0.34, 0.58, and 0.31 kg for CFC-11, CFC-12, and CFC-113, respectively. Multiplied by a U.S. population of 241 million, this results in an aggregate total of 140 million kg of CFC-12.

[22] By this measure, CFC-12 is weighted at 15,800 times CO_2 on a molecule basis, or 5750 times CO_2 on a unit-mass basis. *See* WMO (1990).

[23] Substitutes for CFCs will likely be greenhouse gases, also. Their now unknown effect would have to be included in such future calculations.

something that requires reductions in greenhouse-gas emissions from their agricultural sector? Perhaps this is the real goal of the U.S. policy of pursuing a comprehensive agreement?

Conclusion

This paper compared various estimates of total CH_4 emissions from bovine animals and discussed the relative addition to greenhouse-gas warming from this one source.

Emissions of CH_4 from enteric fermentation in animals have been estimated elsewhere at $\approx$80 Tg/year, 14.8% of global CH_4 releases. This estimate is misleading for two reasons: it ignores the differences in atmospheric residence time between CO_2 and CH_4, and it overlooks the biological and chemical cycling that occurs. The result is an overemphasis on the role of this CH_4 as a greenhouse gas.

This overemphasis has important implications for negotiations on future climate-change accords. By ignoring these two factors, the role of developing countries' total contributions to climate change has been overemphasized. Lashof and Ahuja (1990) conclude that CO_2 emissions alone account for 80% of the contribution to global warming, significantly higher that the oft-cited 50% figure of Ramanathan (1989). On the basis of Lashof and Ahuja's numbers, an agreement aimed solely at reducing future CO_2 emissions would be an important first step. From a practical standpoint, any agreement regulating CH_4 would be exceedingly difficult to develop because of the lack of data available and because of inadequate measurement and monitoring capabilities.

All of this does not imply that bovine CH_4 emissions should be ignored. Policies for reducing CH_4 emissions that follow from the above calculations include improving the quality of animal feed and finding ways to more effectively use animal manure, such as through bio-gas utilization. However, as suggested by the preliminary results of Adams et al. (1992), such reduction strategies may not be economically attractive when compared with CO_2-reduction strategies.

References

Adams, R. M., C.-C. Chang, B. A. McCarl, and John M. Callaway. 1992. "The Role of Agriculture in Climate Change: A Preliminary Evaluation of Emission-Control Strategies," in J. M. Reilly and M. Anderson, eds., *Economic Issues in Global Climate Change: Agriculture, Forestry, and Natural Resources*. Pp. 273–87. Boulder, Colo.: Westview Press.

Abrahamson, D., ed. 1989. *The Challenge of Global Warming.* Washington, D.C.: Island Press.

Chapman, D., and T. Drennen. 1990. "Equity and Effectiveness of Possible CO_2 Treaty Proposals." *Contemporary Policy Issues* 8(3): 16–28.

Cicerone, R. J., and R. S. Oremland. 1988. "Biogeochemical Aspects of Atmospheric Methane." *Global Biogeochemical Cycles* 2(4): 299–327.

Crutzen, P., I. Aselmann, and W. Seiler. 1986. "Methane Production by Domestic Animals, Wild Ruminants, Other Herbivorous Fauna, and Humans." *Tellus* 38B: 271–84.

Lashof, D., and D. Ahuja. 1990. "Relative Contributions of Greenhouse Gas Emissions to Global Warming." *Nature* 344: 529–31.

Miller, T. 1991. "Biogenic Sources of Methane," in W. B. Whitman and J. E. Rogers, eds., *Microbial Production and Consumption of Greenhouse Gases: Methane, Nitrogen Oxides, and Halomethanes.* Washington, D.C.: American Society for Microbiology.

O'Neill, M. 1990. "Cows in Trouble: An Icon of the Good Life Ends Up on a Crowded Planet's Hit Lists." *New York Times,* May 6, section 4, p. 1.

Patterson, J. A. 1989. "Potential Methane Emissions from Animal Manure," in U.S. EPA Intergovernmental Panel on Climate Change, Response Strategies Working Group, Subgroup on Agriculture, Forestry, and Other Human Activities, eds. *Proceedings of the Workshop on Greenhouse-Gas Emissions from Agricultural Systems, Summary Report.* Pp. V-59–V-64. Washington, D.C.: U.S. EPA. (20P-2005.)

Ramanathan, V. 1988. "The Greenhouse Theory of Climate Change: A Test by an Inadvertent Global Experiment." *Science* 240: 293–9.

Rodhe, H. 1990. "A Comparison of the Contribution of Various Gases to the Greenhouse Effect." *Science* 248: 1217–9.

Shea, C. P. 1988. "Protecting Life on Earth: Steps to Save the Ozone Layer." Washington, D.C.: Worldwatch Institute. (Worldwatch Paper 87.)

Tunney, H. 1980. "Agricultural Wastes as Fertilizers," in M. Bewick, ed., *Handbook of Organic Waste Conversion.* Pp. 1–35. New York: Van Nostrand Reinhold.

U.S. Department of Energy. 1991. *International Energy Annual, 1989.* Washington, D.C.: Energy Information Administration, U.S. Department of Energy.

U.S. Department of State. 1990. "Materials for the Informal Seminar on U.S. Experience with 'Comprehensive' and 'Emissions Trading' Approaches to Environmental Policy." Washington, D.C.: U.S. Department of State. (Discussion paper.)

United Nations Environmental Programme (UNEP). 1987. *Montreal Protocol on Substances that Deplete the Ozone Layer, Final Act.* Nairobi: UNEP.

U.S. Environmental Protection Agency. 1990. "Policy Options for Stabilizing Global Climate: Draft Report to Congress." Washington, D.C.: U.S. EPA.

Wolin, M. 1979. "The Rumen Fermentation: A Model for Microbial Interactions in Anaerobic Ecosystems," in M. Alexander, ed., *Advances in Microbial Ecology*, Vol. 3. Pp. 49–77. New York: Plenum Press.

World Meteorological Organization. 1990. *Climate Change, the IPCC Scientific Assessment.* Cambridge, U.K.: Cambridge University Press.

Wuebbles, D., and J. Edmonds, preparers. 1988. "A Primer on Greenhouse Gases." Washington, D.C.: U.S. Department of Energy. (DOE/NBB-0083 TR040.)

PART FOUR

Forestry and Global Change

13

Global Change and Forest Resources: Modeling Multiple Forest Resources and Human Interactions

Michael A. Fosberg, Linda A. Joyce, and Richard A. Birdsey[1]

Introduction

Managing forest resources for sustainable development requires long-range (50 years or so ahead) strategic planning for production of goods and services from the nation's forest lands. In the past, this planning has not been responsive to prospective changes in the climatic influences on natural resources. Because our information base for resource management and for risks associated with management options is empirical, the climatic influences on resources and on interactions among resources, goods, and services have been buried in that information base and have not been clearly identified as controlling variables. Also, interactions among resources will change and will require that we address the problems of strategic planning from a multiple-resource perspective rather than resource by resource.

Timing of global change is based on human population growth and per capita energy consumption. The best estimates for how long it will take to resolve significant impacts of global change are 50–100 years—the same interval used for resource planning. This clearly implies that natural-resource dynamics must become part of the strategic-planning process and that the characteristics of natural resources can no longer be external to the models.

Analyses of future timber resources (Joyce et al., 1990a) and water resources (Frederick and Gleick, 1989) project significant impacts on supplies. Comparable national analyses of recreation, range, wildlife, and

---

[1] From the U.S. Department of Agriculture, Forest Service, Washington, D.C. (M.A.F. and R.A.B.), and Fort Collins, Colo. (L.A.J.).

fisheries resources have not been made. Some inferences about fisheries and wildlife may be made from timber (habitat) and water (stream-flow) assessments. Such spotty inferences tend to focus on threatened and endangered species, but also include impacts on commercial species of anadromous fishes. Significant impacts are identified in each of these analyses. Furthermore, indirect effects on resources may exceed direct effects. The frequency of disturbance from fire, insects, disease, and extreme weather is likely to increase, at least during the transition from the current climate to the changed climate (Fosberg, 1990; Fosberg et al., 1990). Disturbance will be far more important in ecosystem change than the more gradual direct effects of climate change. We can expect all natural resources to be threatened by global change, and we can expect the biological and physical links between resources to change.

Scenarios of global change are based on imperfect computer models of the general atmospheric circulation. Different models do not provide exactly the same projections of global change or of regional depictions of that change (Ephraums et al., 1990). The range of predictions can be used to describe the uncertainties in the global-change scenarios. Also, several resource-impact models have been coupled with the global-change scenarios.

These resource-impact models do not provide the same projections of future resource conditions even for the same climate-change scenario. This variation introduces additional uncertainty into the status of future resources beyond that of climate projection (Joyce et al., 1990a). Research will undoubtedly reduce these uncertainties over time, but uncertainty will still remain in the projections from stochastic factors such as fire, insects, disease, and weather. Past occurrence of these disturbances will not provide an adequate guide to the future. For example, frequency of drought can be expected to increase during climate change, based on long-term historical records derived from tree-ring analysis (Stahle et al., 1988). If these historical drought records are used as a surrogate for other disturbances, then the random component in predicting the state of future resources will increase. This random component is captured in the empirical relationships of current resource models and through simulation of alternative future conditions. If we are to rely on more process-oriented models of resource change, then this random component will need to be explicitly introduced into the projections. Having measures of uncertainty in both the deterministic and the random components of resource projections leads us to consider risk assessment and risk management as a fundamental component of strategic planning.

Strategic planning requires assessment and appraisal of current resources, consideration of dynamics of change in those resources as a result of global change or other physical or biological agents as part of the supply, anticipation of the demand for those resources (in terms of goods and services) with consideration of the international resource base, and attention to demand and price relationships. Within the U.S. Department of Agriculture (USDA) Forest Service, this strategic planning is captured at the global scale in the Resource Planning Act (U.S. Congress, 1974) and at the regional or local scale in forest plans (U.S. Congress, 1976).

At the global scale, we use a collection of models to make projections of future resources and of the economic impacts of forest-policy decisions, under the basic assumption of market equilibrium for timber products (Figure 13.1). Land-use shifts are described by using a two-stage econometric model that projects area by major land-use categories (forest, cropland, urban, etc.) and projects forest area by forest type (Alig, 1986). The dynamics of timber supply and demand are projected with an equilibrium model of the forest-products market coupled with an inventory-accounting model that tracks the movement of forest area and the volume of timber among different age classes, forest types, ownerships,

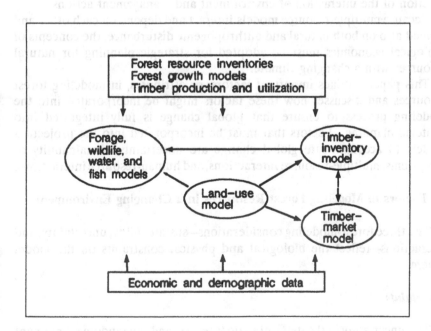

FIGURE 13.1 System of data and models used by the U.S. Forest Service to make multiple-resource projections.

management intensities, stocking classes, and site classes (Adams and Haynes, 1980; Tedder et al., 1987). These models include the assumption that growth of forests is adequately represented by empirical relations estimated from past and current observations. With global change, forest growth will no longer be static, but will become a function of the interaction of environment and management actions.

Models for the resources of forage (grazing), wildlife, fish, and water have been specifically developed to link with the forest-sector system by use of the commonly defined attributes of the land-base (i.e., land area and timber inventory) system (Joyce et al., 1990b). Land-base changes are influenced primarily by economic rather than ecological factors. Some ecological factors known to be significant in resource production and identifiable in the inventory data were incorporated directly into each resource model. The influence of abiotic factors was not included in forest growth, the growth of forage, or water yield. The spatial patterns of land use and cover were not available for consideration. With global change, resource production will no longer be influenced by use of resources alone and, in fact, will become a function of the interaction of environment and management actions.

Because multiple-resource models interact and depend on each other and depend also on both natural and anthropogenic disturbance, the concepts of ecological economics must be adopted for strategic planning for natural resources with a changing climate.

This paper outlines important factors, or attributes, in modeling forest resources and discusses how these factors might be incorporated into the modeling process to ensure that global change is fully integrated into strategic planning. Factors that must be incorporated into the projection models to accommodate global change are uncertainty, sustainability of ecosystems, multiple-resource interactions, and human-resource interactions.

Factors in Modeling Forest Resources in a Changing Environment

The three forest-modeling considerations—sustainability, uncertainty, and interactions—reflect the biological and physical constraints on the model system.

Uncertainty

The uncertainty in the deterministic forecasts and the random component of the forecasts require that we quantify the probable ranges of outcomes through both risk assessment and risk management in strategic planning. Also included, but difficult to quantify, are the surprises introduced by social factors (e.g., the effect of classifying the northern spotted owl as an endangered species on available timber resources). Uncertainty and surprise in the models should match or at least bracket the uncertainties and surprises of the real world. The question to be addressed in this paper is how to incorporate risk assessment and risk management associated with global change into the resource-projection models.

To introduce risk assessment and risk management into strategic-planning models of natural-resource systems, we need first to describe the prediction process (Fosberg, 1987). Prediction of the future must start with an assessment of the current or initial state of each of the resources whose status is to be predicted. Uncertainty exists in the description of this initial state simply because our estimates contain error. Once the initial state is defined, we must determine the rate of change over some time interval.

If our model perfectly mimics the system we are modeling, then the only uncertainty in the prediction will come from the error in the initial conditions and from how that error is propagated in the model. Because models are not perfect, additional uncertainty is introduced into our prediction. Uncertainty in the incomplete model comes about because the deterministic parts of the real system are incompletely modeled, and because random events cannot be predicted at a specific time or place. Random events such as fire, insect or disease outbreak, and extreme weather can be introduced into the forecast system with Monte Carlo techniques. The use of Monte Carlo techniques implies that instead of a single or unique prediction, we will have a probability distribution as our forecast output.

Incorporating global change into strategic planning for natural resources has traditionally been done by constructing a scenario or scenarios of the future climate. If we use a deterministic resource model, then we have a deterministic forecast for each climate scenario. An alternative to this deterministic approach is to construct a composite scenario from the climate models and to use the variations between the climate models to define the climate uncertainty. Again, if this uncertainty is captured in the modeling process through Monte Carlo techniques, we will have a probability distribution as our forecast output. Similarly, the resource models can capture uncertainty in the output.

Linking models, to allow both flow of information from one model to another and feedbacks between models, introduces additional uncertainties. Most models for strategic planning use discrete forms of continuous equations to make forecasts for short time intervals. For climate models, the forecast is for 15-minute intervals, and the forecast state at the end of the 15-minute interval is then the initial state for the next forecast. Time intervals for the resource models vary from 1 month to 1 year. This mismatch of time intervals between different models requires that the models be run asynchronously. Asynchronous solutions provide incomplete feedback between models and therefore introduce additional uncertainty.

Error from estimating initial conditions, from incomplete deterministic models, and from linking models will lead to errors in model prediction that will be revealed during validation. These errors reflect a lack of knowledge that can be addressed through research. The unpredictability of nature is a different kind of uncertainty. Uncertainties from predictions based on slightly different assumptions or from nondeterministic events will bracket the uncertainty in the real world and can be managed. If we accept the model predictions as statements of probability, then we can manage uncertainty in strategic planning for renewable natural resources. The probability statements are, in fact, risk assessment. Decision-analysis techniques allow us to carry these statements of probability into our resource-management models and allow us to begin management of risk.

Sustainability of Ecosystems

Multiple-resource sustainability requires that each generation leave succeeding generations the quality and quantity of the multiple resources, goods, and services that those future generations will need to manage their lives successfully. Resource sustainability in a changing environment must be based on adaptability, resilience, and stability. Genetic diversity within species and diversity of different species within the ecosystem, productivity of the ecosystem, and health of the ecosystem must all be part of a definition of sustainability. Diversity, health, and productivity can be measured and predicted with some certainty, but adaptability, resilience, and stability need quantifiable definitions if they are to be incorporated into strategic-planning models. These quantitative definitions must be valid in a changing environment. Several indicators of sustainability are necessary, and a hierarchy, or structuring, of multiple indicators would be desirable.

Multiple-resource sustainability requires that we concentrate on long-term maintenance of components of forest ecosystems and the interrelations among those components before we focus on the outputs, as goods and services, of forest ecosystems. A definition of sustainability must contain the concepts of adaptability, resilience, and stability. Diversity, health, and productivity are quantifiable components of sustainability, but in themselves do not define sustainability. All too frequently a single attribute or indicator of sustainability of ecosystems has been offered, much as gross national product is used to indicate economic health of a nation. Such single attributes of ecosystem sustainability generally focus on a single resource (e.g., biomass as a surrogate for forest productivity) and fail to capture interrelations between resources (e.g., habitat and abundance of wildlife). Integrating ecosystem dynamics with economic projections in strategic planning requires that there be a quantitative definition of sustainability, because under prospective global change, the metric of potential sustainability can be expected to change.

We define sustainability as a function of adaptability, resilience, and stability. An ecosystem may not have a high value in one or two of these attributes, but it might still be sustainable if it had a high value in the third. Using an analog of conductance of information to capture this multiple-path concept, we propose a functional relation for sustainability (S), adaptability (A), stability (ST), and resilience (R) as

$$1/S = (1/A) + (1/ST) + (1/R) \tag{1}$$

Equating adaptability with diversity (D), and recognizing that diversity is composed of both biodiversity (D_b) and genetic diversity (D_g), we can say

$$1/A = 1/D = (1/D_b) + (1/D_g). \tag{2}$$

This relationship recognizes that monocultures that have high genetic diversity are sustainable, as are ecosystems that have a large number of species. Attainment of stability depends on positive feedback in the ecosystem, whereas maintenance of stability depends on negative feedbacks. Stability does not imply that ecosystems are unchanging. For example, the lodgepole-pine ecosystem in Yellowstone Park is stable. The fires and alteration of the landscape during 1988 are part of that stability. Insect and fire disturbances are a part of the lodgepole pine's life cycle. Stability here means that Yellowstone will remain a lodgepole-pine ecosystem. Stability is defined as a function of biomass, rate of change of biomass (positive or negative feedback), dominance, and contagion (positive feedbacks). This relationship recognizes that low-biomass ecosystems are sustainable (e.g., rangelands vs. temperate rain forests). Contagion, particularly at the landscape scale, provides for nearby external sources of resource renewal. Dominance as a separate measure of diversity captures the grain structure of landscapes.

Resilience is defined as a function of forest health and resistance to disturbance. A healthy ecosystem will have a better chance of resisting a disturbance and recovering from the effects of a major disturbance. Indicators of forest health can be used to compare the state of an ecosystem with a concept of normality. A departure from normality—caused by, for example, climate change—will indicate that an ecosystem is more vulnerable to the effects of a disturbance.

Landscape ecologists (Turner et al., 1989) define dominance and diversity in terms of the proportion of landscape occupied by a given element and the total number of elements. The attributes of equation 1, defined in terms of measurable characteristics defined in equation 2, and the functional statements of stability and resilience provide multiple measures of and paths to sustainability. Use of multiple measures or indices first recognizes that multiple resources, goods, and services exist in ecosystems and, second, that these resources, goods, and services are derived from fundamental biological and physical characteristics of the ecosystem and that those biological and physical characteristics are interrelated through natural laws. Each of the measurable characteristics of sustainability will vary from current values with a changing climate. Because each of these characteristics is predictable within error bounds, given a new climate, sustainability is predictable. A quantitative definition of sustainability provides an integrating measure that will help us achieve sustainability of ecosystems rather than sustainability of one or a few goods and services.

Strategic planning for renewable natural resources is based on models that make projections of future resources and the economic impacts of policy decisions. Currently, resource dynamics are not included in the resource economic models. With global change, we must consider whether resource dynamics should be included in the model analysis, and how changes in sustainability will be considered in strategic planning.

Multiple-Resource Interactions and Human Interactions

Multiple-resource interactions take into account the varying quantities and trade-offs of resources that result from management of natural systems. Traditionally, analyses of resource-production systems have been based primarily on management-related manipulations, and for some single-resource analyses, empirical models are well-established. The increasing demands for resources from forest and rangelands (USDA Forest Service, 1989) has emphasized the need to quantify resource interactions. However, quantitative information on resource interactions, particularly over large geographic areas, is limited (Hof and Baltic, 1988). When environmental factors are included in the analysis of resource production, little quantitative information is available for assessing single-resource production, much less the consequences of climate change on resource interactions.

The role that human interactions might play in affecting single-resource production or the interactions of resource production is unknown. For water resources, the traditional management adjustments to climate variability, based on past human behaviors, may not be the optimal management for water resources under a changing climate (Riebsame, 1988). Yet, the need to engage the involvement of the public in the management of private lands surrounding public lands to preserve biodiversity has been stressed (Westman, 1990). Understanding likely human behaviors and their effects on the future alternatives of resource management will be important in policy decisions affecting public and private forests and rangelands.

Our current understanding of resource production is based on historical climate, previous management results, and historical human reactions to resource change. This understanding, as analyzed in the assessment-appraisal analyses, has recognized the influence of change in the future; however, this influence has been modeled in the economic infrastructure only. Technological improvements and their impact on resource production and use have been recognized. Attempts have been made to quantify the effects of technology changes on the mix of inputs and resource outputs for agriculture and livestock (U.S. Congress, Office of Technology Assessment, 1986) and for timber (Haynes, 1991). The effects that resource and agricultural policy can have on imports and exports, and the interaction with monetary policy, have been recognized (Hahn et al., 1990; Haynes, 1991). Changing consumer preference and its effect on the demand and supply of commodities, such as meat, have been analyzed (Joyce, 1990). Changing consumer preference for increased outdoor recreation has also been analyzed for its effect on supply and demand of recreation facilities (Cordell, 1990). What has rarely been analyzed is the influence of change in the socioeconomic infrastructure on resource interactions and the initiation of conflicts in resource management, such as the spotted owl vs. commercial timber or livestock vs. elk conflicts.

In addition, past assessments and analyses have not recognized the changing nature of the environment—abiotic and biotic. Whereas abiotic changes were seen as unlikely or unaffected by policy, the possibility of biotic changes, such as declining productivity and consequent rising management inputs, has not been included in long-term forecasts in assessments or analyses. Such potentially subtle alterations of resource production, and indirect effects such as changes in management practices, influence long-term future projections in many different ways, including the basic allocation of land to various uses.

The consequences of such changes on wildlife, range forage, water production, recreation, and other goods and services have only been partially analyzed. Resource-production analyses typically quantify only a part of the resource system—that part that currently appears to be important, given the policy question and the current understanding of the system. This premise holds for economic and resource-production analyses; empirically developed models; and mechanistic models. Other parts of the system are assumed to be unchanging or noninfluential. Under a changing environment, these assumptions about other parts of the system-resource interactions may not be valid.

For example, as part of a regional analysis to determine the impacts of land-use change and timber management on the resources of wildlife, forage, fish, and water, one prediction was a decline in the growth of southern pines (USDA Forest Service, 1988). Within this analysis, the land base—land area and timber inventory—served as the integrating factor. Each resource model analyzed production as a function of these commonly defined land-base factors. The scenario of reduced pine growth was implemented by decreasing the growth rate of southern pine 25%, a value within recently recorded declines (Sheffield et al., 1985). Under this scenario, less timber volume was available for harvest and more land area was harvested.

These timber-management changes and land-area shifts were transferred to models for forage, wildlife, water, and fish. Whereas the approach appropriately examined the effect of such a reduction within timber supply and demand, the many possible environmental effects that could cause a reduction in tree growth were not implemented in the timber model; rather, only the rate of tree growth was altered. Thus, the system's perception of reduced tree growth was in the land-area changes that resulted from increased pressure on timber harvesting, not an ecological change. Whereas all resource models projected this harvesting influence, there are many ways for the effect of a reduction in tree growth to occur, and, presumably, some of these ways could influence the production of other resources. The decline in southern softwoods has not been ascribed to a specific cause but to several possible agents: acid deposition, drought, changes in the water table, and changes in the age of stands. An agent such as drought, in addition to reducing annual growth, could have effects on the timber resource, such as changing seedling survivability, root growth, height growth, and quality of wood produced (Miller et al., 1987). Drought would likely affect other parts of the forest system, such as understory vegetation and wildlife. An agent of change, such as drought, implies that there is an identifiable link between resource production and abiotic factors such as precipitation and temperature. Analyzing the effect of one of these agents involves quantifying, empirically or mechanistically, the underlying processes for single resources of interest or for all resources and their interactions in the ecological system.

Finally, these approaches for analyzing multiple-resource interactions assume relatively unchanged human interaction in resource management during the time of the projection (USDA Forest Service, 1989), or the analysis assumes that land managers will optimize their behavior according to some objective, such as minimizing cost or maximizing profit (USDA, 1989). In the first approach, where human interactions are not prescribed, management activities that have been practiced and will be practiced with a certain likelihood on the landscape over the projection period are implicit in both mechanistic and empirical simulation analyses. In the optimization approach, human behavior is assumed to select the most efficient set of management practices, given a set of production costs and output prices.

Alterations in the resource-production system caused by a changing climate may make assumptions in both of these approaches unacceptable.

Within the simulation approaches, human behavior in selecting management activities is usually classified by ownership in timber projections (Haynes, 1991) or by size of farm in agricultural or livestock analyses (Gilliam, 1984). Insufficient information about human behavior restricts the domain of these projections. For example, forage-production analyses assume that tree-planting densities will remain the same or will change in ways that do not increase the sunlight reaching the forest floor. This assumption implies that dramatic shifts in tree-planting density patterns, as advocated in agroforestry management (Lewis et al., 1985), will not occur. How likely is such a shift? We do not know. What classes of owners would be likely to implement such management practices? We do not know. Would such shifts in behavior be more likely in a changing environment? We do not know. Even for the most studied and quantified ownership behaviors in timber management, explanatory factors are missing in our understanding of how these owners select management activities or harvest timber (Rosen et al., 1989; Royer and Convery, 1981).

Optimization approaches, such as analyses with the Center for Agricultural and Rural Development (CARD) model used in support of the Resources Conservation Act (RCA) (USDA, 1989), imply an efficient management of all lands. For example, 6.7×10^{11} m^2 (67 million hectares) of cropland went out of production in the 1985 RCA Appraisal because of prices, environmental-quality constraints, and surplus production of some crops (USDA, 1989). Yet in the real world, nearly all of this land would likely remain in crop production or in cropland but held in large set-aside programs. Whereas this approach might elucidate how government should attempt to optimize production on croplands, it gives little insight into how the behavior of an individual landowner will affect resource production in the future.

Human behavior is motivated by many factors, including economic incentives. Human reactions to the environment are important to understand, because human interactions under a changing climate may be outside previously considered behavioral norms (Miller, 1989). In addition, human interactions are embedded in institutions and cultures (Miller, 1989), and these interactions need to be understood to formulate policy. Climatic variability influences human behavior; for example, the highest densities of cropland entered into the 1950s government set-aside program coincided with the highest variability of biomass production in the Great Plains (Joyce and Skold, 1988). The impact of climatic variability on local and regional human behaviors needs further study.

Analyses of Future Resource Interactions

Policy models, like research models, are changed incrementally to address our changed understanding of how a system works, or are completely replaced—an expensive and often highly resisted option. The decision to modify or replace a model is a function of the policy questions for the moment and of how well the model analyzes these questions. The analysis of a changing environment covers a time of 50–100 years. The Renewable Resources Planning Act (RRPA) requires planning ahead nearly 50 years. The economic infrastructure will undoubtedly change during this time. Long-term policy models attempt to recognize the mechanisms that influence change in infrastructure, including the role of policy. The ecological systems containing and contained within forest and rangelands will likely change over that same period. These changes will be abiotic (the changing climate) and biotic (the response of the ecological systems to the changing climate). Both of these changes will influence resource production, our management responses, and human interactions with forest and rangelands.

We propose that resource production, management responses, and human interactions be examined in a two-step process. The first step involves an examination of the allocation of land and factors underlying that allocation. The second step involves the examination of likely human behavior with respect to choices for management.

We propose using one or more of the following approaches to examine the allocation of land and its effect on resource production:

1. Improve the projections of land use and land cover to include sensitivity to prospective climate changes and human responses to climate change. The land base, including vegetation structure for each land-cover type, provides a common set of factors from which resource production can be estimated. Changes in the vegetation structure, such as return to seral stages, are a function of management activities and ecosystem response to stress and other forces.

2. Establish a spatial assessment and projection of land cover and land use. The spatial land base, including vegetation structure, provides a common set of factors from which resource production can be estimated. Changes in the size, shape, and distribution of land uses and cover should be included. Changes in the vegetation structure as a function of management activities in all resources and other agents of land-use change that can propagate across the landscape should be included. Spatial interactions among resources, management activities, and environmental stress should be included.

3. Capture the processes of ecosystem dynamics. The ecosystem is the common base for all resource models. Ecosystem dynamics and biogeochemical cycles, particularly nutrient cycles, must be quantified at the landscape scale. Process models need links with atmospheric-climate models. Changes in the vegetation structure are a function of natural disturbances, such as fire or insects and disease, and management activities for all resources.

Changes within each ecosystem and between or across ecosystem boundaries also need to be quantified. Once the system states have been identified, resource models describe the likely outputs from these ecosystems and their seral stages.

We propose using one or more of the following approaches to analyze the motivations for human behavior in resource management:

1. Establish the motivations for multiple-resource management. The selection of specific management activities to implement on a particular section of land influences current and future resource production. The decisions influencing that action do not recognize the diversity of ownerships and the diversity of resources produced from forest and rangelands. Although the influence of governmental programs on timber investments has been studied, the influence of governmental programs on other resource management needs to be further quantified.

2. Establish the influence of climatic variability on human motivations in management. Understanding human behavior will become increasingly important as society begins to make choices under a changing environment. Conservation and preservation values will increasingly conflict with production choices. Alternative scenarios of human interactions will be needed to fully assess the effects of management on natural systems. The historical choices by landowners may not adequately describe future choices, because future human interactions with a changing resource base may be outside past behaviors.

Summary

Historically, strategic planning for renewable natural resources has considered climate change as an external factor to which individual resources responded. Internal factors, such as rangeland response to grazing, regulated the behavior of the ecosystem or resource production. In economic models, price has served as an external driving factor to equilibrate the system. The behavior of the resource-economic system may go outside our historic

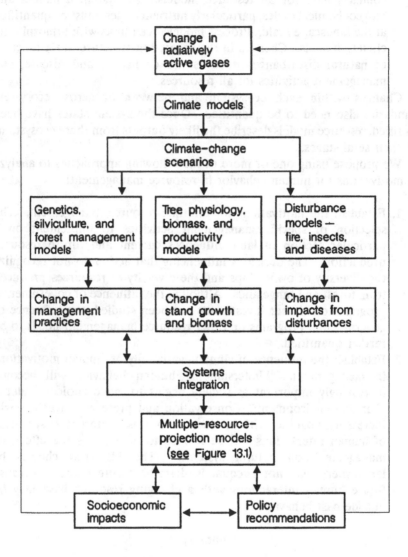

FIGURE 13.2 Example of a process-driven model of forest resources linking global change, forest resources, strategic planning, and policy development (adapted from Wheaton and Thorpe, 1989). Equivalent modules for water, recreation, range, and wildlife are not shown. All feedback loops are not shown.

experience with climate change. If models have been developed empirically, they may not be able to accurately predict the future. Models that account for physical and biological processes are better able to capture the surprises introduced by climate change, but may not be able to capture the natural variability of climate.

Linking climate models to resource-production, ecosystem, and economic models brings climate change into the system as part of the internal dynamics of the resource-econometric models (Figure 13.2). Human factors must also become part of the system. The impact of climate change on human perception and motivation and on resource preferences must be included to minimize surprises in the predictions. Incorporating human influences and climate change into resource-econometric models will increase uncertainty in the predictions. These uncertainties must be captured in the display of the predictions.

Finally, because renewable natural resources are expected to change as a result of climate change, some expression of a desirable future (within the range of possible futures) must be defined. This future must be sustainable. To ensure that we have made the correct choice, we must have a quantifiable definition of sustainability.

References

Adams, D. M., and R. W. Haynes. 1980. "The 1980 Softwood Timber Assessment Market Model: Structure, Projections, and Policy Simulations." *Forestry Science Monographs* 22: 64.

Alig, R. J. 1986. "Econometric Analysis of Forest Acreage Trends in the Southeast." *Forest Science* 32: 119–34.

Cordell, K. 1990. "An Assessment of the Outdoor Recreation and Wilderness Situation in the United States: 1989–2040." Fort Collins, Colo: U.S. Department of Agriculture (USDA) Forest Service, Rocky Mountain Forest and Range Experiment Station. (General Technical Report RM-189.)

Ephraums, J. J., J. T. Houghton, and G. J. Jenkins, eds., 1990. "Climate Change, the IPCC Scientific Assessment. United Nations Intergovernmental Panel on Climate Change, Report of Working Group 1." Cambridge, U.K.: Cambridge University Press.

Fosberg, M. A. 1987. "Forecasting Forecasting," in J. B. Davis and R. E. Martin, eds., *Wildland Fire 2000*. Pp. 105–14. Berkeley, Calif.: USDA. (USDA Forest Service General Technical Report PSW 101.)

Fosberg, M. A. 1990. "Developing Research Programs for Assessing Effects of Global Climate Change on Forests," in *Proceedings of the Society of American Foresters National Convention*. Pp. 134–41. Washington D.C.: Society of American Foresters.

Fosberg, M. A., J. G. Goldammer, C. Price, and D. Rind. 1990. "Global Change: Effects on Forest Ecosystems and Wildfire Severity," in J. G. Goldammer, ed., *Fire in the Tropical Biota*. Pp. 463–86. Heidelberg: Springer-Verlag.

Frederick, K. D., and P. H. Gleick. 1989. "Water Resources and Climate Change," in N. J. Rosenberg, W. E. Easterling, P. R. Crosson, and J. Darmstadter eds., *Greenhouse Warming: Abatement and Adaptation*. Pp. 133–46. Washington, D.C.: Resources for the Future.

Gilliam, H. C. Jr. 1984. "The U. S. Beef Cow-Calf Industry." Washington, D.C.: USDA Economic Research Service, National Economics Division. (Agricultural Economic Report 515.)

Hahn, W. F., T. L. Crawford, L. Bailey, and S. Shagam. 1990. "The World Beef Market—Government Intervention and Multilateral Policy Reform." Washington, D.C.: USDA Economic Research Service, Commodity Economics Division. (Staff Report AGES 9051.)

Haynes, R., ed. 1991. "An Analysis of the Timber Situation in the United States: 1989–2040." Fort Collins, Colo.: USDA Forest Service, Rocky Mountain Forest and Range Experiment Station. (General Technical Report RM-191.)

Hof, J., and T. Baltic. 1988. "Forest and Rangeland Resource Interactions: A Supporting Technical Document for the 1989 RPA Assessment." Fort Collins, Colo.: USDA Forest Service, Rocky Mountain Forest and Range Experiment Station. (General Technical Report RM-156.)

Joyce, L. A. 1990. "An Analysis of the Range Forage Situation in the United States: 1989–2040." Fort Collins, Colo.: USDA Forest Service, Rocky Mountain Forest and Range Experiment Station. (General Technical Report RM-180.)

Joyce, L. A., and M. D. Skold. 1988. "Implications of Changes in the Regional Ecology of the Great Plains," in J. E. Mitchell, ed., *Impacts of the Conservation Reserve Program in the Great Plains*. Pp. 115–27. Fort Collins, Colo.: USDA Forest Service, Rocky Mountain Forest and Range Experiment Station. (General Technical Report RM-158.)

Joyce, L. A., M. A. Fosberg, and J. M. Comanor. 1990a. "Climate Change and America's Forests." Fort Collins, Colo.: USDA Forest Service, Rocky Mountain Forest and Range Experiment Station. (General Technical Report RM-187.)

Joyce, L. A., C. H. Flather, P. A. Flebbe, T. W. Hoekstra, and S. J. Ursic. 1990b. "Integrating Forage, Wildlife, Water, and Fish Projections with Timber Projections at the Regional Level: A Case Study in the Southern United States." *Environmental Management* 14: 489–500.

Lewis, C. E., G. W. Tanner, and W. S. Terry. 1985. "Double vs. Single-Row Pine Plantations for Wood and Forage Production." *Southern Journal of Applied Forestry* 9: 55–61.

Miller, R. B. 1989. "Human Dimensions of Global Environmental Change," in R. S. DeFries and T. F. Malone, eds., *Global Changes and Our Common Future*. Pp. 84–89. Washington, D.C.: National Research Council, Committee on Global Change.

Miller, W. F., P. M. Dougherty, and G. I. Switzer. 1987. "Effect of Rising Carbon Dioxide and Potential Climate Change on Loblolly Pine Distribution, Growth, Survival, and Productivity," in W. E. Shands and J. S. Hoffman, eds., *The Greenhouse Effect, Climate Change, U.S. Forests*. Pp. 157–87. Washington, D.C.: The Conservation Foundation.

Riebsame, W. E. 1988. "Adjusting Water Resources Management to Climate Change." *Climatic Change* 13: 69–97.

Royer, J. P., and F. J. Convery, eds. 1981. *Nonindustrial Private Forests: Data and Information Needs Conference Proceedings*. Durham, N.C.: Duke University.

Rosen, B. N., H. F. Kaiser, and M. Baldeck. 1989. "Nonindustrial Private Forest Landowners as Timber Marketers: A Field Study of Search for Market Information and Decision Quality." *Forest Science* 35: 732–44.

Sheffield, R. M., N. D. Cost, W. A. Bechtold, and J. P. McClure. 1985. "Pine Growth Reductions in the Southeast." Asheville, N.C.: USDA Forest Service, Southeastern Forest Experiment Station. (Resource Bulletin SE-83.)

Stahle, D. W., M. K. Cleveland, and J. G. Hehr. 1988. "North Carolina Climate Changes Reconstructed from Tree Rings: AD 372 to 1985." *Science* 240: 1517–19.

Tedder, P.L., R. M. LaMont, and J. C. Kincaid. 1987. "The Timber Resource Inventory Model (TRIM): A Projection Model for Timber Supply and Policy Analysis." Portland, Ore.: U.S. Department of Agriculture, Forest Service, Pacific Northwest Forest and Range Experiment Station.

Turner, M. G., R. V. O'Neill, R. H. Gardner, and B. T. Milne. 1989. "Effects of Changing Spatial Scale on the Analysis of Landscape Pattern." *Landscape Ecology* 3: 153–62.

U.S. Congress. 1974. Forest and Rangeland Renewable Resources Planning Act of 1974. PL 93-378.

U.S. Congress. 1976. National Forest Management Act of 1976. PL 94-588.

U.S. Congress, Office of Technology Assessment. 1986. "Technology, Public Policy, and the Changing Structure of American Agriculture." Washington, D.C.: U.S. Government Printing Office. (OTA-F-285.)

U.S. Department of Agriculture. 1989. "The Second RCA Appraisal." Washington, D.C.: U.S. Government Printing Office.

U.S. Department of Agriculture, Forest Service. 1988. "The South's Fourth Forest: Alternatives for the Future." Washington, D.C.: U.S. Government Printing Office. (Forest Resource Report 24.)

U.S. Department of Agriculture, Forest Service. 1989. "RPA Assessment of the Forest and Rangeland Situation in the United States, 1989." Washington, D.C.: U.S. Government Printing Office. (Forest Resource Report 26.)

Westman, W. E. 1990. "Managing for Biodiversity." *BioScience* 40: 26–33.

Wheaton, E. E., and J. P. Thorpe. 1989. "Changing Climatic Resources for the Western Canadian Boreal Forest," in D. C. MacIver, H. Auld, and R. Whitewood, eds. *Proceedings of the 10th Conference on Fire and Forest Meteorology*. Pp. 171–76. Ottawa: Environment Canada and Forestry Canada.

14

Climate Change and Forestry in the U.S. Midwest

Michael D. Bowes and Roger A. Sedjo[1]

Introduction

The forest resources of four midwestern states, Missouri, Iowa, Nebraska, and Kansas (the MINK region), were examined as part of a larger project to assess the effect of a warming climate on the resource-based sectors and broader economy of this region. What makes the forests of the region interesting is their likely sensitivity to climate warming. With a position on the fringe of the eastern hardwood forest, bordering the grasslands of the Midwest, these forests might be among the first affected by a changing climate.

The study is presented in four sections. First is a brief description of the current situation. Second, the direct effects of climate change on the forests are estimated. Third is an assessment of currently anticipated trends in the forest sector. Fourth, estimates of the future forest resource under a changed climate, allowing for anticipated trends and climate-driven adaptations in forest activity, are presented.

The Current Base Line

Before settlement, only the eastern half of the MINK region had extensive natural forest cover, and agricultural uses have now taken over much of this potentially forested land. The remaining extensive forests are largely to be found in Missouri, particularly on the rugged Ozarks Plateau

[1] From the Energy and Natural Resources Division, Resources for the Future, Washington, D.C. Support from the U.S. Department of Energy through the Pacific Northwest Laboratory is gratefully acknowledged. We also thank Alan Solomon for providing the FORENA simulation model.

of southern Missouri [U.S. Department of Agriculture (USDA) Forest Service, 1976], an area where soils and slopes are generally not well suited to profitable agriculture. The forests here are dominated by oaks, although some natural pine stands are found in the southeast. These forests are not particularly productive, and they generally yield lower-valued timber products. Still, the Missouri forests are actively harvested and are of considerable importance to local economies.

Elsewhere, significant forest holdings are largely confined to bottomlands and slopes poorly suited to agriculture. In northern Missouri and eastern Kansas and across Iowa, although much of the land is potentially forested, once-wooded lands have been largely converted to agricultural use. In this western edge of the hardwood-forest region, though, the highly valued black walnut is often found, especially in richer, well-drained bottomland soils. Further to the west in Kansas and across Nebraska, little of the land receives enough moisture to sustain significant natural hardwood-forest cover.

The forests of Missouri are not typically under intensive management (USDA Forest Service, 1976). Planting of oaks and other standard hardwood species is not considered a feasible economic practice. Similarly, activities like thinning are rarely undertaken. The hardwood stands regenerate naturally. For the oak-hickory hardwood forest of this region, the method and timing of harvest is very important in management. A harvest should be designed to favor regeneration of the economically preferred oaks. Some of poorer hardwood sites have been converted to pine plantations in areas where pine does well naturally, but, in general, competition from hardwoods makes this impractical. Plantations of black walnut are increasingly found on previously cleared marginal farm lands, but these stands amount to only a tiny component of the total forest area.

The forest industry of the region has become well adapted to the production and use of lower-valued forest products and timber supplies. Most timber goes to lumber production, generally production of lower-grade lumber. Other major uses include the production of charcoal and cooperage, with Missouri leading the nation in output of these two products. The charcoal industry draws much of its supply from the extensive nongrowing stock and from sawmill byproducts. The importance of the primary timber-processing industry for the MINK region is modest, representing only ≈0.8% of total regional value added. The volume of roundwood products from this region amounts to ≈1% of total U.S. volume (USDA Forest Service, 1982, 1988a).

Although the region is not nationally important as a source of wood products, its forests are associated with significant recreational and amenity services. This statement is particularly true for the large Mark Twain National Forest, which is spread across scenic country in southern Missouri. With the exception of this national forest, forests throughout the MINK region are largely in the hands of farmers and small private landowners. The small forest holdings across the agricultural sections of the four states offer wildlife habitat and protection for soils and water. Any decline in these forests could have implications well beyond the foregone harvest value.

Effects of Climate Change

In this section, we assess the direct physical impact of potential climate change on the forests of Missouri. Because of the limited forest cover in Iowa, Kansas, and Nebraska, only the Missouri forests are explicitly considered. The Missouri forests have been simulated by using FORENA (Forests of Eastern North America), a model of forest growth and species succession (Solomon, 1986). Extrapolations from the Missouri simulations will later be used in estimating full regional impacts. A brief discussion of the FORENA model, a description of the data required for the modeling exercise, and selected results from our simulations follow.

Methodology

The analysis focuses on the potential impact of climate change on the supply of timber. The main element in this analysis has been the simulation of growth in the Missouri forests.

FORENA. Forest growth was simulated by using FORENA, a stochastic model of the development of individual trees within a mixed-species forest plot (Solomon, 1986). Ingrowth, growth, and mortality of trees within a plot are modeled as random processes influenced by climate, soils, and the competition for light among trees.

FORENA is one of many similar "forest-gap" models closely derived from an earlier model of Botkin et al. (1972). The forest-gap models are attractive for three reasons. First, past work suggests that these models perform well in simulating the successional development of hardwood forests in the eastern United States (Shugart, 1984). Second, the structure of the model allows an investigation of the effects of climate change on both forest growth and species composition (Solomon, 1986; Botkin et al., 1989). Few other forest-growth models account for climate. Third, it appears that this easily parameterized model (i.e., one that is easy to fit to a new situation by finding that small set of constants that describe that situation) could, if the need arose, be adapted to other forested regions. Shugart (1984) describes model-validation efforts and applications, including applications to forests outside the eastern U.S. hardwood region.

Forest-gap models are now frequently used to simulate the impact of climate changes on the development of natural forest cover. The growth response to climate factors within the model is based on observed limits to the natural geographical range of a species. Modeled growth rates are reduced as climate conditions approach the limits of the natural range. It is this empirical approach that makes the model easy to parameterize. It might be questioned, though, whether the observed range of a tree species is explained solely by climate. Natural forest adaptation to climate change occurs within the model through gradual mortality and ingrowth. Ingrowth of species is modeled as a random draw on a set of eligible species, with eligibility based on soil, climate, and light conditions. This random process can be criticized as perhaps leading to an exaggerated capacity of plots to adapt to climate change, particularly if eligible species include some for which seed sources are not in fact likely to be present.

The Forests. Three areas in Missouri were selected for simulation. The selected locations represent the three major land-resource areas of Missouri that have significant forest cover. The silvicultural constants used to characterize the growth, mortality, and establishment of tree species are from Solomon's FORENA model. Those constants were designed to be broadly applicable to the forests of eastern North America. Forty-nine species likely to be found in the Missouri forest were included in our simulations. Moisture-holding characteristics of the local soils were drawn from USDA Soil Conservation Service documents (USDA SCS, 1978, 1979). Monthly precipitation and temperatures for the control climate were drawn from local climate statistics for the period 1951–1980 [National Climate Center (NCC), 1983].

The existing forest cover was represented by a set of forest plots generated to approximate actual forest inventory (USDA Forest Service, 1976). For each location, the development of 30 plots in each of three age classes was simulated, with initial age classes of 0, 30, and 70 years. The forest plots in each age class were modeled over a period of 200 years. Presented results are averages over the 30 plots. To allow comparison, the same initial plots were used for all climate scenarios.

The Simulations. We ran two groups of simulations. The first simulations were designed to investigate the overall sensitivity of FORENA to climate change in Missouri. The second set of simulations explores the development of the Missouri forests under an analog climate scenario, with climate similar to that of the 1930s (with and without elevated CO_2). An analog climate offers more realistic patterns of climate change, both across the year and across locations, than are possible in the typical sensitivity study. Our focus will be on the results of the analog simulations. The biomass and species mix under the analog climate scenarios are compared with those resulting under control conditions.

Sensitivity of the Missouri Forest to Climate Change

In this section, we present results from simulations designed to investigate the overall sensitivity of FORENA to climate change. These simulations reflect the growth of forest plots on the northwest edge of the Ozarks region, for forest plots initially aged 30 years. The climate changes represented are uniform shifts in monthly temperature or precipitation over the year. A change in climate is assumed to occur immediately.

Figure 14.1 shows average above-ground woody biomass (kg/m^2) after 20 years of simulated growth under different climate scenarios. The asterisk on the surface marks the average biomass under a continuation of current climate conditions. The bottom axes in the figure give deviations in monthly climate from control conditions. For example, after 20 years of growth under a climate scenario with each month 1 °C warmer and 12% drier than at present, simulated average biomass is 3.2 kg/m^2 (32 tons/ha), rather than the 7.4 kg/m^2 resulting under control conditions. It is apparent from Figure 14.1 that the Missouri forests, as simulated, are very vulnerable to climate warming, especially when warming is accompanied by reduced precipitation. At best, modest gains in forest stock might result if limited warming is associated with increased precipitation.

To some extent, the sensitivity to climate in the Missouri forests is not unexpected, given Missouri's position on the edge of the current prairies. Nevertheless, a decline in forest biomass of this degree in a relatively short time period may seem surprising. However, a roughly comparable dieback was observed in Missouri under recent dry conditions. In the early 1980s, 20–30% of mature oaks died, probably from drought-induced stress (USDA Forest Service, 1989). Simulations by others (Solomon, 1986; Botkin et al., 1989) have shown that the forests in drier regions of the Midwest are very vulnerable to climate warming.

Forests under the Analog Climate of the 1930s

Here we report the results of simulations under analog climate conditions like those of the 1930s. Specifically, the results are from simulations with climate years from the 1930s drawn at random and overlaid with random noise. The 1930s in Missouri were on average slightly hotter and drier than the control period, although with some years of high summer temperatures. Although the random noise added to our simulated analog climate does not affect the climate means, it does increase the probability of extended periods of climate stress, thereby raising the associated rate of tree mortality.

It is not yet well known how elevated CO_2 concentrations will affect forest-stand development over time. But, based on seemingly reasonable

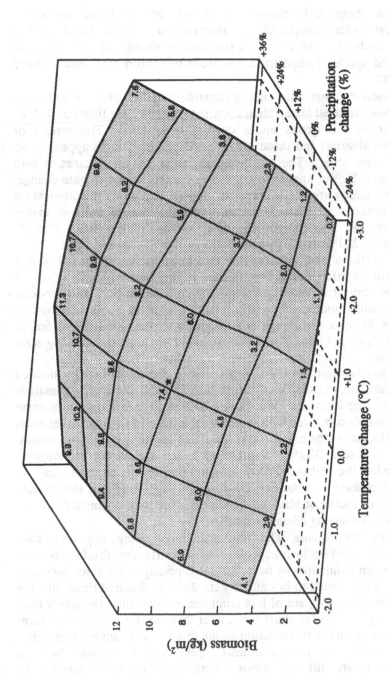

FIGURE 14.1. Biomass response to temperature and precipitation change—for existing 30-year-old stands in Missouri. The climate changes are immediate and permanent adjustments to monthly climate for a central Missouri county. An asterisk marks the biomass resulting under continued control climate conditions.

assumptions about CO_2 effects on growth and moisture efficiency, development under elevated CO_2 concentrations (450 μmol/mol, or 450 ppm) was modeled. The effects of the climate change on average stand biomass and species composition are described, both with and without elevated CO_2.

The Biomass Response. Our analog climate simulation predicts that some areas of Missouri forest lands might eventually support less than one-half of timber biomass that grows under current conditions. The forests of northeastern Missouri and those of the northwestern Ozarks appear to be especially vulnerable. The southeastern areas of the Ozarks, where precipitation is higher, appear to be less vulnerable to the climate change. Shorter-term effects are not as dramatic. Figure 14.2 shows the biomass on simulated forest plots across Missouri, under the control and two analog climate scenarios. The composition of an average plot is shown after 20 years and after 200 years for plots initially in the 70-year age class.

In the northern and west-central locations, increased mortality and declining growth rates lead to forests that, in the long run, have <50% of the biomass expected under current climate conditions. Shorter-term declines are more modest, with $\approx$10% relative decline in biomass on existing stands after 20 years. Although not insignificant, the simulated effects of enhanced CO_2 do not appear sufficient to avoid the potential for long-term forest decline.

In our southeastern location, representing the most heavily forested region of the state, the simulated forest decline under the analog climate is less than at other locations. We show a 25% decline in long-run biomass on existing stands compared with the control forest. In slightly cooler areas of the Ozarks, just to the north of our simulated location, even more modest declines would seem likely. Short-term losses are hardly pronounced. Indeed, under the enhanced-CO_2 assumption, forest biomass may be somewhat increased over the first 20–40 years, suggesting that the primary forest region of southeast Missouri would continue to be a source of timber supply under the analog climate scenario.

The Species-Mix Response. The simulated climate change appears to have relatively little effect on forest species mix. In particular, the forests seem likely to remain dominated by oaks, although perhaps with some increases in those oaks best suited to hot and dry conditions. Some of these, like the post and blackjack oaks, are of less commercial value than the larger oaks they might replace. In the short term, after 20 years, our simulations show almost no change in the composition of the forest. In the longer term, after 200 years, despite the now-pronounced differences in biomass between scenarios, there are still only minor differences in the forest mix for the northern and southeastern forests. Somewhat greater long-run adjustment is apparent for the central Missouri forest.

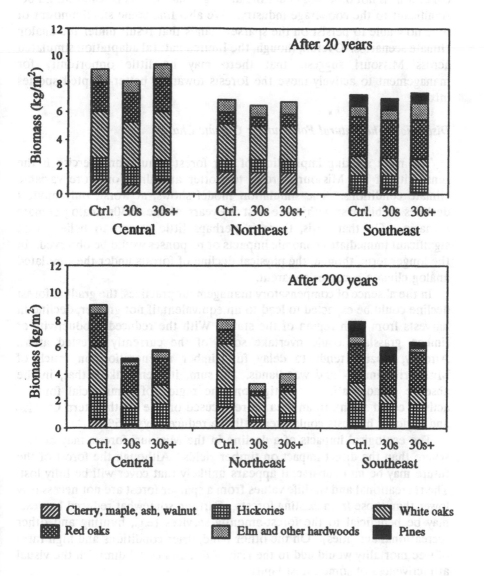

FIGURE 14.2. Impact of 1930s vs. control climate on species composition in central Missouri in 70-year-old stands, comparing values for control, 1930s (30s), and 1930s with elevated CO_2 concentrations (30s +).

Figure 14.2 shows the species composition on the simulated forest plots across Missouri. Most noteworthy is the central Missouri location, where the eventual dominance of stands by white oaks expected under control conditions is not observed under the analog scenarios. This result would be significant to the cooperage industry. We also find some small amount of pine now able to persist on the sparser stands that result under the analog climate scenarios. Overall, though, the limited natural adaptation simulated across Missouri suggests that there may be little opportunity for management to actively move the forests toward a better-adapted species mix.

Discussion: The Natural Forest under Climate Change

The most striking implication of our forest simulation exercise is the sensitivity of the Missouri forests to hotter and drier or more variable climate conditions. The simulation model shows, at worst, only modest declines in biomass within the first 20 years of the 1930s analog climate scenario. On that basis, there is perhaps little reason to believe that significant immediate economic impacts or responses would be observed. In the longer term, though, the physical decline of forests under the simulated analog climate becomes apparent.

In the absence of compensatory management practices, the gradual forest decline could be expected to lead to an equivalent, if not greater, decline in harvests from each region of the state. With the reduced productivity of timber, grassland could overtake some of the currently forested areas. Already, grazing tends to delay full timber regeneration on much of Missouri's timber- and woodlands. In sum, it seems likely that, in the absence of adaptation or mitigation, the region of commercial forestry activity could shrink to an area more focused on the southeastern Ozarks, and regional harvests could eventually be reduced by >25%.

The ecological impacts of a decline in the Missouri forests may be less severe than the direct impact on timber yields. Although the forest of the future may be more sparse, it appears unlikely that cover will be fully lost. The recreational and wildlife values from a sparser forest are not necessarily lower than those from existing forests. Further, somewhat reduced harvests may be beneficial to the forest-amenity services (e.g., hunting and other recreational activities). On the other hand, drier conditions and high rates of tree mortality would add to the risk of fire and could diminish the visual attractiveness of some forest lands.

We now turn to general forestry trends and the possible adaptive responses from forest landholders and industry.

The Future Base Line: Prospects for Change

This section describes general changes that, in the absence of a climate change, are anticipated to occur in forestry and the forestry sector over the period 1990–2030. The trends in forestry technology are examined first. Next, the implications of these trends for forests in the MINK region are considered. Finally, other factors that may influence forestry within the region are discussed, including recreational demands on existing forest and the prospects for biomass plantations on present agricultural lands.

Change in the U.S. Forest Industry

In the United States, the major locations of an active forest industry have been in the Pacific Northwest (PNW) including Northern California, in the South, and increasingly, of late, in the Great Lakes states. Smaller pockets of industry also exist in the Rocky Mountains and in the Northeast, particularly in Maine. Harvests in the PNW consist primarily of logging from old-growth conifer forests for the production of lumber and veneer, with the residue being used as feedstock for pulp mills, both domestic and foreign. In the South, the industry is centered in the coastal plain and also the south-central region. Much of the logging involves southern pine for use both as lumber and as pulpwood. In addition, southern hardwoods are used for lumber and veneer and, increasingly, as feedstock for certain types of pulp. The Great Lakes states have long been a source of specialized paper production. More recently, harvesting has drawn on the abundant aspen forest that developed in the Great Lakes states after the heavy logging in the latter part of the 19th century and first decades of the 20th.

Current Price Trends. Certain wood resources in the United States, unlike most natural resources, have experienced a long-term increase in real price since 1800 (USDA Forest Service, 1988b). The real prices of softwood lumber and saw logs have exhibited this rising trend, whereas pulpwood price has exhibited no trend. However, since 1950, the upward trend in real prices of lumber and saw logs has moderated considerably, even as pulpwood continues to exhibit no price trend (Sedjo and Lyon, 1990). Although long-term real-price trends for hardwood logs are difficult to find, there has been no perceptible long-term trend over the past several decades for either domestic or tropical hardwood logs (de Steiguer et al., 1989). If the real-price trends for wood resources of the past three decades continue, only very modest real-price increases would be anticipated, at best.

Current Technological Trends. In the past several decades, technology in forestry and wood processing has taken on several clearly identifiable forms. The first of these can be characterized as yield-enhancing (Crutchfield, 1988). Probably the largest single example of this is found in the changes that forest management and silviculture have experienced with the advent of the genetic improvement of seedlings used in artificial plantings. Other improvements in technology that enhance yield include improved nursery practices, which have increased survival after planting through such innovations as containerization.

A second effect of technological change is wood extending, increasing supply by allowing the use of what was previously an inferior or unusable wood (Sedjo and Lyon, 1990). There have been several important examples of this type of technology in recent years. For example, wood residues, formerly burned or discarded, now account for >40% of pulp-mill feedstock. Also, an economically viable technology has been developed for making composite wood panels (waferboard and oriented-strand board) with structural properties similar to those of plywood. These new products have resulted in the development of an active market for aspen.

A third type of technological change that has been important in the forest industry is the development of new techniques that can be considered wood-saving, because they use more of the log, thereby reducing the waste. Examples include thinner saws, which reduce sawdust waste, and computer-aided saws that rely on lasers to determine the most efficient cut of the log.

Prospects for Change. In their recent study, Sedjo and Lyon (1990) examined the long-term adequacy of world timber supply over the period 1990–2030, under the assumption that climate would not change. Using a control-theory approach, they examined several scenarios of changes in future demand, alternative sources of future supply, and the gradual introduction of technological innovations. Their findings suggest that although demand can be expected to continue expanding through that period, the combination of factors affecting timber supply—including the timber resources found in old-growth, second-growth, and plantation forests; intensive forest management in regions where productivity is high; and technological change—is able to provide for future demand with only modest increases in real prices.

This and other analyses (e.g., Crutchfield, 1988) see the major sources for meeting additional timber demand as coming from the increased production of plantation forests on high-productivity sites and from the wood-saving and wood-extending effects of technological innovation in wood processing.

Implications for the MINK Forest Industry

If the Sedjo and Lyon perspective is approximately accurate, the implications of technological change in tree growing and wood processing for the MINK region would likely be modest. As noted above, Missouri is the only one of the MINK states with significant forest area and forest-related industrial activity. The rugged forestlands of the Missouri Ozarks are not particularly productive and therefore not well suited for intensive forest-plantation management. Although modest areas of pine plantations have been established in southern Missouri, under current climate conditions, further extension of these plantations is apparently limited by the availability of suitable soils. If real wood-resource prices rise only very modestly as projected, there will not be strong market pressures to utilize marginal lands such as those in Missouri.

A factor in the analysis that could increase the value of the Missouri forest is that of wood-extending technological innovation. The Missouri forest is, by forest-industry standards, marginal. However, these types of forests have lent themselves to innovations that allow the use of relatively low-valued timber. Thus, a newly developed technology might find the Missouri forest an important source of raw material. However, such technologies are usually fiber-oriented and require high-volume, lost-cost material-handling systems. The rough terrain of the Ozarks and the relatively low wood volumes would make low-cost wood collection difficult. In the absence of climate change, it seems likely the forest industry will continue to play the quite modest role that it does now.

Other Factors Influencing the MINK Forests

If America were to face growing problems of energy scarcity, the Ozark forest could be used as a renewable-resource substitute for fossil fuel and other energy sources. Wood-fired electrical-power generation facilities can be constructed to use the low-quality timber of the region. Such a system could probably operate at a moderately high level by drawing fuel wood from the existing forest. Higher levels of operation would require artificial regeneration.

Although tree planting for biomass fuels is possible in this region, neither the biology nor the economics is particularly favorable. The energy crisis of the late 1970s resulted in the Department of Energy's initiation of an experimental woody-biomass program (Geyer and Melichar, 1986). As part of this program, very-short-rotation, intensively managed wood-energy plantations were established on underused alluvial sites in south-central Kansas along the Arkansas River and along the Platte River in Nebraska, as well as in upland areas in the western two-thirds of Kansas and southern Nebraska. The work demonstrated the biological feasibility of short-rotation forestry plantations providing alternative wood-energy supplies. Despite the biological success of the program, the economic returns were poor (Perlack and Geyer, 1987), with the cost of wood produced in this manner higher than that obtained from traditional sources.

Might we see an expansion of timber plantation into the agricultural lands for the production of high-valued lumber rather than biomass? Currently, other than as wind breaks and protection, widespread tree planting is not found in the prairie region. Aside from occasional high-valued black walnut stands and plantings of specialty items, such as Christmas trees, there is little commercial planting. But increasingly, walnut plantations are being established, especially on some flood plains that have limited agricultural use. As the pressures increase to remove the subsidies that keep marginal agricultural lands in production, further expansion in tree plantations might be expected. However, compared with the forest as a whole, the scale of this plantation effort seems certain to remain quite small. The expectation of high future prices could not be sustained if the area of timber in plantations were expanded.

It is then difficult to see the region producing significant volumes of planted industrial wood or fuel wood economically. A very different situation could exist if a major program promoted tree plantations on low-value land to sequester atmospheric carbon (Sedjo, 1989). Because the motivation of such plantations is not financial return but rather, socially desired sequestration of carbon, the decision would be policy driven, and the financing would be provided or subsidized by the government. Under these conditions, the vast areas of submarginal croplands of the western parts of the region could be used for this purpose. Furthermore, in the future, the areas available may increase as some currently irrigated areas are withdrawn from irrigation because of declining supplies of water.

The Future under Current Climate Conditions

In our judgment, the most likely scenario for the MINK region is that, in the absence of climate change, the forest industry will continue to play the quite modest role that it does currently. The forest itself is likely to expand somewhat as agriculture continues to decline in the marginal sites within the mountains and as the existing forest matures. Harvests will continue as an opportunistic use of a largely unmanaged timber resource. This judgment is consistent with the projections by the Forest Service for modestly increasing timber inventories and harvests.

Future Forest under a Warming Climate

This section addresses the development of forests and the forest industry under the assumption that the climate has experienced a permanent change to the analog climate of the 1930s. Projections for future forest inventories and harvests are presented. Our previous simulations suggested that a warmer and drier climate similar to that experienced during the 1930s would lead to a relative decline in forest cover and forest volume as moisture constraints limit the forest. The forest decline would take two forms. First, where forest vegetation continues, increased mortality and slowed growth would be expected. Second, the margin between the prairie and the forest would shift, as grasses gradually replaced the trees that could no longer be sustained by the reduced moisture.

The transition in response to the analog climate is likely to be gradual and probably would not be completed within the 40-year period under consideration, even should warming begin today. Thus, the next 40 years would likely see only a slow decline of the forest and very small expansion of grasslands, should the warmer and drier climate ensue. Over the longer term (200 years), the forest would exhibit some change in the composition of species toward those most highly suited to hotter and drier conditions. In our analysis, CO_2 fertilization was simulated, but the effects only partially offset the results of the warming climate.

Adaptation in Forest Management

The results of the forest simulations described above were based on the development of natural hardwood stands. Might changes in forest management or industry adaptations come into play in response to the negative impacts of climate change? Forest managers could, for example, act through planting, stand improvement, or harvesting practices to alter the species mix or to otherwise reduce the financial risks from potential mortality.

There are several reasons to believe that despite the potential for forest decline, significant mitigating response in this region is unlikely. At present, neither active stand improvement nor planting of hardwoods is observed. Such management does not pay financially. With the possibility of declining forest yields, intensified forest management is likely to become even less financially attractive. Given the rugged terrain and low yields, comparative advantage should shift increasingly to other regions of the country. In addition, it is not altogether apparent what a better-adapted forest for this region might be. Our earlier simulations have shown no obvious transition to a new, commercially viable species mix. The Missouri forests seem likely to remain dominated by those oaks adapted to the rather dry soils of this region. The southern pines do appear to offer one possibility. Shortleaf pine is already a significant component of the southeastern Ozarks forest, and our earlier simulations have indicated that central areas of the state may become somewhat better suited to these pines.

To more carefully investigate the possibilities for adaptation in forest management, FORENA was modified to allow simulation of several artificial regeneration and thinning activities. These simulations of planting and thinning responses confirm that such active management would not be an appropriate response for the private land manager. Although the managed forest simulations do show central areas of the Missouri becoming better suited to shortleaf pine plantations, yields are still below those expected under current climate in southeastern Missouri. There is otherwise little to suggest that a managed transition to any better-adapted species mix would offset the decline in the forest resource. In fact, the simulations show active management to be relatively less attractive financially after a climate change than under the current control climate.

General technological adaptations to climate warming would surely occur, especially over the longer term. In forestry, a major adaptation would likely be found in the changes in species used for plantation forests and also in genetic improvement in seedlings to provide for trees that are more suited to the new climate. In the MINK region, however, the impact of these changes is likely to be modest, because the low productivity of most forest sites has discouraged investments in plantation forests. Thus, there appears to be little role for technology, at least technology driven by financial considerations, in offsetting the effects of climate change on the forest. Should a major warming occur very rapidly, however, thereby severely testing the ability of the natural forest to adapt, the public sector might assist the migration of the forest through the introduction of aerial seeding and other such aids to tree migration and regeneration. In the context of a rapidly changing climate, general improvement in aerial-seeding technology might be expected.

Our judgment is that the region, barring major public intervention, is likely to see only reactive responses to actual forest decline. For example, salvage cutting, along with somewhat earlier harvests, might be undertaken to avoid some of the potential short-term economic losses associated with rising mortality in the standing forest. As in the no-warming scenario, the long-term future of the Ozark forest in a warming would probably be increasingly in providing recreation in the form of vacation homes, woodland experiences, and outdoor recreation. In addition, there is a possible shift to grazing as grasslands replace forest at the intersection of forest and the prairies. Although it is possible that a warming-induced decline in agriculture could free land for trees, for most of the MINK region, we would expect the grasses to dominate. In general, the MINK region would probably experience declines from the warmer and drier climate. With salvage harvesting offsetting some potential early losses, the Missouri forests seem likely to begin a path of gradual decline.

Industry Adaptation to Warming

The forest industry in this region also seems unlikely to make significant positive responses to the changing forest situation. The warmer and drier analog climate is simulated as leading to a relative decline in forest cover and forest volume, albeit a modest decline for the most important forest lands of southern Missouri. In the longer term, under these conditions, it would be expected that forest industries in the region would gradually contract while other land uses would expand.

During transitional periods, which might last several decades, salvaging operations could delay the industry decline. Significant salvage operations have been associated with forest decline elsewhere and might also be expected in the MINK region, especially for the more valuable species. However, because of the already low values of much of the timber in the MINK region, the increased salvaging activities are likely to be modest here. With this an area already well adapted to the use of poorer trees and sawmill byproducts, producing generally lower-valued timber products and a large volume of charcoal, it is unclear how much further such adaptation is possible. The market could not quickly absorb larger supplies of the lower-valued timber products. At best, with salvaged and nongrowing stock timber substituting for some harvests from growing stock, the value of produced output would still decline.

One commercial possibility for a declining forest in the Ozark region is a shift to greater production of charcoal and fuel wood. Early responses to the "energy crisis" of the 1970s included increased use of fuel wood in both individual residences and in certain industrial processes. Should policies provide strong incentives for substituting renewable energy sources, or renewables, for fossil fuels, the Ozark region could become an important supplier, especially to areas of the Midwest with little natural forests of their own. A return to the climate of the 1930s could make such a scenario, while highly speculative, somewhat more likely.

It is possible that over the next several decades, concerns about global warming may result in the imposition of a large carbon tax on the use of fossil fuels. In such an eventuality, the cost structure between fossil fuels and renewable fuels might be altered dramatically in favor of the renewables. In this environment, wood-fired electrical generation plants might become common. Under such circumstances, the forests of the Missouri Ozarks could provide the feedstock for local electrical power generation. The Ozark forests might lend themselves to this task because, for the most part, much wood has relatively few other high-valued commercial uses. However, a couple of caveats remain. For example, as previously noted, the terrain is relatively rough, thereby adding to the logging costs. Fuel demands could draw heavily from the existing stock, but would probably not offer sufficient returns to encourage plantations, with climate leading to reduced yields and increasing mortality. Given the steepness of the terrain, the increased water runoff and erosion associated with the extensive clearcuts likely to attend widespread wood-powered electrical generation might be deemed unacceptable in a society that is becoming increasingly environmentally sensitive.

Our judgment is that under the analog climate, barring major policy changes, timber flows to the forest industry would remain at currently anticipated levels for at least a decade. Salvage operations and market rigidities seem likely to delay the potential impact of the modest, early declines in biomass of forests. A gradual decline in the forest industries is then expected to begin, matching the forest decline.

Estimated Changes in Forest Inventory and Harvests

The FORENA simulations of the Missouri forests under the 1930s analog climate suggest a declining forest resource. Our expectation is that no significant mitigative response to this forest decline will be undertaken. In Table 14.1, the simulated rates of decline for the region's forests are presented, with and without the assumed effects of elevated CO_2. For example, in the year 2000, under the warming scenario, forest biomass is 89% of that resulting under the continuation of current climate conditions. The percentage figures are average rates of decline for the region. The average reflects weightings by the share of timber production accounted for in each simulated region.

Our base-line projections for forest inventory and harvests through the year 2030 for the MINK region are derived from Forest Service projections.

TABLE 14.1 Proportion of Biomass Remaining Under Simulated Analog Climate Relative to Base-Line Climate

Year	Climate Scenario	
	Analog	Analog with CO_2
2000	0.89	0.93
2010	0.89	0.99
2020	0.77	0.88
2030	0.68	0.79

The Forest Service projections for the individual states were combined, with extrapolations made to the year 2030 when necessary. Their figures are based on expected trends in growth, mortality, removals, and land-use change, with climate assumed unchanged over the period. For the region as a whole, under the control climate, slow increases in forest inventory and harvests are anticipated, as shown in Figure 14.3 by the solid lines.

Our projections for inventory and harvests under the analog climate are derived from the base-line projections, adjusted to reflect the simulated forest decline. Forest inventories in each state have been assumed to decline compared with base line, according to the proportions given in Table 14.1 above. Harvests were assumed to decline in the same proportion as inventories—but after a 10-year lag. That lag is intended to reflect both the effects of salvage harvesting and behavioral delays in the supply-and-demand response to declining growth. Ideally, perhaps, the Forest Service projection model would have been rerun to reflect the new simulated rates of forest mortality and stand growth. However, our simple method of adjustment to the base-line Forest Service projections seems adequate as a first approximation of the aggregate forest decline. Note that our projections

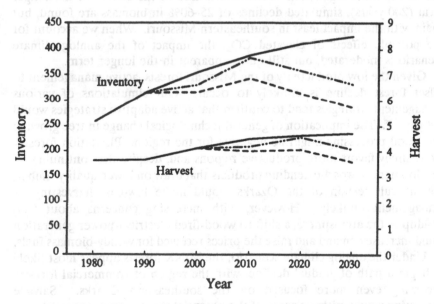

FIGURE 14.3. Projections for timber inventory and harvest (in $m^3 \times 10^6$) under climate change in Missouri, Iowa, Nebraska, and Kansas, 1980–2030.

reflect only the simulated physical decline in the forest and do not allow for any possible change in current land-use trends.

Under the analog climate, a decline in the growth of forest inventories begins immediately (1990), reflecting the advent of the climate change. Harvests start to decline in the year 2000, although with only modest reductions in harvest before the year 2030. These projections are illustrated in Figure 14.3 by the dashed lines. The CO_2-fertilization effect on forest growth offsets some, but not all, of the relative declines associated with the climate change, as illustrated by the dot-dashed lines in the figure.

Summary

Our simulations of the Missouri forests with FORENA show that the forests are extremely sensitive to climate warming. A warming as little as 2 °C is simulated as leading to a severe dieback in the forest, particularly when precipitation is reduced. Under the analog climate of the 1930s, which is only slightly warmer than the baseline climate, our simulated biomass in the forests of Missouri drops by >10% in 10 years. Luckily, the declines are less in the more significant southeastern area of the state. In the longer term (200 years), simulated declines of 25–60% in biomass are found, but again with the impact least in southeastern Missouri. When we account for the possible effects of elevated CO_2, the impact of the analog climate scenario is moderated, but still quite apparent in the longer term.

Given the low productivity of the Missouri forests, active management to offset forest decline is unlikely to occur. Our simulations of various management strategies tend to confirm that active adaptive strategies would not pay off. The implication of general technological change in tree growing and wood processing might be negative for the region. Plantation forestry increasingly favors more productive regions and, despite the continuing development of wood-extending products that draw on lower-quality timber, the difficult terrain of the Ozarks would make low-cost harvesting or management unlikely. However, with increasing concerns about CO_2 buildup in the atmosphere, a shift to wood-fired electrical-power generation could increase demand and raise the prices received for woody-biomass fuels.

Under the analog climate scenario, the MINK forests appear most likely to begin a path of gradual decline, with the region of commercial forestry becoming even more focused on the southeastern Ozarks. Salvage harvesting would mitigate some of the potential short-term economic losses that could be associated with rising mortality. However, our projections for the MINK region show a long-term decline in both inventory and harvest compared with baseline projections, with a rising CO_2 concentration offsetting some, but not all, of this decline.

References

Botkin, D. B., J. F. Janak, and J. R. Wallis. 1972. "Some Ecological Consequences of a Computer Model of Forest Growth." *Journal of Ecology* 60: 849–72.

Botkin, D. B., R. A. Nisbet, and T. E. Reynales. 1989. "Effects of Climate Change on Forests of the Great Lake States," in J. B. Smith and D. A. Tirpak, eds., *The Potential Effects of Global Climate Change on the United States*, Appendix D: Forests. Pp. 2-1–2-31. Washington, D.C.: U.S. Environmental Protection Agency Office of Policy, Planning and Evaluation. (Report PM-221.)

Crutchfield, D. M. 1988. "The 2030 Forest: The Industrial Direction," in *Proceedings of the Fifth Biennial Southern Silvicultural Research Conference*. Pp. 5–10. New Orleans: U.S. Department of Agriculture (USDA) Forest Service. (General Technical Report SO-74.)

de Steiguer, J. E., L. W. Hayden, D. L. Holly Jr. et al. 1989. "Southern Appalachian Timber Study." Asheville, N.C.: Southeastern Forest Experiment Station. (General Technical Report ES-56.)

Geyer, W. A., and M. W. Melichar. 1986. "Short-Rotation Forest Research in the United States." *Biomass* 9: 125–33.

National Climate Center (NCC). 1983. "Climate Normals for the U.S., Base 1951–80." Detroit, Mich.: Environmental Data and Information Service, National Oceanic and Atmospheric Administration.

Perlack, R. D., and W. A. Geyer. 1987. "Wood Energy Plantation Economics in the Great Plains." *Journal of Energy Engineering* 113: 92–100.

Sedjo, R. A. 1989. "Forests: A Tool to Moderate Global Warming?" *Environment* 31: 15–20.

Sedjo, R. A., and K. S. Lyon. 1990. *The Long-Term Adequacy of World Timber Supply*. Washington, D.C.: Resources for the Future.

Shugart, H. H. 1984. *A Theory of Forest Dynamics: The Ecological Implications of Forest Succession Models*. New York: Springer-Verlag.

Solomon, A. M. 1986. "Transient Response of Forests to CO_2-Induced Climate Change: Simulation Modeling Experiments in Eastern North America." *Oecologia* 68: 567–79.

USDA Forest Service. 1976. "Timber in Missouri, 1972." St. Paul, Minn.: U.S. Department of Agriculture Forest Service. (Resource Bulletin NC-30.)

USDA Forest Service. 1980. "Iowa Forest Resources 1974." St. Paul, Minn.: U.S. Department of Agriculture Forest Service. (Resource Bulletin NC-52.)

USDA Forest Service. 1982. "An Analysis of the Timber Situation in the United States 1952–2030." Washington, D.C.: U.S. Department of Agriculture Forest Service. (Resource Report 23.)

USDA Forest Service. 1984. "Kansas Forest Inventory 1981." St. Paul, Minn.: U.S. Department of Agriculture Forest Service. (Resource Bulletin NC-83.)

USDA Forest Service. 1986. "Nebraska's Second Forest Inventory." St. Paul, Minn.: U.S. Department of Agriculture Forest Service. (Research Bulletin NC-83.)

USDA Forest Service. 1988a. "An Analysis of the Timber Situation in the United States 1989-2040." Washington, D.C.: U.S. Department of Agriculture Forest Service (draft).

USDA Forest Service. 1988b. "U.S. Timber Production, Trade, Consumption, and Price Statistics 1950-1986." Washington, D.C.: U.S. Department of Agriculture Forest Service. (Miscellaneous Publication 1460.)

USDA Forest Service. 1989. "History of Hardwood Decline in the Eastern United States." Broomall, Pa.: U.S. Department of Agriculture Forest Service. (General Technical Report NE-126.)

USDA Soil Conservation Service (SCS). 1978. "Land Resource Regions and Major Land Resources Areas of the United States." Washington, D.C.: U.S. Department of Agriculture Soil Conservation Service. (Agricultural Handbook 296.)

USDA SCS. 1979. "Missouri General Soil Map and Soil Association Descriptions." Columbia, Mo.: U.S. Department of Agriculture Soil Conservation Service State Office.

15

The Role of Agriculture in Climate Change: A Preliminary Evaluation of Emission-Control Strategies

Richard M. Adams, Ching-Cheng Chang, Bruce A. McCarl, and John M. Callaway[1]

Introduction

Agricultural productivity is of obvious importance to human welfare. Climate is a major determinant of both the location and the productivity of agricultural enterprises. It is thus not surprising that agriculture has been identified as an area of concern in the current public debate over the causes and effects of climatic change. Indeed, agriculture has been the central focus of several studies of potential effects of climatic change (Decker et al., 1986; Sonka and Lamb, 1987; Rosenzweig, 1988; Smith and Tirpak, 1989).

Most studies have evaluated the sensitivity of several dimensions of agricultural activity to climate change, including yields, input use, and locational (geographic) patterns. The economic implications of such potential sensitivities have also been explored (Dudek, 1988; Adams et al., 1990; Kane et al., 1990). The results of these economic evaluations, while preliminary, suggest that climate change does not threaten the U.S. food supply, although substantial regional adjustments are likely.

When considered globally, the agricultural sector is more than a receptor of possible climatic changes arising from anthropogenic trace-gas emissions; it is also a source of trace gases, including CO_2, methane (CH_4), and nitrous

---

[1] From the Department of Agricultural and Resource Economics, Oregon State University, Corvallis (R.M.A.); the Department of Agricultural Economics, Texas A&M University, College Station (C.-C.C. and B.A.M.); and RCG/Hagler, Bailly, Inc., Boulder, Colo. (J.M.C.).

oxide (N_2O). The understanding of agriculture's contributions to trace-gas emissions has increased considerably over the past decade, leading to several potential strategies for reducing such emissions. In addition, interest is growing in the use of agricultural land as a potential sink for carbon through the establishment of forest plantations (Marland, 1988; Lashof and Tirpak, 1989; Moulton and Richards, 1990; Sedjo and Solomon, 1989; Dudek and Leblanc, 1990). Several studies have suggested that tree plantations on marginal agricultural lands may be a relatively cost-effective way to slow the build-up of greenhouse gases. Such tree-planting or "carbon-growing" activities have also received considerable political attention as an alternative to carbon taxes or "command-and-control" strategies.

Objectives

There are few economic evaluations of the costs to agricultural producers and consumers of strategies for reducing CH_4, CO_2, and N_2O emissions from agriculture. To the extent that reduction of these residuals requires changes in management, including reduced use of current inputs such as nitrogen fertilizer and certain types of feed rations for livestock, such strategies imply rising per-unit costs, at least in the short run. Whereas forest plantations on agricultural lands to sequester carbon have been the subject of economic analyses (Sedjo, 1983; Sedjo and Solomon, 1989; Dudek and Leblanc, 1990), most studies to date have not included the opportunity cost of converting large areas of agricultural lands to tree plantations. Large shifts in agricultural land use suggest reduced agricultural output and rising commodity prices. The associated welfare losses to consumers need to be considered.

The overall objective of the research reported here is to perform a preliminary economic evaluation of the social costs of selected strategies to reduce trace-gas emissions from agriculture and to sequester carbon in tree plantations. These strategies are intended to reduce CH_4 emissions from livestock and rice production, reduce N_2O emissions from crop production, and increase through reforestation the amount of carbon sequestered in agricultural lands. In presenting this information, we include (1) a review of the role of agriculture in trace-gas emissions; (2) a discussion of strategies for reducing such emissions; (3) a quantification (in dollars per 10^3 kg) of the costs of reducing agricultural trace gases with specific strategies; and (4) an evaluation of the costs per 10^3 kg of sequestering carbon in forest plantations on agricultural lands.

Two caveats for this evaluation need to be noted. First, the empirical focus is limited to U.S. agriculture. The costs of similar strategies in the rest of the world are ignored. Second, the analysis is a comparative static evaluation of selected control strategies. The dynamic or long-run adjustments to reduce emissions are likely to be different from those provided here. Thus, the results are preliminary and should be viewed more as sensitivity analyses, given the uncertainty in some key data, as noted below. However, because the data used here are comparable to those in other studies, the results can be compared with existing cost estimates and suggest the importance of including some economic aspects not explicitly addressed elsewhere.

Background

The contribution of agricultural sources to total trace-gas emissions varies by the individual gas and location. For example, U.S. agriculture consumes $\approx 3\%$ of total U.S. fossil fuel for on-farm production activities. If storage, dehydration, and other food- and fiber-processing activities are added, agriculture still is responsible for $<10\%$ of total U.S. fossil-fuel use. Thus, emissions of CO_2 from agricultural production are a relatively small share of the $\approx 1 \times 10^{12}$ kg (≈ 1 billion metric tons) of carbon produced from fossil-fuel combustion in the United States per year (Darmstadter and Edmonds, 1989). However, CO_2 released by agricultural activity globally is much higher, primarily because of the clearing of tropical forests for agriculture. Tropical deforestation may account for 1.8×10^{12} kg (1.8 billion metric tons) of carbon, or as much as 33% of total annual carbon emissions from all sources (Woodwell et al., 1983; Houghton et al., 1985; Palm et al., 1986).

For other trace gases, such as CH_4 and N_2O, agriculture's role in total U.S. emissions is higher. Worldwide, agriculture is believed to account for $\approx 55\%$ of total N_2O emissions, primarily because of deforestation and the use of nitrogenous fertilizers such as anhydrous ammonia (Cates and Keeney, 1987; Hutchinson, 1988). As much as 7% of total nitrogen applied as anhydrous ammonia may be lost as N_2O. Agriculture also plays a role in CH_4 emissions, primarily from rice production and livestock. Worldwide, $\approx 60\%$ of total CH_4 is believed to come from agriculture, with 30% from rice production and 15% from livestock (Gibbs et al., 1989; Rasmussen and Khalil, 1981). In addition, rice area and livestock numbers, although stable in the U.S., are increasing worldwide (Gibbs et al., 1989).

Both CH_4 and N_2O are much smaller constituents of the earth's atmosphere than is CO_2. For example, in the atmosphere, CH_4 concentrations are currently 1.7 μmol/mol (1.7 ppm) and N_2O concentrations are 0.3 μmol/mol, whereas CO_2 concentrations are 350 μmol/mol. However, each molecule of CH_4 and N_2O is more radiatively active than a molecule of CO_2 (N_2O is >200 times as active as CO_2 as a greenhouse gas).

Strategies for reducing CO_2 emissions from U.S. agriculture are generally similar to those for other sectors—increase fuel efficiency or seek alternative energy sources. Strategies for reducing CH_4 include changing feed rations (to rations lower in CH_4) in the short run and genetic and dietary improvements in the long run (Lashof and Tirpak, 1989; Gibbs et al., 1989). In the short run, use of some feed rations lower in CH_4 will mean higher finished livestock costs. Methane emissions from rice production result from several factors. Nitrogen fertilization is believed to be a contributing factor, because CH_4 emissions from fertilized rice fields have been found to be three to five times as high as those of unfertilized fields (Cicerone and Shetter, 1981), although data from other research have not confirmed the size of this differential (Matthews et al., in press; Yagi and Minami, 1990). Thus, one potential strategy is to reduce nitrogen fertilization of rice. Emissions of N_2O are directly related to nitrogen-fertilizer applications and global deforestation. In the United States, reduced use of nitrogen fertilizer, particularly those easily volatilized forms such as anhydrous ammonia, could reduce N_2O emissions.

In the short term, reduced use of nitrogen fertilizer and lower CH_4 feeding systems for livestock are expected to reduce yields and/or increase costs. Such effects, in turn, suggest higher food costs and, hence, losses to consumers. In the long run, improved breeding programs for livestock, better management of nitrogen in rice and other crop production, and improved crop breeding to reduce fertilizer dependence are needed to reduce emissions.

Procedure and Data

The preceding discussion highlights some simple strategies that could be used both to reduce agriculture's contribution to the worldwide increase in CH_4, N_2O, and other trace gases and to help offset increasing CO_2 emissions. This section discusses the procedure used to quantify the social costs of some of these strategies and the data and assumptions that underlie this quantification effort.

The procedure used here imposes the yield and cost implications of each strategy on a spatial-equilibrium model of U.S. agriculture. This model is discussed in detail in Chang et al. (1990) and has been used in several related evaluations of environmental issues, including the recent Adams et al. (1990) analysis of climate-change effects on agriculture. The features of this model that apply to the current effort are (1) the model represents the range of crop and livestock activities found in the major agricultural regions of the United States; (2) the model allows for changes in yields, cost, input, and other relevant factors; (3) the model captures the effects of changes in these factors through economic-welfare measures (consumers' and producers' surpluses, in total and by commodity); and (4) the model includes arable cropland as well as pasture and rangelands in the land inventory, with

1.43 $\times$ 10^{12} m^2 (353 million acres) of cropland and 3.52 $\times$ 10^{12} m^2 (869 million acres) of pasture and rangeland. This land inventory is important to the tree-planting, or carbon-sequestering, analysis. For these analyses, the model was expanded to include forest-planting activities in each region, an issue that will be discussed in more detail later. As with any programming-based exercise of this type, the resulting solution generated for each analysis represents the minimum social cost of achieving the desired objective or constraint within the context of the production possibilities included in the model.

To address the cost of reducing emissions of trace gases, we evaluated strategies for reducing CH$_4$ (from rice and livestock production) as well as N$_2$O (from fertilizer applications). The first CH$_4$ analysis simulates reductions in the use of nitrogen fertilizer in rice production (arbitrarily set at 50% and 100%). The implications for yield of this reduced use of nitrogen in rice production were estimated with nitrogen-response functions for rice (Hexem and Heady, 1981), which suggests 25% and 45% yield reductions, respectively. The changes in CH$_4$ emissions associated with this reduction in rice fertilization were estimated to be 35% and 70% (Cicerone and Shetter, 1981). The costs of reducing CH$_4$ emissions from livestock are evaluated in two ways. In the first case, the use of certain high-energy feed rations is assumed to be reduced 10%, resulting in a 5% reduction in meat and dairy-product yields. In the second case, beef consumption is reduced by a simulated reduction in the demand for beef products.

A reduction in the use of nitrogen in crop production is assumed in the N$_2$O evaluations. Specifically, anhydrous ammonia is reduced 50% and other nitrogenous fertilizers are assumed to be reduced 10%. The percent reduction in yields associated with this reduced use of nitrogen is derived from a series of nitrogen-fertilizer response functions taken from the agronomic literature (Hexem and Heady, 1978; Andersen and Køie, 1975; Arnold et al., 1974; Brinkman and Rho, 1984; Cochran et al., 1978; Constable and Rochester, 1988; Eck, 1984; Køie and Morrill, 1976; Perry and Olson, 1975; Reneau et al., 1983; Singh et al., 1979; Sorenson and Penas, 1978). Estimated reductions in yields, which range from 3% to 15% across crops and regions, are then used to adjust the yields in the sector model to evaluate the costs of these strategies.

The evaluation of a policy of converting agricultural lands to forest plantations for growing carbon is the most complex of the evaluations performed here. Assumptions about total amount of carbon sequestered, the extent of land available for such purposes, and the use of timber from such plantations will affect the social costs of such a policy. In this analysis, we use the cost, yield, and carbon-sequestering information from Moulton and Richards (1990). Establishment costs are converted to annual costs by discounting at 10% for an assumed rotation of 50 years. We also adopt carbon-fixation goals that are similar to Moulton and Richards' (1990). These cost and yield data are then used to develop the forest-plantation activities for each of 10 regions in the economic model. In this analysis, the resulting stumpage was assumed not to be sold and thus not to affect existing timber markets.

An important issue in evaluating agricultural land as a potential site for forest plantations is the inventory of lands capable of growing trees (without the need for major investments, such as irrigation) (USDA, 1989). Several studies have noted the apparent large areas of "marginal" farmland in the United States, such as the inventory of $>1.6 \times 10^{11}$ m^2 (40 million acres) of lands currently enrolled in the Conservation Reserve Program (CRP) (USDA, 1989). CRP and similar land areas are frequently suggested as candidates for tree-planting activities. However, $>50\%$ of CRP-enrolled area is in the semiarid West [with rainfall of <450 mm/year (<18 inches/year)], where the dominant forms of natural vegetation are drought-tolerant grasses and shrubs, not trees. This suggests that any realistic tree-planting program should emphasize more easterly regions of the United States, which involves competition with higher-valued crops at the margin. The importance of the land-inventory assumption is explored in the analysis below.

Results

This section presents quantitative estimates of the social costs of reducing trace-gas emissions from U.S. agriculture, as well as the costs of sequestering carbon through planting trees on agricultural land. These costs are in terms of net economic-surplus changes associated with each policy option or strategy. Thus, in some cases, consumer losses are partially offset by gains to producers (associated with rising commodity prices). Further, these are aggregate effects; the regional implications will not be explored in detail. All these estimates are preliminary, given the nature of the physical and natural science assumptions underlying each analysis.

Methane Reduction

Worldwide, agricultural activities are estimated to contribute $\approx 50\%$ of total CH$_4$ emissions. In the United States, the two agricultural sources are rice production and livestock (Shültz et al., 1989). As evaluated here, CH$_4$ reductions are achieved by reducing nitrogen applications to rice in the United States (by 50% and 100%) and by adopting low-CH$_4$ rations for livestock (cattle, including beef and dairy, and sheep). In the latter case, the use of low-CH$_4$ feed is treated in two ways: a shift in supply to reflect a decrease in total livestock meat production and a shift (reduction) in livestock demand to reflect reduced consumption, analogous to a tax on finished meat products. These CH_4 results are reported in Table 15.1.

As is evident from Table 15.1, the average costs of CH_4-emission reductions range from ≈$500 to nearly $4,000 per 10^3 kg of CH_4, depending on the strategy (the reductions for rice are 0.5 and 1.1 ×10^9 kg of CH_4 for the 50% and 100% nitrogen reduction, respectively, and 1.1 ×10^9 kg for the low-CH_4 feed-ration policies). For rice, the costs are the result of reduced yields associated with reduced nitrogen applications. These reductions in nitrogen increase the total area in rice production. For livestock, the costs are the result of either (assumed) changes in per-unit-product yields for all livestock or reduced consumption of fed beef. Costs are borne primarily by consumers; producers in some cases gain because of rising prices associated with reduced supply. Whether these social costs are large depends on the benefits per 10^3 kg associated with reductions of these gases. However, the United States is a relatively small contributor to total world CH_4 emissions, given that the U.S. shares of world rice and livestock production are ≈1% and ≈8%, respectively.

TABLE 15.1 Costs of Methane and Nitrous Oxide Emission Reductions (per 10^3 kg)

Strategy	Changes in Economic Surplus ($ millions)	Cost per 10^3 kg ($)
Methane		
Reduce rice fertilization by		
50%	−268	590
100%	−724	663
Lower CH_4 rations		
5% yield reduction (supply shift)	−1282	1166
5% demand shift for fed beef (tax)	−4590	4180
Nitrous oxide		
Reduce anhydrous ammonia and nitrogen applications	−643	4708

Nitrous Oxide Reductions

Agricultural N_2O emissions come from two sources: deforestation and nitrogenous fertilizers. The United States accounts for $\approx 10\%$ of global nitrogenous-fertilizer consumption. The United States also has a higher proportion of anhydrous ammonia in the mix of nitrogenous fertilizers than the rest of the world. Because anhydrous ammonia releases more N_2O than other forms of nitrogenous fertilizers (as much as 7% of the total nitrogen is lost as N_2O), the U.S. contribution to worldwide N_2O emissions (from fertilizers) is probably somewhat higher than 10%. In this analysis, it is assumed that the United States can reduce its total annual N_2O emissions by 50% (1.3×10^8 kg, or 130 thousand metric tons) by reducing the use of anhydrous ammonia by 50% and reducing the use of all other nitrogenous fertilizers by 10%. The net costs of this strategy, as presented in Table 15.1, are $\approx$\$650 million, or $\approx$\$4,000 per 10^3 kg of N_2O. These costs, as modeled here, are borne entirely by consumers; in the aggregate, producers gain through increased commodity prices. Again, whether these costs are large depends on the benefits associated with the reduced N_2O emissions.

There are obvious sources of error in both the CH_4 and N_2O cost estimates of emission reductions. For example, relatively small adjustments in some inputs (i.e., 10% reduction in total nitrogen applications) are assumed here to result in major reductions in emissions, suggesting that these cost estimates may be understated. Given the lack of good data on CH_4 and N_2O emissions under alternative management systems for actual field conditions, it is difficult to know whether the extent of emission reductions reported here could be achieved. Conversely, the substitution of other inputs or technologies (which are not explicitly modeled here) in response to control strategies could reduce social costs.

Forest Plantations

As noted earlier, several studies have focused on the technical feasibility of large-scale forest plantations as a way to slow atmospheric-CO_2 increases. The use of marginal agricultural land to "grow carbon" has appeal for several reasons. These reasons include the benefits of reduced erosion, reduced use of agricultural chemicals, increased wildlife habitat, and, perhaps, aesthetic value arising from forested landscapes. Tree plantations also would appear to be a less politically painful alternative to other potential strategies, such as carbon taxes.

The social costs of such tree-planting and -growing operations include the establishment and management costs, plus the opportunity costs of tying up potentially large areas of agricultural land for long periods. Whereas the United States has substantial areas of marginal agricultural land or pastureland available for such purposes, the climatic and edaphic characteristics of some regions are not well suited to tree plantations. Thus, successful tree growing needs to be confined to a subset of the available agricultural land inventory in the United States. Obviously, the more carbon sequestered, the greater the effects in those regions that are suited for tree growing.

In the analysis reported here, the sensitivity of the economic-cost estimates is tested under several assumptions about the carbon-fixing goals (total amount) and area of pastureland vs. cropland used. In addition, carbon-growing activities are restricted to regions with enough rainfall to grow trees. As discussed in the Procedure and Data section, these sensitivities are explored by changes in a sector model of the U.S. agricultural sector. The changes include addition of regional tree-growing activities in regions with sufficient rainfall and the specification of carbon-fixing goals. The total carbon-fixing goals are 1.27, 2.55, 3.82, and 6.36×10^{11} kg (140, 280, 420, and 700 million short tons) of carbon per year. They represent $\approx 10\%$, $\approx 20\%$, $\approx 30\%$, and $\approx 50\%$ of total annual carbon emissions from the United States. These carbon-fixing goals are similar to amounts analyzed in Moulton and Richards (1990). The highest amount (6.36×10^{11} kg) corresponds to amounts found in Sedjo and Solomon (1989).

The results of these evaluations are reported in Table 15.2. The results include estimates for the four carbon goals under several assumptions about location of carbon-fixing activities and the use of pastureland vs. cropland within each region. Analysis 1 allows carbon to be grown in most regions of the United States; the exceptions are arid portions of the Southern Plains, Southwest, and parts of the Northern Plains, where rainfall is <450 mm/year. The analysis also allows unrestricted use of the pastureland inventory for fixing carbon. As is evident from the table, the marginal costs (averaged across all regions) per 10^3 kg increase as the carbon-fixing goals are increased, from a low of $16.30/10^3$ kg for 1.27×10^{11} kg (140×10^6 tons) to $\approx 62.00/10^3$ kg for a goal of 6.36×10^{11} kg (700×10^6 tons), about half of total U.S. carbon emissions. The rising costs per 10^3 kg reflect the rising opportunity costs of agricultural land as more land and land of higher quality is diverted from crops to trees. This diversion increases the losses to consumers caused by rising commodity prices.

The range of estimates from this analysis can be compared with other recent studies, but caution is needed because of differences in assumptions about annual or cumulative carbon fixation and discount rates and in whether the analyses report average or marginal costs. For example, the marginal costs per 10^3 kg reported here are slightly to moderately higher than the marginal-cost estimates in Moulton and Richards (1990). Specifically, Moulton and Richards (1990) report costs of $12.00, $15.70, and $18.00 per short ton (i.e., per 907 kg) as least-cost estimates to sequester

TABLE 15.2 Marginal Costs and Agricultural Land Needed to Sequester Carbon by Tree Plantations

	Total Carbon Sequestered [a]							
	140 (1.27)		280 (2.55)		420 (3.82)		700 (6.36)	
Evaluation	Cost ($/$10^3$ kg)	Area (× 10^{10} m^2)	Cost ($/$10^3$ kg)	Area (× 10^{10} m^2)	Cost ($/$10^3$ kg)	Area (× 10^{10} m^2)	Cost ($/$10^3$ kg)	Area (× 10^{10} m^2)
Analysis 1[b]	16.30	22.18	23.30	43.85	30.25	65.70	61.90	112.00
Analysis 2[c]	17.50	21.49	26.40	42.04	34.65	63.41	118.70	109.52
Analysis 3[d]	22.30	19.38	41.10	41.11	66.55	62.21	218.35	108.38

[a] Short tons × 10^6; × 10^{11} kg in parentheses.
[b] Restricts tree planting to regions where rainfall exceeds 450 mm/year.
[c] Same as analysis 1 but also restricts pastureland to <50% of total area in tree plantations.
[d] Same as analysis 1 but removes Southern Plains from the solution.

140, 280, and 420 million short tons (1.27, 2.55, 3.82, and 6.36 × 10^{11} kg) of carbon. The current estimates also appear similar to the average costs reported by Dudek and Leblanc (1990) (of $3.50 to $11.00 per short ton, or per 907 kg) for sequestering ≈140 million short tons (1.27 × 10^{11} kg) of CO_2 [or 38 million short tons (3.5 × 10^{10} kg) of carbon]. However, the costs here are considerably higher than Sedjo and Solomon's (1989) average cost range of $12.00 to $19.00 per short ton (907 kg) to achieve a larger carbon sequestration (2.9 billion short tons, or 2.6 × 10^{12} kg). The somewhat higher costs generated by the spatial optimization used here reflect the costs to consumers of foregone agricultural production and the elimination of selected arid areas from the land inventory (as well as the differences between marginal and average costs).

Analysis 2 from Table 15.2 is similar to analysis 1 except that a restriction is imposed on the proportion of pastureland converted to trees. Specifically, pastureland is limited to 50% of total tree production (in analysis 1, it is ≈60%). The purpose of this analysis is to test the effect of the land inventory and productivity assumptions on such estimates (some "pastureland" in the inventory is of low productivity). As the numbers in Table 15.2 suggest, use of slightly more cropland to achieve the carbon goals translates into slightly higher costs for the modest goals (e.g., 1.27 × 10^{11} kg, or 140 × 10^6 short tons, of carbon) but results in dramatically higher costs at more extreme goals (6.36 × 10^{11} kg, or 700 × 10^6 short tons), because of increasing pressure on the cropland-resource base.

Analysis 3 further investigates the effects of changes in the land-inventory assumption on estimated carbon-fixing costs. In this case, the Southern Plains (Texas and Oklahoma) are removed from the potential land inventory for tree planting. While primarily a sensitivity exercise, some analyses suggest that climate change will adversely affect the ability of this region to produce trees in the future. Specifically, forecasts from two general circulation models (Goddard Institute of Space Studies, or GISS, and Princeton's Geophysical Fluids Dynamics Laboratory, or GFDL) indicate that the Southern Plains will experience a hotter and drier climate under a doubled-CO_2 environment (or the equivalent amount of climate forcing from all trace gases). If the production of trees in the Southern Plains is eliminated from the model solution, then the costs per 10^3 kg of carbon increase by 40% to almost 100%. The rise in costs across each carbon goal is caused by increased diversion of land in regions of higher agricultural productivity.

The results of all analyses suggest that a modest program of carbon sequestering can be achieved without major impacts on agricultural land area and commodity production. For example, 1.27 × 10^{11} kg (140 × 10^6 short tons) of carbon can be sequestered by diverting ≈2 × 10^{11} m^2 (50 million acres) from agricultural use to tree plantations. However, the results also suggest that costs per 10^3 kg are likely to rise rapidly as more ambitious goals are attempted, because of increased land diversions from traditional agricultural activities. At 636×10^{11} kg sequestered, $>1.1 \times 10^{12}$ m^2 (270 million acres) are diverted from agriculture. As the costs per 10^3 kg rise, the use of tree plantations as a "greenhouse" strategy becomes less appealing. Indeed, purchase of tropical forests as carbon sinks appears to be less costly as an "offset" strategy than continued conversion of U.S. cropland and pastureland.

Conclusions

The economic consequences of reducing trace-gas emissions from agriculture as estimated here are subject to considerable uncertainty but indicate potentially high social costs to agricultural constituents. Most of these costs are borne by consumers in rising commodity prices. Strategies to reduce inputs, such as reducing nitrogenous fertilizers, also have some other undesirable side effects, including pressure to increase land use for affected commodities. This would conflict with any policy of converting agricultural lands to forest plantations.

The costs of sequestering carbon through tree plantations vary with the carbon-fixing goals and the land inventory available for such conversion. Average costs per 10^3 kg of carbon rise as more land is diverted from agriculture, suggesting that modest levels of carbon sequestration may be feasible. However, these analyses indicate that tree plantations are only one component of an overall strategy to control trace-gas buildups; other solutions are needed.

These analyses suggest the importance of specific biological and physical data in performing economic assessments. The analyses of changes in input use depend on the accuracy of a broad set of biophysical relationships, such as fertilizer response functions and trace-gas-emission rates. The evaluation of costs of land conversion to tree plantations reflects the importance of the land-inventory assumptions, including differences in land productivity within and across regions. Because of these data uncertainties, the results reported here are best viewed as sensitivity analyses. The estimates can, however, provide some general sense about whether the costs of these agricultural strategies are large or small relative to nonagricultural options for reducing trace-gas emissions.

References

Adams, R. M., C. Rosenzweig, R. M. Peart et al. 1990. "Global Climate Change and U.S. Agriculture." *Nature* 345: 219–24.

Andersen, A. J., and B. Køie. 1975. "N Fertilization and Yield Response of High Lysine and Normal Barley." *Agronomy Journal* 67: 695–8.

Arnold, J. M., L. M. Josephson, W. L. Parks, and H. C. Kincer. 1974. "Influence of Nitrogen, Phosphorus, and Potassium Applications on Stalk Quality Characteristics and Yield of Corn." *Agronomy Journal* 66: 605–8.

Brinkman, M. A., and Y. D. Rho. 1984. "Response of Three Oat Cultivars to N Fertilizer." *Crop Science* 24: 973–7.

Cates, R. L. Jr., and D. R. Keeney. 1987. "Nitrous Oxide Production Throughout the Year from Fertilized and Manured Maize Fields." *Journal of Environmental Quality* 16: 443–7.

Chang, C. C., B. A. McCarl, J. W. Mjelde, and J. W. Richardson. 1990. "Sectoral Implications of Farm Program Modifications." Texas Agricultural Experiment Station technical article. College Station, Texas: Texas A&M University.

Cicerone, R. J., and J. D. Shetter. 1981. "Sources of Atmospheric Methane: Measurements in Rice Paddies and a Discussion." *Journal of Geophysical Research* 86: 7203–9.

Cochran, V. L., R. L. Warner, and R. I. Papendick. 1978. "Effect of N Depth and Application Rate on Yield, Protein Content, and Quality of Winter Wheat." *Agronomy Journal* 70: 964–8.

Constable, G. A., and I. J. Rochester. 1988. "Nitrogen Application to Cotton on Clay Soil: Timing and Soil Testing." *Agronomy Journal* 80: 498–502.

Darmstadter, J., and J. Edmonds. 1989. "Human Development and Carbon Dioxide Emissions: The Current Picture and Long-Term Prospects," in N. Rosenberg, W. Easterling, P. Crosson, and J. Darmstadter, eds., *Greenhouse Warming: Abatement and Adaptation*. Pp. 35–52. Washington, D.C.: Resources for the Future.

Decker, W. L., V. K. Jones,. and R. Achuntuni. 1986. "The Impact of Climate Change from Increased CO_2 on American Agriculture." Washington, D.C.: U.S. Department of Energy. (DOE/NBB-0077.)

Dudek, D. J. 1988. "Climate Change Impacts Upon Agriculture and Resources: The Case of California," in J. B. Smith and D. A. Tirpak, eds., *The Potential Effects of Global Climate Change in the United States*. Pp. 5-1–5-38. Washington, D.C.: U.S. Environmental Protection Agency.

Dudek, D. J., and A. Leblanc. 1990. "Offsetting New CO_2 Emissions: A Rational First Greenhouse Policy Step." *Contemporary Policy Issues* VIII: 29–42.

Eck, H. V. 1984. "Irrigated Corn Yield Response to Nitrogen and Water." *Agronomy Journal* 76: 421–8.

Gibbs, M. L., L. Lewis, and J. S. Hoffman. 1989. "Reducing Methane Emissions From Livestock: Opportunities and Issues." Washington, D.C.: U.S. Environmental Protection Agency. (400/1-89/002.)

Hexem, R. W., and E. O. Heady. 1978. *Water Production Functions for Irrigated Agriculture*. Ames, Iowa: The Iowa State University Press.

Houghton, R. A., R. D. Boone, J. M. Melillo et al. 1985. "Net Flux of Carbon Dioxide from Tropical Forests in 1980." *Nature* 316: 617–20.

Houghton, R. A., R. D. Boone, J. R. Fruci et al. 1987. "The Flux of Carbon from Terrestrial Ecosystems to the Atmosphere in 1980 Due to Changes in Land Use: Geographic Distribution of the Global Flux." *Tellus* 39B: 122–39.

Hutchinson, G. L. 1988. "Nitrogen Losses From Soil Calculated." *Agricultural Research* 36: 5–6.

Kane, S., J. Reilly, and J. Tobey. 1990. "An Empirical Study of the Economic Effects of Climate Change on World Agriculture." Paper presented at the American Agricultural Economic Association Annual Meeting, August 3–5. Vancouver, B.C.

Køie, S. E., and L. G. Morrill. 1976. "Influence of Nitrogen, Narrow Rows, and Plant Population on Cotton Yield and Growth." *Agronomy Journal* 68: 897–901.

Lashof, D. A., and D. A. Tirpak, eds. 1989. "Policy Options for Stabilizing Global Climate." Washington, D.C.: United States Environmental Protection Agency Draft Report to Congress. Volume I: chapters I–VI.

Marland, G. 1988. "The Prospect of Solving the CO_2 Problem through Global Reforestation." Oak Ridge, Tenn.: U.S. Department of Energy. (NBB-0082.)

Matthews, E., I. Fung, and J. Lerner. In press. "Methane Emission from Rice Cultivation: Geographic and Seasonal Distribution of Cultivated Areas and Emissions." *Global Biogeochemical Cycles*.

Moulton, R. J., and K. B. Richards. 1990. "Costs of Sequestering Carbon Through Tree Planting and Forest Management in the U.S." Washington, D.C.: USDA Forest Service. (General Technical Report WO-58.)

Palm, C. A., R. A. Houghton, J. M. Melillo, and D. L. Skole. 1986. "Atmospheric Carbon Dioxide from Deforestation in Southeast Asia." *Biotropica* 18: 177–88.

Perry, L. J. Jr., and R. A. Olson. 1975. "Yield and Quality of Corn and Grain Sorghum Grain and Residues as Influenced by N Fertilization." *Agronomy Journal* 67: 816–8.

Rasmussen, R. A., and M. A. Khalil. 1981. "Atmospheric Methane (CH_4): Trends and Seasonal Cycles." *Journal of Geophysical Research* 86: 9826–32.

Reneau, R. B. Jr., G. D. Jones, and J. B. Friedericks. 1983. "Effect of P and K on Yield and Chemical Composition of Forage Sorghum." *Agronomy Journal* 75: 5–8.

Rosenzweig, C. 1988. "Potential Effects of Climate Change on Agricultural Production in the Great Plains: A Simulation Study. Report to Congress on the Effects of Global Climate Change." Washington, D.C.: U.S. Environmental Protection Agency.

Schültz, H., W. Seiler, and R. Conrad. 1989. "Processes Involved in Formation and Emission of Methane in Rice Paddies." *Biogeochemistry* 7: 33–53.

Sedjo, R. A. 1983. "The Comparative Economics of Plantation Forests." Washington, D.C.: Resources for the Future.

Sedjo, R. A., and A. M. Solomon. 1989. "Climate and Forests," in N. J. Rosenberg, W. E. Easterling, P. R. Crosson, and J. Darmstadter, eds., *Greenhouse Warming: Abatement and Adaptation*. Pp. 105–20. Washington, D.C.: Resources for the Future.

Singh, N. T., A. C. Vig, R. Singh, and M. R. Chaudhary. 1979. "Influence of Different Levels of Irrigation and Nitrogen on Yield and Nutrient Uptake by Wheat." *Agronomy Journal* 71: 401–4.

Smith, J. B., and D. A. Tirpak. 1989. "The Potential Effects of Global Climate Change in the United States." *Agriculture*. Vol. 1, Appendix C. Washington, D.C.: U.S. Environmental Protection Agency.

Sonka, S. T., and P. J. Lamb. 1987. "On Climate Change and Economic Analysis." *Climate Change* 11: 291–311.

Sorensen, R. C., and E. J. Penas. 1978. "Nitrogen Fertilization of Soybeans." *Agronomy Journal* 70: 213–6.

U.S. Department of Agriculture (USDA). 1989. "Conservation Reserve Program: Progress Report and Preliminary Evaluation of the First Two Years." Washington, D.C.: U.S. Government Printing Office.

Woodwell, G. M., J. E. Hobbie, R. A. Houghton et al. 1983. "Global Deforestation: Contribution to Atmospheric Carbon Dioxide." *Science* 222: 1081–6.

Yagi, K., and K. Minami. 1990. "Effects of Organic Matter Applications on Methane Emission from Japanese Paddy Fields," in A. F. Bouwman, ed., *Soils and the Greenhouse Effect*. Pp. 467–73. Chichester, U.K.: John Wiley & Sons.

16

Policy and Research Implications of Recent Carbon-Sequestering Analysis

Kenneth R. Richards[1]

Introduction

Recent research developed marginal and total cost curves for the use of tree planting and modified forestry practices to capture atmospheric carbon on marginal agricultural land and forestland in the United States (Moulton and Richards, 1990). The analysis considered the type of land (private cropland, pastureland, or forestland) and the region of the United States in which the forestry activities take place, and indicated that the costs per unit of carbon sequestered range widely as a function of these two factors. The report is data-intensive, which makes it possible to test for the effects of factors such as land availability, rental costs, and discount rates in ways not possible with previously available reports of carbon-sequestering costs. Because the data, methodology, and assumptions are clearly stated, other researchers can develop alternative analyses using variations in either the data or the assumptions.

Because of the U.S. government's interest in potential technologies for reducing net CO_2 emissions, this study has received considerable attention. The U.S. Department of Energy, the U.S. Environmental Protection Agency (EPA), the U.S. Department of Agriculture (USDA), the congressional Office of Technology Assessment, and the National Academy of Science have all found the figures useful. Analysis of the president's proposed tree-planting program was facilitated by the study.

This report will describe the results of Moulton and Richards (1990), examine its potential use in forming policy, and suggest economic and scientific research that would be useful in further developing the analysis.

---

[1] From the Graduate Program, Wharton School and Law School, University of Pennsylvania, Philadelphia.

The second section describes the results of the study and some of the implications for policy. The third section examines some of the limitations of the analysis that suggest the cost figures should be used with caution until further study is completed. That section also explores some of the questions about implementation that the study raises. To test the magnitude of the uncertainty in the model, the fourth section describes the results of some simple sensitivity analyses. The fifth section discusses the importance of the shape of the forestry growth curve and its relation to dynamic analysis and policy formulation. Finally, the last section draws some conclusions about future research that will carry this analysis forward.

Model, Data, and Results

The Model [2]

The model used to develop the cost curves was simple. The basic notation and calculations are presented here, followed by a summary of the procedure by which the cost curves were derived. The variables of the model are defined as follows.

LA^{ij} = area of land type i in farm production region j, i=1...x, j=1...y
R^{ij} = annual rental cost/unit area of land type i in region j
P^{ij} = tree planting/treatment cost (i.e., capital cost) on land type i in region j per unit area
Y^{ij} = annual incremental yield (volume/unit area) of merchantable wood on land type i in region j
K^{i} = conversion factor for the ratio of incremental increase in carbon in forest ecosystem (entire tree, soil, surface litter, and understory growth) to incremental increase in carbon in merchantable wood on land type i
D^{ij} = density of carbon in the merchantable wood grown on land type i in region j.

The calculation of annual incremental carbon uptake in mass per unit area on land type i in region j, G^{ij}, is simply

$$G^{ij} = Y^{ij}K^{i}D^{ij}.$$

---

[2] This section is adapted from Moulton and Richards (1990).

The potential total national carbon uptake is calculated as

$$TG = \sum_i \sum_j [LA^{ij}][CO_2^{ij}].$$

The cost associated with reaching the total national CO_2 uptake is

$$TC = \sum_i \sum_j LA^{ij}[R^{ij} + A(r,n)P^{ij}]$$

where $A(r,n)$ is the annualized cost of the capital investment, P, spread over the n years of the project life, at a discount rate of r. The average cost per unit mass of carbon is simply

$$AC = TC/TG.$$

The model was fitted with data from each of the 10 USDA farm-production regions (Figure 16.1). The marginal agricultural land in each region was identified and categorized into one of seven classifications.[3] This created

Figure 16.1 USDA farm-production regions.

---

[3] The seven economically marginal types of land were wet and dry cropland, wet and dry pastureland, and forestlands requiring planting, active management, or passive management.

70 distinct region–land type combinations.   Each region–land type combination was matched with an appropriate forestry treatment, such as planting or regeneration, and with the appropriate mix of species. For each land type in each region, the planting cost and rental cost per 10^4 m^2 (or, per hectare) was determined. The product of the incremental merchantable wood yield, the specific carbon density of the wood, and a factor relating carbon in merchantable wood to total forest-ecosystem carbon yielded the carbon-fixation rate for each type of land in each region. The gross carbon-fixation costs were calculated as

$$\text{(Capitalized planting costs } + \text{ rent)/carbon-fixation rate}$$

The cost curves were derived by arranging each type of land in each region (LA^{ij}) in ascending order according to its associated carbon-fixation cost, with the land areas that capture carbon most cheaply at the top of the list. The marginal cost curve was derived simply by plotting the dollars per 10^3 kg in the ascending list against the cumulative amount sequestered. The total program cost associated with a given amount of annual carbon sequestering was derived by plotting the cumulative cost figures against the cumulative amount sequestered.

Data Collection

The data were largely derived from historical figures from past USDA programs. Although this source helps avoid the problems of overly optimistic approximations, for several types of data, either the data were not available or were collected over such a small sample that they would be inappropriate in a large program such as this (e.g., land-rental cost data). In these cases, the assumptions that were necessary are explicitly stated in the detailed description of data sources provided in Moulton and Richards (1990). The complete data set is also provided in that report.

Results[4]

The total and marginal cost curves, shown in Figures 16.2 and 16.3, lead to several interesting observations. On the basis of an estimated current annual U.S. net emissions of carbon in the form of CO_2 of 1.3×10^{12} kg/year (1.4 billion short tons/year), a tree-planting and -management program limited to economically marginal agricultural land and forestland could achieve as much as a 56.4% decrease in net emissions.

---

[4] The Results section is largely excerpted from Moulton and Richards (1990).

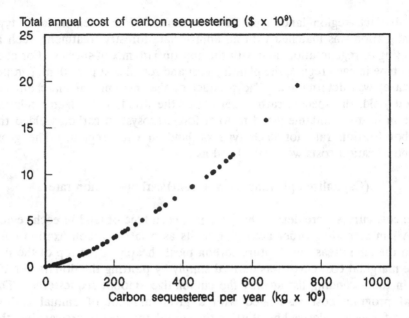

Figure 16.2 Total cost curve.

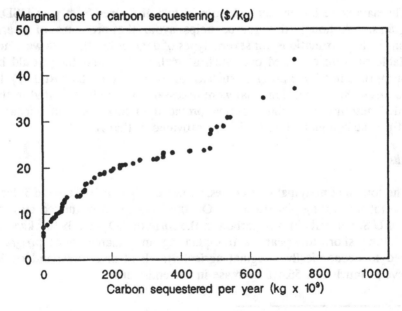

Figure 16.3 Marginal cost curve.

The cost of a program to achieve a 56.4% reduction would be ≈$19.5 billion/year. As shown in Table 16.1, the annual cost of achieving 10%, 20%, and 30% reductions would be ≈$1.7, $4.5, and $7.7 billion, respectively. The marginal costs of carbon captured in programs designed to reduce net CO_2 emissions by 10%, 20%, 30% and 56% are $18.6, $23.0, $26.0, and $47.7 per 10^3 kg, respectively.

As shown in Table 16.1, a least-cost program to reduce net emissions of CO_2 by 10% would require ≈2.9×10^{11} m^2 (71 million acres). Of this, 0.9×10^{11} m^2 (31%) are pastureland, 1.5×10^{11} m^2 (52%) are forestland, and 0.5×10^{11} m^2 (17%) are cropland (Figure 16.4). The costs of the program are dominated by land-rental costs, with the establishment or planting costs generally constituting <40% of total annualized costs on the cropland and pastureland.

One concern is the extent to which a tree-planting program would compete with other productive uses of the land, particularly crop production. This analysis has been limited to economically marginal and environmentally sensitive croplands, pasturelands, and forestlands on which growth rates of trees could be enhanced. As Figure 16.4 indicates, the first 10% (1.30×10^{11} kg/year) offset would use relatively little cropland. As indicated in Figure 16.5, however, the relatively small amount of cropland included at the 1.80×10^{11} kg/year level provides a disproportionate share of the carbon sequestration because of the contribution of relatively inexpensive but productive cropland in the Mountain Region. Above 1.80×10^{11} kg/year, virtually all of the significant capacity is on cropland.

TABLE 16.1 Program Statistics by Percentage Reduction from 1.3×10^{12} kg/year

Annual Carbon Offset (%/billions of kg)	Land Requirement (billions of m^2)	Total Annual Cost ($ billion)	Average Cost ($/$10^3$ kg carbon)
5/65	149	0.7	10.72
10/129	287	1.7	13.25
20/259	560	4.5	17.34
30/389	800	7.7	19.75

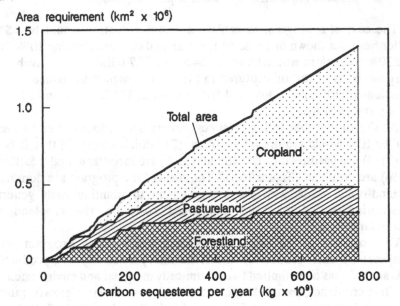

Figure 16.4 Area requirements by land type.

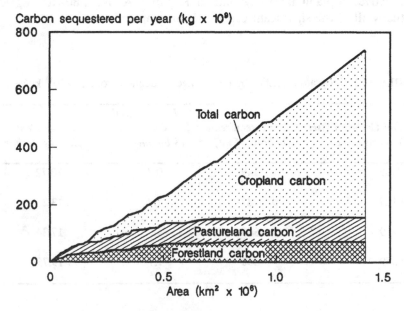

Figure 16.5 Carbon sequestration by land type.

Table 16.2 illustrates that the relative land area used in each region of the country depends on the size of the tree-planting program. For example, a least-cost program designed to offset 5% of total U.S. CO_2 emissions would use no land in Appalachia, whereas in a 30%-reduction program, that region would have the highest share of land area. In contrast, the Pacific region would have more land used in a 5% program than any other region, but the total amount of its contribution would increase only slightly in a 30% program, making it one of the least significant regions. At no level is a large land area from the Northeast involved.

Limitations of the Analysis

The purpose of the research on carbon sequestering was to develop a range of least-cost estimates for carbon sequestration as expressed in marginal and total cost curves. These estimates would allow preliminary comparison with least-cost estimates for alternative technologies that could reduce net carbon emissions. Because of the limited scope of the study, many questions arise, some dealing with the data and methodology used, some related to corollary issues such as implementation and social effects.

Methodology and Data Limitations

Although the Moulton and Richards analysis used the best data available at the time it was conducted, there is room for further refinement. In particular, the carbon-sequestering rates were average rates over the 40–120-

TABLE 16.2 Regional Land Areas by Percentage Reduction from 1.3×10^{12} kg/year

	Area ($\times 10^4$ m^2) Included in the First			
	5%	10%	20%	30%
Northeast	0	0	691	1,672
Appalachia	0	2,876	9,899	15,462
Southeast	324	3,113	10,556	10,556
Lake states	2,213	3,187	3,911	5,660
Corn belt	1,603	3,088	5,099	7,217
Delta states	2,193	2,843	7,127	14,506
North Plains	1,046	1,046	1,046	3,945
South Plains	3,201	3,201	6,468	6,468
Mountain	736	5,581	5,581	5,920
Pacific	3,639	3,639	4,012	4,014

year expected life of the trees, disaggregated by species and region. But the rate of accumulation of tree and ecosystem carbon is not constant; rather, the rate of sequestration is initially quite low and increases rapidly for the first 15–35 years, after which it gradually decreases for the remainder of the tree stand's existence. Because this pattern affects the flow of CO_2 out of the atmosphere, it also affects the timing of benefits associated with carbon fixing—a crucial consideration in an analysis in which discounting can make such a difference. The fourth section presents a rough sensitivity analysis to indicate the size of the effect that dynamic considerations could have on the analysis. The fifth section presents examples of some preliminary growth curves and addresses potential methods for dealing with the dynamics problem.

In addition to the dynamic considerations, the analysis is limited because it is based on a partial-equilibrium concept. The quantity of land removed from the market represents a nonmarginal change, but the rental rates in the basic analysis are based on prevailing market prices (or marginal costs), albeit conservatively estimated. The sensitivity analysis described in the next section uses rental rates three times as high as current private-market prices. But a more detailed analysis is needed to carefully examine the supply curve for long-term cropland and pastureland rentals. Similarly, the effect of removing $5–100 \times 10^{10}$ m^2 (10–200 million acres) of the least-productive cropland from agricultural production on food prices and national security should be examined. At the same time, the technology-driven decrease in demand for agricultural land must be considered.

As a partial response to this problem, Adams et al. (1992) used the cost, yield, and carbon-sequestering data from the Moulton and Richards analysis in a model of the U.S. agricultural sector to estimate secondary effects such as rising land prices and food prices, as measured by consumer-welfare losses. Although there are differences between the two studies in the definition of the land inventory, the data are similar enough to warrant a comparison of results. Examination of the base-case (analysis 1) marginal costs in Adams et al. reveals that the estimated social costs are slightly lower than those shown in Figure 16.3 of this report at the 1.25×10^{11} kg (140 million tons) annual rate of sequestering, but climbs more rapidly as annual carbon-sequestration rates rise. Consequently, the spatial-equilibrium model results in figures nearly equivalent to the partial-equilibrium model of Moulton and Richards up to $\approx 3.80 \times 10^{11}$ kg/year (420 million tons/year) sequestration rates, but that diverge dramatically at 6.35×10^{11} kg/year (700 million tons/year). The wide variance between the spatial-equilibrium model and the partial-equilibrium model at high land-area use—at $\approx 1.20 \times 10^{12}$ m^2—is not surprising. It simply indicates that removing large quantities of even marginal land from agricultural production will have a significant affect on land rents and food prices. What is more interesting is the observation that at 3.8×10^{11} kg/year, or nearly 30% of the U.S. carbon-emissions rate, with nearly 8.0×10^{11} m^2 (200 million acres), there is relatively little effect on food and agricultural land prices. When Adams et al. imposed additional constraints—restricting pastureland to no more than 50% of total area in their analysis 2 and removing Southern Plains land area in analysis 3—the estimated marginal costs increased dramatically, exceeding the estimates of the Moulton and Richards analysis by $\approx25\%$ at 1.3×10^{11} kg/year (140 million tons/year) and 250% at 3.8×10^{11} kg/year (420 million tons/year). The admittedly preliminary nature of the Adams et al. analysis highlights the need for additional research in this area.

The estimates of increases in carbon-sequestration rates through improved practices on forestlands of low productivity are obviously based on the difference between two factors: the rate of net carbon uptake after treatment and the rate of sequestration in the absence of treatment. Richard Birdsey, whose work provides the basis for most of the yield data used in this study (Birdsey, 1990a), has suggested that the Moulton and Richards analysis may have underestimated the rate of carbon capture in a portion of the marginal forestlands in the absence of treatment (personal communication, 1990). If so, the estimates of gains from treatment would be higher than warranted; correction of the data would lead to a slight rise in the low end of the marginal cost curve in Figure 16.3 and an overall shortening of the length of the curve (the total amount that can be captured).

The Moulton and Richards analysis made no attempt to capture external effects of a large-scale tree-planting program. These effects might include soil-erosion improvements, lower water tables, and changes in wildlife habitat, as well as many unforeseen effects, suggesting the need for a separate detailed study.

Finally, this treatment of carbon-sequestering costs is limited to the United States, where land ownership, infrastructure, climate, soil types, and administrative and technical expertise make such a program at least possible. Land-tenure patterns in many other countries may suggest the need for a separate analysis of global tree-planting potential that starts not with science and economics, but rather with law and anthropology.

Implementation Considerations

The limitations of the study methodology suggest many questions about the implementation of a tree-planting program. Although the analysis is very rich in data, using U.S. Forest Service records for land-type areas and historic yield and cost figures that reflect actual species mix in current planting practices, it is purposefully simple in its assumptions. In particular, in an effort to develop a "least-cost curve," the model assumes that all privately owned marginal agricultural land and forestland will be available for expanded tree-planting and forest-management activities. However, several constraints, both political and social, may limit the availability of some lands. For example, to avoid significant disruptions to local economies, the USDA's Conservation Reserve Program (CRP) has limited the percentage of land in any county that may be enrolled in the program. Also, there may be resistance from landowners, particularly those holding land for speculative purposes, to the idea of leasing land for 40 years at a time. These and other social and political factors suggest the need for a separate detailed study of the feasibility of a large-scale land-leasing and -acquisition program.

Many other factors related to the more general issue of implementation should be addressed before committing to a large domestic carbon-sequestration program. In particular, the relation of the program to other government policies must be understood. Presumably, much of the land would be drawn from areas already enrolled in crop price-support programs. It appears likely that if the tree-planting program could simply be substituted for price-support and land-area-removal programs, then the savings of federal funds from those transfer programs would nearly pay for the costs of carbon sequestration on a very large scale. It will also be necessary to work out the relation between the tree-planting program and a government carbon-tax program or carbon-emissions-permit program. Similarly, it will likely be necessary to develop a tax or permit program for the cutting of trees; otherwise, large landowners will have an incentive to cut all standing timber to clear areas on which they can plant "new" tree stands for which they will be paid.

This study has not addressed the administration of a large-scale tree-planting program and the associated costs. The obvious federal agency to control the program is the U.S. Forest Service, acting through its state counterparts, which are more familiar with local land conditions and suitability for participation. Because much of the foreseeable infrastructure for the program is already in place, particularly for monitoring and compliance, the additional cost of administration could be as low as 15% of total costs. However, it is at least imaginable that if a large-scale program is implemented in which landholder participation is not voluntary, because of the property rights involved, the program could become mired in crippling litigation costs.

Additional issues about the administration of the program include the governance of landowner withdrawal from the program, the number of years over which the planting activities will be phased in, the rate at which seedling nurseries can be expanded, and the marketing question of how quickly large percentages of landowners can be induced to participate in a voluntary program.

A recurring issue is the ultimate disposal of the timber. There are several ways to approach this problem. First, a carbon-sequestration program may be viewed as temporary, an effort to buy time for 40–100 years while society searches for other solutions, such as economic alternatives to fossil fuels. In this case, the timber can either be left standing indefinitely or cut once the alternatives are developed and the land is in greater demand. Evidence from previous federal planting programs suggests that once tree stands are established, a large percentage of land owners will not remove the trees, even in the absence of continued government payments and control. Alig et al. (1980) indicate that nearly 20 years after the end of the program, trees were still growing on 89% of the area planted through the Conservation Reserve phase (1956–1961) of the USDA's Soil Bank Program. Alternatively, the timber could be used as a substitute for fossil fuels. Although the burning of biomass results in a direct release of CO_2 when a tree-planting and wood-fuel program is viewed as a whole, the net carbon release is relatively low compared with the burning of fossil fuels. Finally, the trees might be harvested for incorporation in semipermanent timber uses such as houses and bridges. In all likelihood, the final resolution will include some combination of these approaches. The government should not need to intervene in that decision if, as mentioned earlier, it institutes taxes or permits that provide incentives not to cut timber unless the sum of the timber's value plus the value of the newly freed land exceeds the cost of the permit or tax plus the cost of the harvest. The absence of such a program would lead to either under- or overdeterrence to cutting.

Sensitivity Analysis

Land-Rental Costs

The land-rental costs were the most difficult numbers to estimate. The numbers chosen for the basic analysis were derived from historic market rates and government rental rates in the CRP, but were altered modestly to reflect the nonmarginal amount of land rental involved in a large-scale planting program. An alternate run was conducted in which rental rates were set at three times the 1987 and 1988 average market rates. The results of that run are shown in Figures 16.6 and 16.7. The change in rental rates shifted the carbon-sequestering cost range from $5.80–$47.77 per 10^3 kg to $7.57–$96.13 per 10^3 kg. Although the cost of sequestering the full 7.25×10^{11} kg/year nearly doubled, the cost of capturing 1.27×10^{11} kg (10% of current U.S. emissions) increased by only 35%.

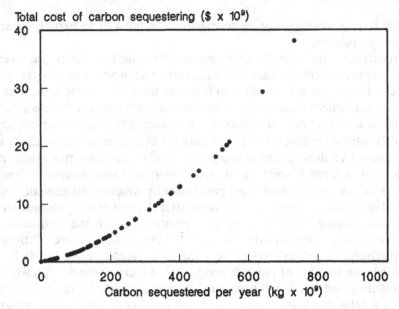

Figure 16.6 Total cost curve when land-rental cost is three times market price.

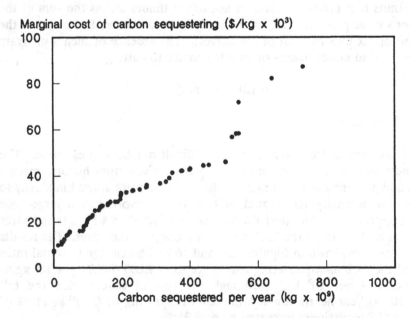

Figure 16.7 Marginal cost curve when land-rental cost is three times market price.

Delayed Benefits

One premise of the basic analysis was that the trees sequestered carbon at the average rate throughout their life cycle. In fact, there are several years as the trees mature during which the carbon-fixation rates are much lower than the lifetime average. This effect is intensified because relative increases in litter and soil carbon lag in time behind the increases in tree carbon. To examine the importance of this effect, an analysis was conducted in which the carbon-fixation rates were delayed by 10 years by bringing the initial treatment cost in the first year and the annual rental costs for the first 10 years forward at appropriate interest rates to the beginning of the 10th year. These costs were then capitalized over the next 30 years and added to the annual rental rates to derive a cost per square meter. The result, shown in Figures 16.8 and 16.9, was a relatively consistent increase in both the marginal and total cost curves by a factor of 2.6. This result suggests that the shape of the carbon-uptake curves over time is an important consideration in future research. A preliminary discussion of the effect of the growth curves of the trees on the cost of carbon sequestration is provided in the fifth section of this report.

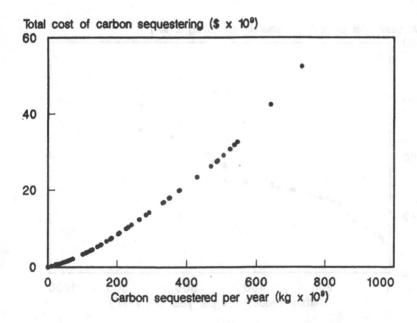

Figure 16.8 Total cost curve when carbon sequestering is delayed until year 10.

Discount Rate

Because most of the costs associated with a large-scale planting program are in the rental or land-opportunity costs, the effect of the discount rate is not strong. As shown in Figure 16.10, at 7.25×10^{11} kg/year, lowering the discount rate from 10% (which was used for this analysis) to 4% decreases total costs by $\approx$10%. It should be expected that the effect of the discount rate would be much stronger when delays in the benefits are considered, as described in the previous paragraph. That effect is illustrated in the next section.

Forestry Growth Curves and Dynamic Analysis

As mentioned in the previous two sections, the carbon-yield figures in the carbon-sequestration analysis were based on average rates of sequestration over the expected lifetime of the trees. The cost per square meter was calculated by capitalizing the preparation and planting costs, incurred at the commencement of the program, at a given discount rate, 10%, over the 40-year planning horizon of the analysis, and adding that cost to the rental

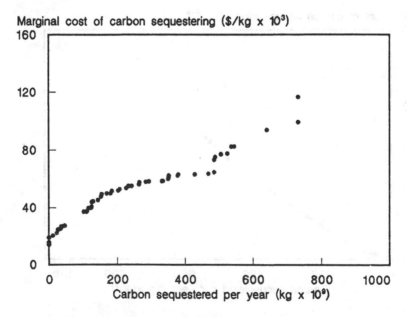

Figure 16.9 Marginal cost curve when carbon fixing is delayed until year 10.

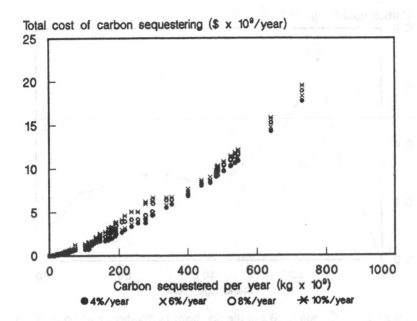

Figure 16.10 Effect of discount rate on total costs.

cost in each year. The cost per kilogram of carbon on a given type of land is the same for each year, derived by dividing the annual cost per square meter by the annual kilograms per square meter. Simple discounting permits carrying forward a proportionate amount of the capital expenses to each year in which sequestering takes place, and expressing that cost as if the expenditure had been made in the same year as the sequestering.

However, the carbon-uptake rates are not constant, as illustrated by the examples shown in Figures 16.11 and 16.12. These examples of cropland conversion in the Southeast and Pacific Coast regions show that the carbon-uptake rate is quite varied over the life of the tree stand and that maximum rates of sequestration are not reached until after 15 and 35 years, respectively. This uneven pattern of sequestration means that it is very difficult to distribute the capital costs over the years in any way proportionate to each year's sequestration.

A tempting solution to this problem is simply to bring all annual costs (land rents) and all benefits (carbon sequestered) back to an equivalent discounted present value. By adding the present value of the 40-year stream of rental payments to the capital costs of the planting and preparation and then dividing by the present value of the carbon sequestering, a figure for the cost-to-benefit ratio is derived, just as in the analysis based on annual

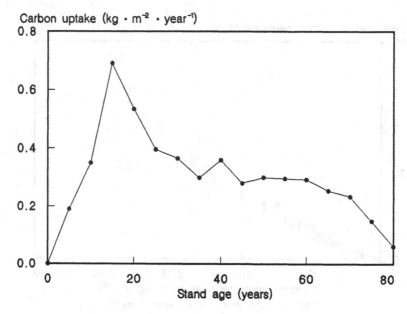

Figure 16.11 Carbon uptake in the Southeast by Southern pine on cropland.

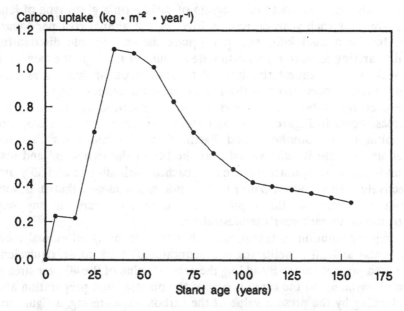

Figure 16.12 Carbon uptake in the Pacific Coast by Douglas fir on cropland.

average carbon-uptake curves. But there is a conceptual problem here. The meaning of discounting dollars is clear. Adopting a 10% interest rate means that $1.00 today is valued as much as a promise of $1.10 next year, given the risks involved in the promise and alternative opportunities to use the dollar this year. But it is not as simple to discount the carbon sequestering. If the dollar value of the benefits involved in net emissions reductions were known, it would be simple to discount the expected benefits flow. However, the benefits are extremely uncertain, a function of innumerable scientific, social, and political factors.

On the other hand, one could simply ask, "Given the state of knowledge about alternative technologies, how much do we value a kilogram of reductions in net emissions this year vs. next year?" The answer to this question is largely a function of the progress of technology and of whether the costs of emissions reductions are expected to decrease over time. Not much value is placed on the promise made today of 1 kg of carbon reductions next year if it is believed that, in the interim, a technology will emerge that will reduce carbon emissions at almost no cost. This belief would lead to adoption of an infinite discount rate. Even if no technological advance is expected, the resources that would go to 1 kg of reductions this year could be invested, thereby earning a return that could be used to make reductions next year. For example, the $1.00 could be invested this year to receive $1.10 next year, which would buy 10% more reductions. If the investment option is preferred, it suggests that the trade-off rate is at least 10% and that future reductions decrease in value rather rapidly as the planning horizon is stretched.

The importance of this calculation is illustrated in Table 16.3, which lists the "equivalent present kilograms" of carbon associated with Figures 16.11 and 16.12 for various physical-discount values. Before considering these figures, note that the yields for the Southern pine are projected over 80 years, whereas those for the Douglas fir are spread over 155 years. The figure for equivalent present kilograms per square meter at a 0% discount rate is simply the total area under the growth curves, implying that the average yield of the Southern pine and the Douglas fir are 0.30 and 0.53 $kg \cdot m^{-2} \cdot year^{-1}$ (3.0 and 5.3 $tons \cdot ha^{-1} \cdot year^{-1}$), respectively. This result suggests that if all kilograms of carbon sequestration are valued equally, no matter when they are captured, and the costs of establishment and rent are equivalent for the two stand types, then planting Douglas fir in the Pacific Coast region is strongly preferred.

However, when the timing of the yield of carbon is considered, the two systems become much more similar, largely because at high discount rates, such as 10–20%, the value of any sequestration after 45 years rapidly approaches zero. Note in Figures 16.11 and 16.12 that the difference in yields between the two stand types is greatest at that time.

TABLE 16.3 Discounting Carbon Flows (in equivalent present kg/m^2)

	Discount Rate		
	0%	10%	20%
Southeast Southern pine	25.24	32.3	1.50
Pacific Coast, Douglas fir	82.84	3.67	1.30

This analysis leads to the observation that if the timing of carbon sequestration is considered, species and types of land that can capture carbon rapidly will have a distinct advantage over those with higher average yields but longer life cycles. This result in turn suggests that although the initial analysis using average yields was not particularly sensitive to the choice of discount rates (Figure 16.10), under the new analysis using carbon-uptake curves, discount rates will be a critical factor.

Conclusions about Future Research

The results of the Moulton and Richards (1990) analysis indicate that the use of tree planting to sequester carbon in the United States bears further consideration. When compared with the results of studies such as Manne and Richels (1990) and Nordhaus (1990), it suggests that the costs of reducing carbon emissions may have been overestimated. At the same time, the discussion in this paper indicates that the cost curves are preliminary, leaving much additional research to be conducted.

The most intriguing area for development is the dynamics of carbon emissions and reductions. This work falls in two general categories. In the first, Reilly (1992) recently presented preliminary results of an effort to use an optimal-control model to develop a greenhouse-gas index that expresses the relative economic effect of different gases over time. Nordhaus (1982) used an optimal-growth model that estimated shadow prices for carbon, given assumptions of growth at a specified rate for a specified number of years, after which the economy is in a steady state. For the carbon-sequestration analysis, the useful output of the optimization models would be an estimate of the shadow prices of reductions in carbon emissions in any given future year. Development of these shadow prices would serve as an input to the second category of research, use of the carbon growth curves developed by Birdsey (1990b). If present-value shadow prices were substituted for the kilograms of carbon sequestered, then it would be relatively easy to develop cost-benefit ratios for tree planting.

The second area of research suggested by this report has already been started by Adams et al. (1992). Use of general-equilibrium or related models should provide important information on land-use costs. It should also facilitate examination of the possible government savings from replacing those farm-subsidy programs whose essential purpose is still served by a tree-planting program. Finally, a multisector model would permit investigation of the effects of large-scale tree planting on the timber industry.

Additional research is needed to understand the external social and environmental effects of massive tree-planting programs. It is quite possible that limited planting could have quite beneficial side effects, whereas extensive reforestation could negatively affect such natural systems as watershed hydrology and wildlife habitat.

That fast-growing trees may provide significant advantages when considered in a dynamic framework suggests that more attention should be paid to the potential for reforestation in tropical countries. Establishment and rental costs are apt to be much lower in developing countries.[5] EPA is considering development of a global database to track land types and areas, possible types of treatment for sequestering carbon, and the potential yield of the land type–treatment type combinations. One lesson from the Moulton and Richards research is that to be useful, such a database should be developed in conjunction with the economic model that will ultimately transform the information into policy analysis.

Development of these research areas will eventually make it easy to compare carbon sequestration with other greenhouse gas–reducing options and assist with the effort to estimate the costs of emissions limitations.

References

Adams, R. M., C.-C. Chang, B. A. McCarl, and J. M. Callaway. 1992. "The Role of Agriculture in Climate Change: A Preliminary Evaluation of Emission-Control Strategies," in J. M. Reilly and M. Anderson, eds., *Economic Issues in Global Climate Change: Agriculture, Forestry, and Natural Resources.* Pp. 237–87. Boulder, Colo.: Westview Press.

---

[5] A logical extension of this observation is that all efforts should be made to discontinue uneconomical deforestation.

Alig, R. J., T. J. Mills, and R. L. Shakelford. 1980. "Most Soil Bank Plantings in the South Have Been Retained; Some Need Follow-up Treatments." *Southern Journal of Applied Forestry* 4: 60–4.

Birdsey, R. 1990a. "The Carbon Cycle Impacts of Forests and Forestry Changes," in *Proceedings of the North American Conference on Forestry Responses to Climate Change.* May 15–17, 1990. Washington, D.C.: The Climate Change Institute.

Birdsey, R. 1990b. "Estimation of Regional Carbon Yields for Forest Types in the United States." Washington, D.C.: U.S. Department of Agriculture, Forest Service (draft).

Manne, A. S., and R. G. Richels. 1990. "CO$_2$ Emission Limits: An Economic Cost Analysis for the USA." *The Energy Journal* 11: 51–74.

Moulton, R. J., and K. R. Richards. 1990. "Costs of Sequestering Carbon through Tree Planting and Forest Management in the United States." Washington, D.C.: U.S. Department of Agriculture, Forest Service. (General Technical Report WO-58.)

Nordhaus, W. D. 1990. "To Slow or Not To Slow: The Economics of the Greenhouse Effect." New Haven, Conn.: Economics Department, Yale University. (Discussion paper).

Nordhaus, W. D. 1982. "The Global Commons I: Costs and Climatic Effects; How Fast Should We Graze the Global Commons?" *American Economics Association Papers and Proceedings* 72: 242–6.

Reilly, J. 1992. "Climate-Change Damage and the Trace-Gas-Index Issue," in J. M. Reilly and M. Anderson, eds., *Economic Issues in Global Climate Change: Agriculture, Forestry, and Natural Resources.* Pp. 72–88. Boulder, Colo.: Westview Press.

PART FIVE

International Perspectives of Global Change

17

The Enhanced Greenhouse Effect and Australian Agriculture

David Godden and Philip D. Adams[1]

Introduction

Investigations of the potential impact of the enhanced greenhouse effect (EGE) on agriculture have generally emphasized production-capacity effects. These effects are a CO_2 fertilization effect; yield effects resulting from climate changes, with lesser effects from pests, diseases, and weeds; and effects on the variability of agricultural production. Second-order adaptive effects include changes in agricultural technology and management responses (e.g., Parry et al., 1990).

Parry and Carter (1988) presented a more complete, but essentially linear, analysis of climatic change operating through agriculture: Climate affects yield, which affects farm output, with consequent regional and/or national output effects, with ultimate farm, regional, and national adjustments to output changes (*see also* Parry and Carter, 1989). This model was mirrored in the draft report of Section A (Agriculture and Forestry) of Working Group II of the Intergovernmental Panel on Climate Change (IPCC, 1990). Crosson (1989a, 1989b) added environmental costs in agriculture and, summarizing earlier authors, changes in comparative advantage.[2] Easterling et al. (1989) located the problem of global warming for Midwestern farmers

---

[1] From the Department of Agricultural Economics, University of Sydney, New South Wales (D.G.), and the Centre of Policy Studies, Monash University, Clayton, Victoria, Australia (P.D.A.). This paper was written while the first author was employed by NSW (New South Wales) Agriculture & Fisheries. Comments on earlier versions of this paper from attendees at seminar presentations and from Fredoun Ahmadi-Esfahani are gratefully acknowledged.

[2] In this context, "comparative advantage" appears to have been solely related to soil-moisture status.

in the United States in the context of an interconnected world agricultural model. S. Kane, J. M. Reilly, and R. Bucklin (reported in Walker et al., 1989) investigated possible world-trade effects of the EGE, based on assumed crop-yield changes in different countries.

A production-oriented focus is far too narrow for investigating the potential impact of an EGE, especially for Australian agriculture. So, too, is Parry and Carter's (1988) "linear" model, in which the direction of causation is from climate change, through agricultural-production effects, to higher-order effects. A more comprehensive investigation of EGE-agriculture relationships in Australia requires analysis of (1) EGE's potential direct effects on natural resources and agricultural production; (2) the potential indirect effects of EGE on natural resources and, thus, on agricultural production; and (3) socioeconomic effects of the predicted EGE, including management responses to the direct and indirect effects of EGE on natural resources (e.g., on production costs); management responses to changing demands for goods and services as a consequence of EGE; and collective policy responses—both of agriculture and of the wider economy—to EGE or its consequences.

Because the individual effects of EGE are likely to be interrelated, the direction of these effects can only be modeled in a general-equilibrium framework. Evaluation of the size of these responses requires quantitative general-equilibrium modeling. Incorporation of EGE effects in quantitative general-equilibrium models requires a systematic examination of the economic structure of the agriculture in which these effects operate.

In this paper, therefore, a brief description of the key characteristics of Australian agriculture is followed by a similarly brief survey of likely direct, indirect, and socioeconomic effects of EGE on Australian agriculture. A systematic analytic framework for incorporating EGE effects on Australian agriculture is then described, including a discussion of possible welfare effects in a partial-equilibrium setting. Because of the effect of macroeconomic variables on Australian agriculture, aggregative variables potentially directly affected by EGE—which are likely to interact with purely agricultural production-oriented EGE effects—are described in the next section. In the final section, a simple experiment is described that uses a computable general-equilibrium model of the Australian economy to compare two classes of economic effects of EGE on Australian agriculture.

Australian Agriculture

Approximately half of Australia's land area receives < 400 mm (16 inches) of rain per year (Davidson, 1981). Rainfall variability is high, and more than three-quarters of the land mass has a growing season (defined in terms of a ratio of rainfall to evaporation) of < 5 months (Davidson, 1981). Broad land-use regions depicted in Davidson (1981) are briefly described below. Two separate areas are suited to moderately intensive agriculture; both are, except for some irrigated areas, within ≈400 km of the coast. The first area runs from near Townsville, Queensland, in the northeast, down the east coast to midway along the southern coastline. The second area suited to moderately intensive agriculture lies in the southwest corner of the continent.

Areas of more intensive agriculture are generally structured as mixed farms—principally sheep (for wool and meat), cattle, and crops (predominantly wheat, with barley and oats as the other main winter crops; summer cropping in summer-dominant rainfall areas includes dryland corn, sorghum, sunflower seed, and soybeans). Enterprise substitution in mixed farming, which is constrained in the short run by investment in cropping machinery and the build-up phase of livestock enterprises, is generally high. Irrigated areas of the South include rice in a mixed-farming system. There are some areas where agriculture tends to be more specialized: sugar cane on the northeast coast; sheep or cattle grazing on eastern uplands; dairy farming in well-watered or irrigated areas of the South; cotton on irrigated areas in the inland part of the central eastern seaboard; wine grapes in many areas of the Southeast and Southwest and sultana grapes in the South; intensive horticulture (stone and pome fruits, citrus fruits, and vegetables) in many, especially irrigated, areas; and intensive livestock (pigs and poultry) in the grain belt or close to urban areas.

Rural exports (including small amounts of forest and fish products) constitute ≈30% of Australia's total exports of goods and services (ABARE, 1990). Many agricultural industries are heavily export-oriented. In 1988–1989, all crops constituted 20% of the total value of farm production (TVFP); wheat alone was 12% of TVFP. Approximately 40% of all crops by value (and 70% of wheat) was exported. Corresponding figures for other groups in 1988–1989 were as follows: meat and live sheep, 23% of TVFP and 40% exported; wool, 26% of TVFP and >90% exported; and dairy, 8% of TVFP and 30% exported. (High proportions of cotton and sugar are also exported, but their degree of processing makes it difficult to estimate a proportion of farm value. Ratios for other product groups were roughly reduced to reflect processing value added in the value of exports.)

Even in more intensively farmed areas, agriculture is very extensive compared with that in the Northern Hemisphere. Agriculture is relatively extensive because of high ratios of land to labor (Hayami and Ruttan, 1985), a traditional reliance on export markets, and a general reluctance or inability of Australian governments to subsidize Australian farmers sufficiently for more intensive agriculture to be profitable.

Framework of EGE Effects on Australian Agriculture

The potential effects of a possible EGE on natural resources include CO_2 availability, ambient temperature (level and seasonal distribution), rainfall (amount, intensity, and distribution), and wind and their effects on farm-production characteristics such as yield, land degradation (erosion, salinity, nutrient loss, and leaching), and externalities [e.g., pests, diseases, and weeds; compare with Randall (1987), chapter 9].

Effects of EGE on natural resources would partly be directly environmental, but would also be substantially conditioned by management response. For example, possible effects of EGE on erosion include the direct effects of rainfall amount, intensity, and distribution; the indirect ecological effects through changes in associated flora (both indigenous and exotic species) and fauna (indigenous, and exotic species both husbanded and feral); and a management response (conditioned by both the natural and socioeconomic environments; *see* below)[3]. Management response would include changes in husbandry practices and enterprise mix as a consequence of EGE's effect on direct-production costs as well as changes in farm structure, externalities, and product demands. Of major importance to management would be possible changes in production variability.

Because Australian agriculture is heavily export-oriented, changes in the demand for Australian agricultural produce may be as important as, or even more important than, domestic-production effects. These demand effects would be principally generated externally. For example, in an EGE-warmed world, substituting clothing warmth for fuel warmth would benefit Australian wool production if fiber-fuel substitution into wool outweighs the reduced demand for warmth protection. These effects for Australia would arise not only from final demand, but also as a consequence of production changes in other regions. For example, demand for Australian grains could rise if grain production in the rest of the world were detrimentally affected compared with Australia, or the demand for Australian grass-fed beef could increase compared with that for grain-fed beef in the rest of the world if EGE were followed by rising international grain prices. Similar changes may also arise from macroeconomic effects (*see* below).

---

[3] IPCC (1990) noted in its discussion of EGE effects on pests and diseases that cattle ticks threatened the profitability of the Australian beef industry because of increased insecticide resistance and high costs of dipping. This threat was confined to northern Australia (which currently produces a small proportion by value of Australia's beef production). The threat posed by cattle ticks was also a direct result of management failures to adopt the biologically sound response of increasing the proportion of *Bos indicus* in the threatened herds.

Potential socioeconomic consequences of EGE also include collective policy responses to EGE. Both domestic and external policy responses are likely to affect Australian agriculture. Domestic policy responses to EGE include allocative-efficiency responses (e.g., changing investment and disinvestment decisions in publicly provided infrastructure—roads, railways, grain-handling facilities, and agricultural research and development—in response to the demand for, and costs of providing, these services) and distributive responses (e.g., changing direct subsidies, or providing subsidized infrastructure, in response to the effects of EGE on particular groups). A second type of domestic policy response includes economic-management decisions, such as the direct and indirect effects of EGE-mitigating policies (e.g., carbon taxes or emission quotas) and macroeconomic responses that might affect the level of domestic demand, interest rates, and exchange rates.

Possible external policy responses mirror domestic responses. Of most direct consequence—from Australia's perspective—are other countries' policies that affect the excess demand for Australian agricultural produce[4]. If EGE detrimentally affects farm incomes in developed countries in the Northern Hemisphere, increased farm protection in response to EGE would further exacerbate existing negative effects on Australian agriculture of these countries' existing agricultural protection (Crosson, 1989b). A public-choice, rather than a public-interest, theory of agricultural policy—for example, Ruttan's (1978) "induced institutional innovation"—suggests that, if EGE obviously and adversely affects developed countries in the Northern Hemi-

---

[4] IPCC (1990) expressed concern that 77% of all traded cereals in 1987 originated in just three countries (United States, Canada, and France), three regions (United States, Canada, and the European Community) held one-third of world stocks of wheat and coarse grains, and some of these countries may be vulnerable to production decreases with an EGE. However, some of the quantitative importance of these countries to cereal production arises from their agricultural-protection policies, not from their crop-production capacity. There has been much focus in the past decade on the most protectionist countries reducing their agricultural support levels and, thus, reducing their importance to agricultural production and trade, because this result would provide incentives to other countries to increase their agricultural production. Current patterns of world agricultural production cannot be evaluated solely by physical production characteristics, but also by the economic incentives underlying this production pattern. Similar comments apply to IPCC's comments about the possible disadvantage that sugar cane may suffer compared with temperate sugar beet because of the differential CO_2 effect in C3 and C4 plants. Sugar beet is produced as extensively as it is because of the protection afforded its producers (e.g., WCED 1987). An EGE may induce a (relative) decline in sugarcane production, but—if protection of beet sugar was substantially reduced—this reduction would occur from a much higher output and income base.

sphere, then these governments are likely to respond by increasing agricultural protection.[5] Imposition of direct or indirect constraints on greenhouse-gas emissions outside Australia may indirectly affect Australian agriculture through exchange rates (*see* below).

Analyses of potential effects of EGE on Australian agriculture have concentrated on yield and output effects (e.g., Walker et al., 1989; Pittock, 1989; Nulsen, 1989; Landsberg, 1989), climate and agricultural variability (e.g., Hobbs et al., 1988), and the effects of pests (e.g., Sutherst, 1990), although it has been noted in passing that export demand will play a major role (e.g., Pittock, 1989; MPE, 1989).

Microeconomic Effects of EGE on Production

The production-economics framework outlined below is conventionally neoclassical. We do not model production discontinuities—for example, changes in production sets as a consequence of extreme events—even though, in the more fragile ecosystems of Australian agriculture, such discontinuities have been observed (e.g., the long-run degradation of arid pastoral lands in western New South Wales after overgrazing by introduced sheep and rabbits was capped off by a calamitous drought from 1899 to 1902). Ignoring discontinuities in production sets was justified because the better-watered Australian agricultural areas appear more resilient to extreme events, and most Australian agricultural output comes from these areas.

Production Function—Single-Product Firm

For the single-output production function, effects of EGE may be modeled as changes in the input-output relationship. If EGE is uniformly favorable (unfavorable) to production, then EGE will be comparable to disembodied technological progress (regress) and may be modeled by shifts in the location of production isoquants. These changes may be "Hicks-neutral," if the isoquants migrate homothetically, or "Hicks-biased" (nonhomothetic migration). Several possible biological interpretations of such changes exist:

Hicks-neutral changes. If EGE simply resulted in uniformly more (or less) growth of all desirable plant species because of higher rainfall with no other

---

[5] Parry and Carter (1989) categorized the elements of national agricultural policy likely to be affected by EGE as agricultural self-sufficiency, regional equity, farm-income support, and agricultural extension (e.g., resource conservation, water management, and pest control). Their characterization of agricultural policy was firmly in the context of a public-interest theory of regulation.

effects, then output would increase (or decrease) in a Hicks-neutral way. Agricultural production changes in response to input-price changes would be similar to the non-EGE regime, but at different levels.

Hicks-biased changes. Continuing the previous example, if EGE brought greater growth of desirable plant species because of higher rainfall, but fertilizer was leached more rapidly from soils, then EGE would result in a Hicks-biased form of technological regress and would be fertilizer-saving. Thus, for unchanged relative prices of inputs, the production system would use relatively more (less) of the inputs advantaged (disadvantaged) by EGE.

As well as affecting the location of the isoquants, EGE may also change the elasticity of input substitution. Such changes are important because they would affect the flexibility of farm response to changes in input prices. If the elasticity of input substitutability increased (decreased) with EGE, the production technology would become more (less) responsive to changes in input prices. Consider the biological underpinnings of the substitutability of machinery and labor: suppose substitution of machinery for labor occurs through the use of larger or heavier machinery. If EGE is accompanied by higher rainfall, then, on heavier soils, larger machinery may become relatively less suitable because of increased compaction in generally wetter soil conditions, with accompanying yield reductions. The elasticity of substitution of machinery for labor would be reduced, indicating fewer opportunities for substituting machinery for labor as the price of machinery falls relative to the price of labor. Alternatively, for similar reasons, wetter conditions may change the substitutability between land-based machinery and aerial machinery: if generally wetter conditions favor plant growth, but land-based machinery leads to increased soil compaction, then an inward shift of isoquants primarily at relatively high levels of use of aerial inputs would increase the elasticity of substitution between land- and air-based machinery.

If EGE leads to biased technological change that, for a particular input, is input-saving for some input combinations but input-using for other combinations, then the enterprise effects of EGE will be considerably more complex. The post-EGE isoquant map will intersect pre-EGE isoquants, and the consequences for relative input use for constant relative input prices will be more difficult to determine. These problems will be exacerbated if elasticities of substitution among inputs also change with EGE.

Multiproduct Firms: Regional Impacts

Because Australian farms usually produce more than one product, it is preferable to model them as multiple-output, multiple-input firms. This framework may also be used to examine regional production effects. Analogous to previous arguments, the potential effect of EGE on the mix of outputs and inputs in multiproduct firms or regions may be represented by Hicks-neutral or Hicks-biased technological progress or regress, and/or by changes in the elasticity of substitution among inputs and outputs.

The potential effect of EGE on a multiproduct firm's input demands and output supplies could be examined analytically through its effect on production parameters. EGE's quantitative impact could be examined in a model where EGE effects were represented numerically (e.g., econometric or programming models of production systems). With appropriate aggregation conditions, or assuming profit maximization at the regional level, the regional effects of EGE could be investigated analytically through the regions' input-demand and output-supply functions, and the quantitative impact examined by appropriate models.

Aggregate Supply

The possible agricultural effects of an EGE were characterized in the first two sections as firm- and regional-level production functions. By using profit maximization, simple analytical representations of firm-, regional-, or industry-level production functions may be converted into their corresponding supply-response analogues. With duality, supply-response equations may be derived from cost or profit functions. The effect of EGE may then be represented in supply-response or input-demand equations. EGE may be introduced as an additional variable (most simply, as a time shifter representing the effect of EGE), or, alternatively, the parameters (or even functional form) of the supply-response or input-demand equations may be modified directly to represent EGE.

Conversion of the production effects of an EGE into a supply-response framework permits consideration of the welfare effects of production changes induced by EGE. In the single-output framework in a closed economy, with EGE causing a simple parallel shift in a linear supply schedule, the welfare effects of EGE can be represented by the area between the supply schedules and bounded by the demand schedule. The direction of welfare effects for simple types of EGE impacts can be summarized (Table 17.1). Without such simple and unlikely assumptions, numerical modeling would be required to estimate the likely welfare effects of EGE, even on small regions. If the size of the EGE effect on production can be estimated, the welfare effects of EGE may be estimated in a partial equilibrium context with models used for investigating the benefits of research, like that of Davis et al. (1987).

Table 17.1 Direction of Partial Equilibrium Aggregate Welfare
Effects of the Enhanced Greenhouse Effect (EGE)[a]

| | Direction of EGE Shift for Enterprises | |
EGE as Techno-logical-Shift Type	Unidirectional	Mixed
Hicks-neutral		
Progress	Positive	
		Undefined
Regress	Negative	
Hicks-biased		
Progress	Positive	
		Ambiguous
Regress	Negative	

[a] Table considers only aggregate benefits of shifts; distributional aspects and shifts in demand from EGE are ignored. How different groups (e.g. farmers vs. consumers) fare under EGE change would depend on demand elasticities and other factors.

Production Dynamics

In the preceding discussion, it was implicitly assumed that EGE causes an instantaneous, once-and-for-all change in production conditions. However, EGE is likely to occur incrementally, perhaps even unnoticeably, from year to year, and possibly with significant lagged effects. The process of EGE-induced production effects, and the nature of the emergent responses in agriculture, may be as important as the effect of EGE itself. Also, EGE may not simply affect environmental variables unidimensionally, but may also have more complex effects. For example, there may not just be increases or decreases in rainfall or temperature, but there may also be changes in the variability or distribution of these variables.

Asset Fixity. Physical and human capital is specialized in varying degrees to particular forms of agricultural production. If EGE becomes progressively evident, there will be a shift away from those enterprises that are relatively less favored, or detrimentally affected, by EGE. However, the speed and extent of this shift will be modified by the reactions of decision makers to these changes. As EGE changes the relative profitability of enterprises, the value of physical and human capital specialized in those enterprises of declining profitability will decrease. Some of these assets may become immobilized in the declining enterprises as the value of this capital declines. If the cost of investment in human capital for new enterprises is high (e.g., significant new knowledge and skills are required for the preferred new enterprises) and some farmers cannot afford the necessary investment (e.g., they cannot afford the opportunity cost of investing in new human capital), then these farmers may become locked in to the increasingly less-favored enterprises. Alternatively—and perhaps relatedly—if capital markets are imperfect and the cost of borrowing to invest in new physical capital is high, then some farmers may be locked in to the less-favored enterprises because they cannot acquire the capital necessary for the more-favored enterprises. This asset immobilization will be exacerbated if the average age of farmers is high, because off-farm employment opportunities will be increasingly restricted for older farmers, and farmers approaching retirement are likely to be less willing to change occupation and more able to survive decreased family income because they are likely to have fewer dependents.

The slower the rate at which EGE affects agricultural production, the lower the likelihood that asset fixity will be a serious problem. Further, for assets with short asset lives, there will be frequent opportunities to consider whether such assets should be replaced, or alternative investments should be undertaken. Similarly, the slower the rate of expression of EGE, the lower the opportunity cost of acquiring new knowledge and skills, and the greater the possibility that new capital can be acquired by a new generation of farmers at the same cost that would have been required for investment in the previously optimal set of knowledge and skills.

New technologies. If EGE has a detrimental effect on the production process, new technologies may be developed to completely or partially offset the impact of EGE. For example, suppose EGE in eastern Australia results in wetter summers and drier and more variable winters. The consequent deleterious effects on winter-cereal production could be offset if new winter-cereal varieties are developed that are more efficient in their use of water and more tolerant of water shortage and that could be harvested successfully in wetter conditions. If harvesting occurs in wetter conditions with increasing frequency, grain drying might be increasingly adopted and improved.

It is not immediately obvious whether or not the allocation of resources should vary because of EGE. However, the general principles concerning the allocation of resources to research are the same whether or not there is an EGE. Research resources should be allocated among enterprises so that the marginal benefit from research is identical regardless of the enterprise additional resources are allocated to, and total research resources should be allocated until the marginal cost of research is identical to the present value of the marginal benefits of this research.

The research-resource-allocation problem with EGE is therefore similar to the maintenance-research problem (in which new plant varieties become increasingly susceptible to diseases over time). With EGE, however, the problem is to evaluate production systems for all enterprises in terms of a possible EGE and then to assess how EGE may interact with possible new technologies.

The importance of formally considering allocation of agricultural research resources in the context of EGE is that, should EGE detrimentally affect particular enterprises, there will be the temptation to attempt to counteract these effects by developing new technologies. In an institutional setting in which research resources are allocated by public-choice rather than public-interest mechanisms, there may be strong political pressures to develop counteractive technologies as a short-run response to EGE problems in an attempt to protect existing farm investments in enterprises detrimentally affected by EGE. Although it may be feasible to counteract EGE in this way, it may not necessarily be socially optimal. If EGE affects enterprises monotonically—that is, once a beneficial or detrimental effect occurs, it intensifies over time—the shorter the perspective taken in allocating research resources, the more likely it is that research resources will be misallocated. Unfortunately, this short-run perspective is likely to characterize a public-research effort that is responsive to political pressure.

The pattern of agriculture's response to a possible EGE will depend on the particular sequence of decisions by public and private decision makers in response to this EGE. Response through time will partly be governed by the current stocks of physical and human capital in the farming sector and by stocks of physical capital in the downstream industries that transport, store, and process agricultural products. The responses to EGE of public agricultural agencies—for example, those agencies that provide rural infrastructure; undertake research, extension, and regulatory activities; or redistribute income to the agricultural sector—will also affect the way agriculture responds. If these agencies see their roles as Canute's courtiers, agricultural adjustment to EGE will be impeded.

Aggregate Effects

Because EGE, or policy responses to EGE, may affect the supply of inputs to agriculture or the demand for agricultural outputs—for example, through the imposition of carbon or methane taxes—examination of the possible impact of EGE cannot appropriately be effected assuming constant input or output prices.

If carbon taxes are imposed, the price of hydrocarbon-based energy (e.g., petroleum-based fuels and coal-derived electricity) or inputs whose production requires high amounts of these inputs (e.g., petroleum-based chemicals and steel) would become relatively more expensive. In agriculture, the prices of inputs like diesel fuel, gasoline, electricity, agricultural chemicals, and fertilizers would rise compared with the prices of inputs like labor. It is difficult, however, to determine how the relative prices of inputs within the "high-greenhouse-gas" group would change, because the degree to which input prices may rise depends on the greenhouse-gas intensity of their production processes. The effect of these relative price changes on relative input use in agriculture depends not only on the degree of the change in relative prices, but also on the elasticity of substitution of these inputs in agricultural production. If this elasticity is low, the consequent change in relative input use will also be low, even if there is a marked change in relative input prices.

The effect of EGE on agricultural-input prices will not only arise from an EGE-induced policy response affecting production costs for greenhouse-gas–intensive industries. Competing demands in other sectors for resources used in producing agricultural inputs may also be affected by EGE or EGE-induced policy responses. If an optimal EGE-induced policy response increases the production of energy from nonhydrocarbon sources (e.g., nuclear, hydroelectric, solar, or wind), major construction works would be required to harness energy from these sources, and there would be a consequent large increase in demand for capital, material, and labor inputs required for the necessary investment. Unless the supply of these inputs is infinitely elastic, the increased nonagricultural demand for these inputs would increase the price of these inputs in agriculture.

EGE-induced changes in the demand for Australian agricultural produce may emerge from a variety of sources, including (1) direct changes in patterns of final demand for agricultural products—either in Australia or overseas—thus affecting the demand for Australian agricultural production, and (2) the excess demand for Australian agricultural output, a function not only of final demand, but also of the production of similar products, substitutes, or complements in other countries.

The emergence of EGE may affect the relative prices of agricultural and nonagricultural products. For example, if the external cost of greenhouse gases is internalized in the price of goods, then the price of energy for heating may rise relative to the price of using clothing for warmth. Thus, the demand for clothing for warmth may rise and, consequently, the demand for fibers such as wool and cotton. This effect would be reinforced by the potential effect of carbon taxes on the price of synthetic fibers. This substitution effect may conceivably be large enough to offset the reduced demand for clothing fibers because of general atmospheric warming (Pittock, 1989).

Three broad macroeconomic mechanisms that may transmit EGE effects external to agriculture into the Australian agricultural sector are exchange rates, aggregate demand, and savings and wealth.

If overseas countries' reactions to EGE lead to reductions in the export demand for Australian energy exports and their complements, or energy-intensive products, then a significant downward pressure on the Australian exchange rate is likely, other things being equal. This effect would raise the Australian dollar price of non-energy-related exports and increase the price of all Australian imports, thus increasing the profitability of non-energy-intensive export and import-competing industries. Agricultural exports, and some agricultural commodities that face (or may face in the future) import competition, would benefit significantly from exchange-rate changes arising from adverse changes in energy-related exports.

The aggregate effect of changes affecting the Australian exchange rate can only be derived in a quantitative general-equilibrium framework, but the direction of change in increasing the relative profitability of agricultural production seems plausible. This increased profitability will increase agricultural production, drawing factors (i.e., land, labor, and capital) and other production inputs into the agricultural sector. Thus, an EGE-induced policy response—generated either within or outside Australia—that reduced Australian exports of energy-intensive commodities would increase the relative prosperity of the Australian agricultural sector and encourage output expansion in Australian agriculture.

Policy responses to EGE that increase the direct costs of activities producing greenhouse gases will reduce the consumption of goods and services produced by these activities. Conversely, EGE-induced policy responses will increase consumption of some goods and services, especially under the general heading of environmental amenity. In particular, individuals' private consumption expenditures are likely to be reduced because of increased prices of many conventional goods and services that are greenhouse-gas intensive. Similarly, private wealth is likely to be substantially reduced because the profitability of private production activities producing high amounts of greenhouse gases will also have been reduced. The size of these effects is unlikely to have a major impact on agriculture because the income elasticity of demand for agricultural products is low and wealth effects on demand for agricultural products is similarly likely to be low.

Addressing the potential problem of EGE brings up the fundamental issue of intergenerational transfers of productive assets. Continued production activities that store up greenhouse problems for future generations are a form of dissaving, or borrowing the income of future generations. Conversely, current policy responses that reduce future potential greenhouse problems will force savings on present generations to the potential benefit of future generations. The additional savings by present generations to reduce potential future greenhouse problems could come from two possible sources: diversion of current investment from other uses, and increasing the current amount of savings to enable both investment in greenhouse mitigation and continuation of existing investment. If greenhouse-gas–mitigating strategies were financed by reduced investment outside agriculture, then agricultural investment would only be affected if agriculture itself were directly affected by the mitigation strategy—for example, increased input prices through carbon taxes or methane taxes. If greenhouse-mitigating strategies were financed by diversions of government investment away from present infrastructure investment or by reductions in transfer payments, then agriculture would be affected to the extent that government infrastructure investment reductions reduced agricultural profitability or that reductions in transfer payments reduced farmers' incomes. If greenhouse-mitigating strategies were financed by greater savings and investment, then the effects on agriculture would be a higher interest rate on savings for net lenders and a higher interest on borrowings for net borrowers, and increased demands for investment goods, leading to higher prices for these investment goods to the agricultural sector. The likely impact of these effects on Australian agriculture is small.

EGE-Effects Experiment

The possible agricultural effects of EGE in Australia are likely to be so pervasive, potentially reinforcing or self-canceling, and often subtle that it is impossible to use yield effects alone as a proxy for welfare effects. In addition, potential EGE effects on Australian agriculture could arise either from direct impacts in the agricultural sector itself, or possibly from quite remote effects elsewhere in the domestic or international economies. As indicated in preceding sections, a general-equilibrium framework is required to derive satisfactory conclusions about the possible effects of EGE on Australian agriculture.

A simple general-equilibrium experiment for investigating possible EGE effects on Australian agriculture is summarized in Table 17.2. In this experiment, one possible detrimental effect of EGE on Australian agricultural production—a reduction in wheat production in the wheat-sheep zone[6]—is compared with one possible beneficial demand effect (*see* below). There is much conjecture about the possible effect of EGE on Australian wheat production: there are possible beneficial or detrimental yield effects

---

[6] We originally intended to model the impact of EGE on agriculture as a southward migration of the wheat belt. However, ORANI does not disaggregate the wheat-sheep zone geographically, so it is not possible to directly model this effect.

Table 17.2 Experiment Examining EGE Effects in Australian Agriculture

	Agricultural Effects of EGE	
Carbon Tax	No EGE Effects	*Wheat Production Decreases in Wheat-Sheep Zone*
No carbon tax	Control	EGE only
Carbon tax	Policy response only	EGE + policy response

(from changing rainfall patterns, pests, and diseases) and possible expansions or contractions in wheat area from movement of the wheat belt's current boundaries (compare with Walker, et al., 1989; Pittock, 1989; Nix, 1990; NSW Agriculture & Fisheries, 1990). We assume a detrimental effect of an EGE on wheat production in the wheat-sheep zone as a possible worst-case scenario.

The assumed demand-side effect was the imposition of carbon taxes overseas, which reduces the demand for Australian coal. With a trade-balance constraint, this reduction in coal exports will lead to increased exports of Australian agricultural products. This demand effect operating through the exchange rate is not exclusively beneficial to agriculture: exchange-rate depreciation increases the overseas competitiveness of all traded goods industries.

This experiment was run using the ORANI computable general-equilibrium model of the Australian economy (Dixon et al., 1982). The effect of EGE was modeled as a 1%/year decline in wheat yield in the wheat-sheep zone over the period 1988–1989 to 1994–1995 [in physical terms, 1.5 $g \cdot m^{-2} \cdot year^{-1}$ (≈ 15 $kg \cdot ha^{-1} \cdot year^{-1}$), and the policy response as a 1%/year decline in world demand for Australian black-coal exports over the same period (a price reduction of ≈ 0.05–0.06 U.S. cents/kg, or ≈ 50–60 U.S. cents/metric ton, per year at current export levels).[7]

The principal assumption of the base-case scenario is that Australia's foreign debt as a proportion of gross domestic product (GDP) is stabilized by 1994–1995 (compare with Dixon et al., in press). The other main exogenous assumptions are an average real interest rate on foreign debt of

---

[7] The macroeconomic settings are similar to those reported in Dixon et al. (in press), but the version of ORANI used in the present experiment has a changed representation of the export of nontraditional commodities.

4%, inflation of 5%/year, real government consumption increasing at 2.25%/year, decline in Australia's terms of trade of 1%/year, aggregate employment increasing at 1.8%/year, and labor-saving technological change of 1.0%/year. This scenario requires GDP to rise at 2.6%/year with a rapid export growth of 8.1%/year, accompanied by a 1.6%/year devaluation of the real exchange rate (Base column in Table 17.3). This rapid export growth is achieved in part by rapid increases in the volume of agricultural production (2.4–7.6%/year of the major agricultural enterprises), with consequent increases in farm incomes ranging from 3.8 to 5.7%/year.

The columns of Table 17.3 labeled Agric and Macro report projections of the effects of a lower wheat yield and of a decline in world demand for Australian black coal, respectively. Values in these columns are to be interpreted as deviations from the base projections. These deviations were computed by assuming that none of the model's exogenous variables (e.g., aggregate employment or the change in the debt-GDP ratio at the end of the period) are affected by the shocks.

At the aggregate level, the assumed 1%/year wheat-yield-depressing effect of EGE reduces the GDP growth rate by 0.01%/year compared with the base case, depreciates the real exchange rate by a further 0.03%/year, reduces the growth in export volume by 0.02%/year, and marginally reduces the fall in the terms of trade by 0.01%/year (Agric column in Table 17.3). Extra real depreciation is required to eliminate the initial deleterious effects on the balance of trade of the Agric shock. Without this depreciation, the debt-GDP ratio would not be stabilized by 1994–1995; this stabilization is one of the exogenously imposed requirements of the projections. In the agricultural sector, the effect of EGE-induced wheat-yield decline is to reduce the rate of growth of wheat output by 1.7%/year, with substitution mainly by sheep and other grains (increasing their volume growth rates by 0.07 and 0.25%/year, respectively). Farm-operating surplus grows faster in zones growing little wheat (pastoral and high rainfall, 0.19%/year and 0.23%/year, respectively), and the growth of farm-operating surplus is reduced by 0.38%/year in the wheat-sheep zone. Agriculture- and forestry-sector-output growth are reduced 0.08%/year by the wheat-yield decline.

The overseas imposition of carbon taxes, leading to a reduction in export demand for Australian black coal of 1%/year, has a GDP effect similar to that of declining wheat yield (Macro column in Table 17.3). The real exchange-rate effect is, however, much stronger: there is an additional 0.12%/year depreciation of the real exchange rate, with an associated increase in the rate of growth of export volumes of 0.09%/year. Because export demand for black coal is assumed to fall, the decline in the terms of trade is accentuated (an additional 0.12%/year). In the agricultural sector, the effects of the decline in export demand for coal are—except for wheat and other grains—much more dramatic than those caused by the wheat-yield

Table 17.3 Estimated Direct and Indirect EGE Effects in Australian Agriculture[a]

| | *Scenario* | | | |
Variable	*Base*	*Agric*	*Macro*	*Agric+Macro*
Macro summary				
GDP	2.62	−0.01	−0.01	−0.02
Real exchange rate[b]	1.56	0.03	0.12	0.15
Exports (volume)	8.09	−0.02	0.09	0.07
Terms of trade	−1.55	0.01	−0.12	−0.11
Agriculture output volume				
Wool	4.22	0.01	0.22	0.23
Sheep	7.61	0.07	0.45	0.52
Wheat	3.84	−1.67	0.23	−1.44
Other grains	4.14	0.25	0.22	0.47
Beef	2.40	0.02	0.14	0.16
Operating surplus				
Pastoral zone	3.81	0.19	0.15	0.34
Wheat-sheep zone	3.78	−0.38	0.22	−0.16
High-rainfall zone	5.72	0.23	0.34	0.57
Industry output				
Agriculture and forestry	4.07	−0.08	0.21	0.13
Non-metal mining	−0.53	−0.02	−0.03	−0.05

[a]All variables reported as changes in percent per year.
[b]Positive sign means devaluation.

decline. The growth rate of wool output increases by 0.22%/year over the base case (compared with 0.01%/year in the wheat-yield-decline scenario); sheep-output volume grows an additional 0.45%/year (cf. 0.07%/year), and beef-output volume grows an additional 0.14%/year (cf. 0.02%/year). The rate of growth of wheat-output volume is an additional 0.23%/year (compared with a reduction of 1.67%/year with wheat-yield decline), and the change in the growth rate of other grains is similar to that with wheat-yield decline. The growth rate of total agriculture- and forestry-sector output is increased 0.21%/year over the base case by the reduced export demand for black coal.

Because the ORANI model is linear, the combined effects of the simulated wheat-yield decline and carbon tax (Agric+Macro in Table 17.3) are estimated by adding corresponding values in the Agric and Macro columns. The combination of EGE effects through wheat-yield decline and carbon tax is favorable for the agriculture and forestry sector as a whole, and only has a net negative effect for that industry (wheat) and zone (wheat-sheep) directly disadvantaged by the wheat-yield decline. Because of ORANI's linearity, the results presented in Table 17.3 can also be used to estimate the consequences of any change that is a multiple of the assumed 1%/year change (all corresponding values in Table 17.3 scaled by relevant multiple), including beneficial effects (all corresponding values in Table 17.3 multiplied by −1).

Conclusions

In a multiproduct, multi-input production system, the key economic impacts of EGE to be identified are whether the result is equivalent to technological progress or regress, whether this technological effect is product- and/or input-neutral or biased, whether there is an effect on the substitution elasticities, or whether there are changes in functional form, leading to changes in the degree or elasticities of output supply or input demand. The agricultural impacts of EGE may be exacerbated by capital fixity and labor immobility, and exacerbated or ameliorated by decisions about the development of new technologies.

The agricultural impacts of EGE are not, however, solely confined to the physical environment of agricultural production. EGE or greenhouse-induced policies may change the price of agricultural inputs, especially those that are energy-intensive. Aggregate supply effects within Australia, or in Australia's export competitors or importing countries, may affect the prices at which Australia's agricultural produce may be sold. Aggregate effects of EGE or greenhouse-induced policies in Australia or overseas may affect the exchange rate and, thus, border prices for agricultural commodities. Other macroeconomic variables, such as aggregate demand and wealth effects on consumption, are less likely to have a significant impact on demand for Australian agricultural products. However, greenhouse-ameliorating policies may have substantial effects on investment-goods industries and the capital market and may thus have significant impacts on agricultural investment. The potential effects of EGE on agriculture, taking into account these nonagricultural effects, can only be modeled in a general-equilibrium framework.

Clearly, it is impossible to draw general conclusions from the simple EGE experiment reported in this paper. The results of the general-equilibrium modeling reported here, however, demonstrate that EGEs' damage to agriculture may be sufficiently offset by EGE-induced impacts elsewhere in the economy and that agriculture may be a net gainer from EGE even if it is directly damaged and even if the economy as a whole is also damaged. A parenthetical caution might also be added: it is possible for agriculture to suffer net damage from EGE even if the biological effects of EGE are favorable for agriculture.

EGE can have many possible direct and indirect effects on agriculture other than wheat-yield decline and carbon taxes. It is therefore likely to be more profitable to refine estimates of EGE impacts on wheat yields and coal prices while modeling a much wider range of possible impacts of EGE on agriculture and variables affecting agriculture. We plan to expand this general-equilibrium modeling. It may be desirable to link national general-equilibrium models of EGE directly to world-trade models, to assess simultaneously international and national effects of EGE. ORANI has already been linked in this way to examine agricultural trade reform (Horridge et al., 1990).

References

Australian Bureau of Agricultural and Resource Economics (ABARE). 1990. "Statistical Tables." *Agriculture and Resources Quarterly* 2: 337–71.

Crosson, P. 1989a. "Greenhouse Warming and Climate Change: Why Should We Care?" *Food Policy* 14: 107–18.

Crosson, P. 1989b. "Climate Change and Mid-Latitudes Agriculture: Perspectives on Consequences and Policy Responses." *Climatic Change* 15: 51–73.

Davidson, B. R. 1981. *European Farming in Australia: An Economic History of Australian Farming.* Amsterdam: Elsevier Scientific.

Davis, J. S., P. A. Oram, and J. G. Ryan. 1987. *Assessment of Agricultural Research Priorities: An International Perspective.* Canberra: Australian Centre for International Agricultural Research.

Dixon, P. B., B. R. Parmenter, J. Sutton, and D. P. Vincent. 1982. *ORANI: A Multisectoral Model of the Australian Economy,* Amsterdam: North-Holland.

Dixon, P. B., M. Horridge, and D. T. Johnson. In press. "ORANI Projections for the Australian Economy for 1988/9 to 2019/20 with Special Reference to the Land Freight Industry." *Empirical Economics.*

Easterling, W. E., M. L. Parry, and P. Crosson, P. 1989. "Adapting Future Agriculture to Changes in Climate," in N. J. Rosenberg, W. E. Easterling, P. R. Crosson, and J. Darmstadter, eds., *Greenhouse Warming: Abatement and Adaptation.* Pp. 91–104. Washington, D.C.: Resources for the Future.

Hayami, Y., and V. W. Ruttan. 1985. *Agricultural Development: An International Perspective*, revised and expanded edition. Baltimore: Johns Hopkins University Press.

Hobbs, J., J. R. Anderson, J. L. Dillon, and H. Harris. 1988. "The Effects of Climatic Variations on Agriculture in the Australian Wheatbelt," in M. L. Parry, T. R. Carter, and N. T. Konijn, eds., *The Impact of Climatic Variations on Agriculture. Volume 2: Assessments in Semi-Arid Regions*. Pp. 665–57. Dordrecht, The Netherlands: Kluwer Academic.

Horridge, M., D. Pearce, and A. Walker. 1990. "World Agricultural Trade Reform: Implications for Australia." *Economic Record* 66: 235–48.

Intergovernmental Panel on Climate Change (IPCC), Working Group II (Impacts). 1990. "Section A: Report on Agriculture and Forestry" (draft).

Landsberg, J. J. 1989. "The Greenhouse Effect: Issues and Directions for Australia; An Assessment and Policy Position Statement by CSIRO." Canberra: Commonwealth Scientific and Industrial Research Organisation (CSIRO). (Occasional Paper 4.)

Ministry for Planning and Environment. 1989. "The Greenhouse Challenge: The Victorian Government's Response. A Draft Strategy for Public Comment." Melbourne: MPE.

NSW (New South Wales) Agriculture & Fisheries. 1990. *Scenario Development Exercise*. Sydney: Roneo.

Nix, H. 1990. "Global Change—a Southern Hemisphere Perspective,", Global Change and the Southwest Pacific, 59th Australian and New Zealand Association for the Advancement of Science (ANZAAS) Congress, Hobart, Feb. 14–16.

Nulsen, R. A. 1989. "Agriculture in South-western Australia in a Greenhouse Climate," in *Proceedings of the Fifth Agronomy Conference*. Pp. 304–11. (Conference held in 1990.) Parkville: Australian Society of Agronomy.

Parry, M. L., and T. R. Carter. 1988. "The Assessment of the Effects of Climatic Variations on Agriculture: Aims, Methods and Summary of Results," in M. L. Parry, T. R. Carter, and N. T. Konijn, eds., *The Impact of Climatic Variations on Agriculture. Volume 1: Assessments in Cool Temperate and Cold Regions*, Pp. Pp.11–95. Dordrecht, The Netherlands: Kluwer Academic.

Parry, M. L., and T. R. Carter. 1989. "An Assessment of the Effects of Climatic Change on Agriculture." *Climatic Change* 15: 95–116.

Parry, M. L., J. H. Porter, and T. R. Carter. 1990. "Climatic Change and its Implications for Agriculture." *Outlook on Agriculture* 19: 9–15.

Pittock, A. B. 1989. "The Greenhouse Effect, Regional Climate Change and Australian Agriculture," in *Proceedings of the Fifth Agronomy Conference, Australian Society of Agronomy*. Pp. 289–303. (Conference held in 1990.) Parkville: Australian Society of Agronomy.

Randall, A. 1987. *Resource Economics: An Economic Approach to Natural Resource and Environmental Policy*. 2nd ed. New York: John Wiley & Sons.

Ruttan, V. W. 1978. "Induced Institutional Change," in Binswanger, H. P., V. W. Ruttan, et al., *Induced Innovation: Technology, Institutions, and Development*. Pp. 327–57. Baltimore: Johns Hopkins University Press.

Sutherst, R. W. 1990. "Impact of Climate Change on Pests and Diseases in Australia." *Search* 21: 230–32.

Walker, B. H., M. D. Young, J. S. Parslow et al. 1989. "Global Climate Change and Australia: Effects on Renewable Natural Resources," First Meeting, Prime Minister's Science Council, Canberra, Oct. 6.

World Commission on Environment and Development (WCED). 1987. *Our Common Future*. London: Oxford University Press.

18

Global Warming and Mexican Agriculture: Some Preliminary Results

Diana M. Liverman[1]

Introduction

This paper discusses some of the potential impacts of global warming on the agriculture of Mexico. The discussion is based on the results of several general circulation models (GCMs), which project what climate may be like if CO_2 levels in the atmosphere double, and uses their output to estimate changes in evaporation and crop yields at selected sites in Mexico. Most studies that use GCMs to assess the agricultural effects of global warming have focused on developed rather than developing countries. Although climate models project greater temperature increases in higher than in lower latitudes, tropical regions are nevertheless expected to experience significant temperature increases. Moreover, developing countries may be relatively more vulnerable to any climatic shift because of their economic situation and great reliance on both rain-fed and irrigated agriculture (Jodha, 1989; Gleick, 1989).

The crop-modeling work discussed in this paper will contribute to the U.S. Environmental Protection Agency (EPA) study on the international agricultural impacts of climate change. The EPA study uses GCM output to project the impact of changes in climate on crop yields at sites in 22

[1] From the Department of Geography and the Earth System Science Center, Pennsylvania State University, University Park. I would like to thank Karen O'Brien, Leticia Menchaca, Maxx Dilley, and Greg Edmeades for their help with the research for this study. The work with the CERES model has been partly supported by the International Climate Change Impacts Project at the U.S. Environmental Protection Agency, with particular advice from Cynthia Rosenzweig and Bruce Curry.

countries by using the crop models of the International Benchmark Sites Network for Agrotechnology Transfer (IBSNAT).

GCM runs with doubled CO_2 ($2 \times CO_2$) indicate significant increases in temperature and changes in precipitation in Mexico. Mexico is the 13th largest contributor to global greenhouse gases, producing 1.4% of the total net emissions (World Resources Institute, 1990). For Mexico, continually striving to support a growing population with an agricultural system that relies on relatively low and variable rainfall, any warmer, drier conditions could bring nutritional and economic disaster. More than one-third of Mexico's rapidly growing population works in agriculture, a sector whose prosperity is critical to the nation's debt-burdened economy. Although only one-fifth of Mexico's cropland is irrigated, this area accounts for half the value of the country's agricultural production, including many export crops. Many irrigation districts rely on small reservoirs or wells, which deplete rapidly in dry years. The remaining rain-fed cropland supports many subsistence farmers and provides much of the domestic food supply. Frequent droughts already reduce harvests and increase hunger and poverty in much of Mexico (Liverman, 1990).

There is a significant geographic mismatch between water and population in Mexico (Enge and Whiteford, 1989). Seven percent of the land, lying in the extreme southeast of the country (states such as Veracruz and Tabasco), receives 40% of the rainfall (Figure 18.1). Only 12% of the nation's water is on the central plateau (states such as Puebla, Mexico, and Guanajuato), where 60% of the population and 51% of the cropland are located. In arid northwest Mexico (Sonora and Sinaloa), much of the agriculture depends on irrigation development. Water-supply infrastructure has tended to lag behind rapid urban and industrial development in regions around Mexico City and along the U.S. border. Cities, industry, and hydroelectric power compete with agriculture, which consumes more than 80% of total water supplies.

How might global warming alter climate conditions in different parts of Mexico? Mexico lies between latitudes 15° and 32° north, with a climate influenced by three seasonally shifting features of the general circulation: the westerly winds, which bring winter rain to the extreme northwest of Mexico; the subtropical high-pressure belt, which brings stable, dry conditions to most of the country in winter and to the north in summer; and the intertropical convergence zone and trade winds, which bring substantial rains as they move north in the summer. In winter, cold polar air can surge southward over Mexico in "*nortes*," bringing rain and damaging frost. In midsummer, high pressure sometimes develops over the central plateau with a disruption of easterly flow, bringing a period of drought conditions called the "*canicula*" (Garcia, 1965; Metcalfe, 1987; Mosino, 1975).

We have obtained results from five major GCMs for the grid squares covering Mexico. The models are as follows:

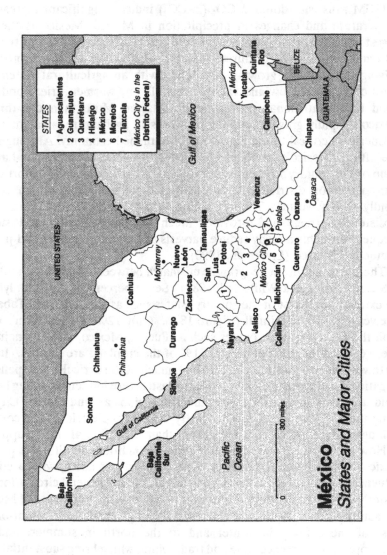

FIGURE 18.1 Map of Mexico showing states.

1. The Geophysical Fluid Dynamics Laboratory (GFDL) model developed at Princeton and described by Manabe and Wetherald (1987). The results used in this study are from the "Q" run, which includes variable clouds and has a 7.5°-by-4.44° geographic grid.
2. The Goddard Institute for Space Studies (GISS, or Goddard) model described by Hansen et al. (1983). We use the "Model 1" equilibrium run results for a 100°-by-7.83° grid.
3. The National Center for Atmospheric Research (NCAR) model described by Washington and Meehl (1986). We obtained the mixed-layer ocean version and seasonal, rather than monthly, results on a 7.5°-by-4.44° grid.
4. The Oregon State University (OSU, or Oregon) model discussed by Schlesinger and Zhao (1989) with a mixed-layer ocean and a 5°-by-4° grid.
5. The United Kingdom Meteorological Office (UKMO, or British) model described by Wilson and Mitchell (1987) with a 7.5°-by-5° grid.

The models vary in their assumptions about clouds, ice feedbacks, and oceans and in their characterization of topography, geography, and soil depth, among other factors. Despite these differences, there are many similarities in the models' structures and the ways complex processes are represented by statistical relationships (i.e., parameterizations). The model characteristics and results have been compared by Bach et al. (1985), Schlesinger and Mitchell (1987), and Jenne (unpublished observations).

GCM Ability to Replicate Current Climate

Our research shows that the models perform rather poorly in simulating the current climate of Mexico. Comparisons between model results for "current," or "1×CO$_2$," concentrations of CO$_2$ ($\approx$300 μmol/mol, or $\approx$300 ppm) and observed climates are complicated by differences in, and the large size of, the spatial grids of the models (Figure 18.2), as well as the uneven distribution of meteorological stations. One of our methods of comparison uses individual meteorological stations, comparing the most recent climatic normals (e.g., 1951–1980 average temperature and total precipitation) to the annual average temperature and total precipitation predicted by the 1×CO$_2$ simulation of the models for the grid square in which the station is located. Table 18.1 shows the summary results of such a comparison for a representative station, Puebla, located in an important corn-growing region in the central highlands of Mexico.

All models, except for the UKMO model, overestimate temperatures at Puebla—by almost 10 °C in the case of the OSU model. This error probably results from the lack of topographical detail in the models, because Puebla's

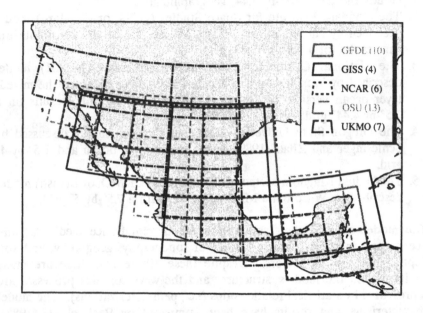

FIGURE 18.2 General circulation model (GCM) grid coverage for Mexico.

TABLE 18.1 Comparison of Observed, Rand, and $1\times CO_2$ Climate at Puebla, Mexico[a]

| | Observed | Projected | | | | | |
		Rand	GFDL	GISS	NCAR	OSU	UKMO
Average temperature (°C)	16.7	20.3	19.6	22.1	21.9	26.2	16.7
Annual precipitation (mm)	833	1016	3198	1710	444	467	2462

[a] Rand, Rand Corporation 5°×4° grid average of actual measurements; GFDL, Geophysical Fluid Dynamics Laboratory model; GISS, Goddard Institute for Space Studies model; NCAR, National Center for Atmospheric Research model; OSU, Oregon State University model; UKMO, U.K. Meteorological Office model.

cooler observed temperatures result from its 2166-m altitude. The UKMO model may have more accurate temperatures because it sets the Puebla grid square at 1351 m, whereas the Goddard and GFDL models have much lower elevations at 785 m and 616 m, respectively, with correspondingly warmer temperatures. For rainfall, the NCAR and OSU models only produce about half the observed annual amount at Puebla, whereas the other models overestimate rainfall by as much as four times. Monthly comparisons are shown in Figure 18.3.

Individual meteorological-station conditions may not be representative of the regional climates that correspond to the scale of the GCM grids. The heterogeneity of the Mexican environment means that meteorological observations within a model grid square may vary widely. To investigate this, we use regional averages of observed climates developed by the Rand Corporation (Santa Monica, Calif.), in which temperature and precipitation are averaged for meteorological stations within a 5°-by-4° geographic grid. As can be seen from Table 18.1, the Rand conditions are a little wetter and slightly warmer than the individual station value at Puebla. Figure 18.4 shows model results compared with Rand climatology for January and July temperatures and annual precipitation, across an east-west transect at latitude 30° north. Observed conditions in January show a general increase in temperatures from east to west, but with a decline around 105° west longitude, which represents the cooling influence of higher-elevation stations. The NCAR and OSU models most closely follow this climatology; the UKMO and GFDL models produce temperatures that are too low, even in the highland region. The Rand annual rainfall declines from east to west across northern Mexico, from the more humid Gulf Coast to the deserts along the Pacific. There is a slight increase in observed rainfall as air starts to rise over the highlands at longitude 110° west. The Goddard and GFDL models show rainfall decreasing from east to west, but the OSU model produces an opposite trend. The UKMO model seems to best replicate the orographic rainfall increase over north-central Mexico.

These results demonstrate some of the inadequacies in the ability of climate models to reproduce observed climates and can lead to distrust of the changes projected by the models, particularly for precipitation. Further caution in using model results for climate-impact studies is indicated in the chapter on climate variability in a recent EPA report (U.S. EPA, 1989), which suggests that the models are unable to reproduce the variability of observed climate.

Projected Climate Changes in Mexico

Despite the models' problems in reproducing observed climates, it may still be possible to use their output to construct relative estimates of how

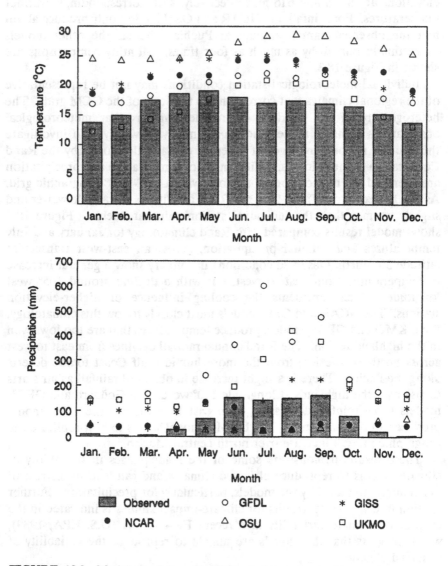

FIGURE 18.3 Model simulation (1×CO₂) of observed monthly temperature and precipitation in Puebla, Mexico, based on climate-model estimates. (NCAR, National Center for Atmospheric Research model; GFDL, Geophysical Fluid Dynamics Laboratory model; OSU, Oregon State University model; GISS, Goddard Institute for Space Studies model; UKMO, U.K. Meteorological Office model.)

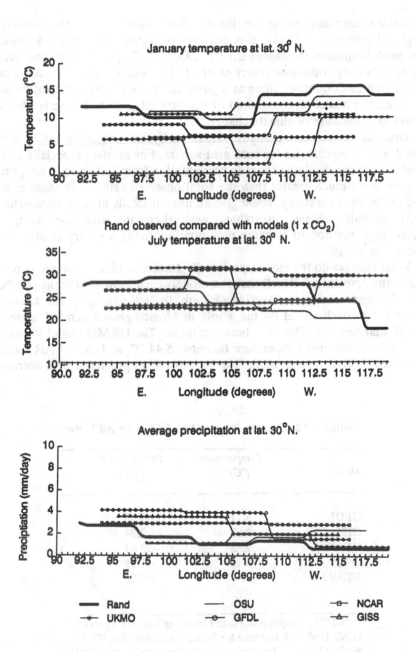

FIGURE 18.4 Comparison of model simulations of current climate (1×CO₂) with the Rand grid averages of observed climate at latitude 30° north, and average precipitation. (*See* Figure 18.3 caption for definition of terms.)

global warming may change the climate. Some authors claim that although regional estimates of $1 \times CO_2$ climates may be unrealistic, the way in which the model estimates the change from $1 \times CO_2$ to $2 \times CO_2$ conditions is assumed to be relatively consistent (Parry et al., 1988; Cohen, 1986). Rather than relying on $2 \times CO_2$ results alone as a guide to future climates, the difference between the model representation of the present and the future is used to generate a measure of climate change.

Thus, the temperature change is calculated by subtracting the $1 \times CO_2$ from the $2 \times CO_2$ temperatures for each grid square. For rainfall, especially in an arid region, the absolute change in precipitation from $1 \times CO_2$ to $2 \times CO_2$ runs is sometimes much greater than the total observed rainfall. In these cases, use of the absolute change could give an unrealistically high, or even below-zero, rainfall. Many researchers have therefore used the ratio, or percentage, method to estimate precipitation changes (Parry et al., 1988; Adams et al., 1990).

What changes do the models project for Mexico, and do these projections show any agreement? Recall that the models each have a different number of grid squares covering Mexico (Figure 18.2). Averaging the changes projected by each model for the whole of Mexico gives a general sense of the similarities and differences between them. The UKMO model forecasts the greatest regional temperature increase, 5.44 °C, and the NCAR model the least, 2.38 °C (Table 18.2). The NCAR model predicts a 23% decrease

TABLE 18.2 Changes Projected by GCMs for All Mexico

Model	Temperature (°C)	Precipitation (%)
GFDL	3.11	−1.75
GISS	3.92	2.74
NCAR	2.38	−23.01
OSU	3.15	−1.05
UKMO	5.44	−0.09

[a]GFDL, Geophysical Fluid Dynamics Laboratory model; GISS, Goddard Institute for Space Studies model; NCAR, National Center for Atmospheric Research model; OSU, Oregon State University model; UKMO, U.K. Meteorological Office model.

in rainfall, and the GISS model, a 3% increase. The size and, in the case of rainfall, the direction of change, varies depending on the spatial and temporal scale of analysis. When we look at average changes for the whole of Mexico but seasonally and monthly rather than annually, we see much greater differences between the models (Figure 18.5). For example, we find that the OSU model has its greatest temperature increases in the summer, whereas the GISS and NCAR models have their greatest increases in the fall and winter. The differences between the model predictions for precipitation are very dramatic seasonally. For example, in the NCAR model, fall rainfall decreases by almost 50% with doubled CO_2, and that for the OSU model increases by 15%. Monthly comparisons show even more extreme differences in the size and direction of projected changes.

We chose to use the monthly results for individual grid squares to investigate climate change in Mexico. This choice highlights the differences between the models and maximizes the representation of seasonal variability. Daily data, not available for this study, would permit an even better examination and representation of variability within and between the models. We also chose to use the results from all five models in generating climate-change scenarios for Mexico to capture some of the uncertainties in the projections for global warming.

The most common approach to analyzing the regional impacts of global warming is to take the type of changes estimated by the models, as discussed above, and modify current climates to represent doubled-CO_2 conditions. Temperature changes are added or subtracted from observed conditions, and observed precipitation is multiplied by the estimated percentage change.

Figures 18.6 and 18.7 show how climate may change at some meteorological stations in Mexico. At all stations, all of the models project increases in temperature in every month of the year, resulting in more months of extremely high summer temperatures. For example, mean monthly temperatures of ≈25 °C in Chihuahua could increase to almost 30 °C, with some models predicting at least 1 month above 35 °C. In Ciudad Obregon, which already experiences summer temperatures near 30 °C, several months could approach 35 °C. These temperature increases in the already hot summers of northern Mexico are likely to bring great stress to humans, plants, and animals, especially if these monthly averages imply even higher daily extremes. On the other hand, warmer temperatures may extend the growing season in areas at higher elevations, such as Chihuahua, which sometimes has cold temperatures and frosts in winter months. In Ciudad Obregon, located in one of the most important irrigated agricultural regions of Mexico, the GISS and UKMO models indicate that rainfall may increase in July and August, whereas other models bring lower rainfall throughout the year. Chihuahua's climate scenarios follow a similar pattern, but with distinct drops in OSU and GFDL estimates for summer rainfall.

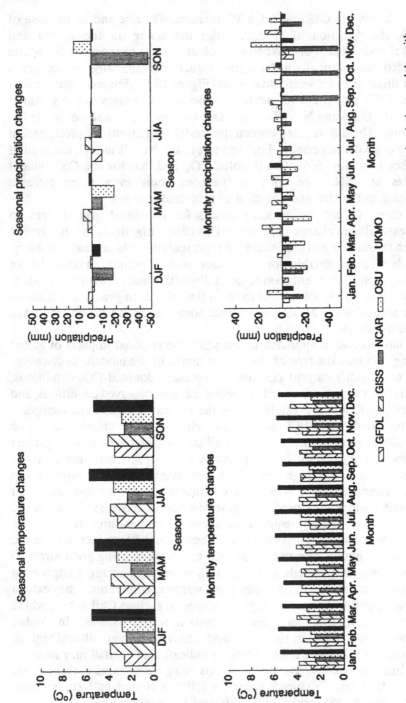

FIGURE 18.5 Seasonal and monthly changes in temperature and rainfall projected by five general circulation models, with a doubling of CO_2, averaged for the whole of Mexico. (DJF, Dec, Jan, and Feb; MAM, Mar, Apr., and May; JJA, June, July, and Aug.; SON, Sept., Oct., and Nov.)

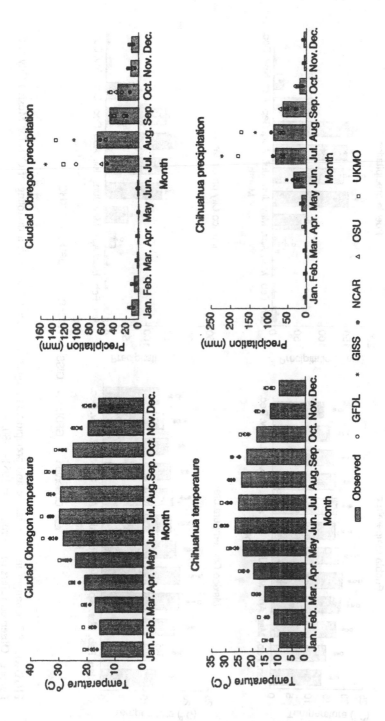

FIGURE 18.6 Changes in temperature and precipitation projected by five GCMs with a doubling of CO_2 for Ciudad Obregon and Chihuahua. Observed refers to averages for 1951–1980.

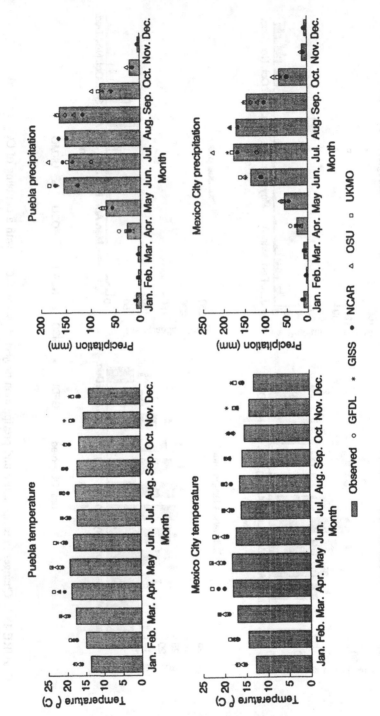

FIGURE 18.7 Changes in temperature and precipitation projected by five GCMs with a doubling of CO_2 for Mexico City and Puebla. Observed refers to averages for 1951–1980.

In central Mexico, temperature increases are not quite as high as in the north. For example, Mexico City's current winter temperatures of ≈12 °C could increase to as high as 18 °C. Warmer temperatures could reduce frost risks in the highland agricultural regions around Mexico City and Puebla.

Water Availability

For Mexican agriculture, the most important effects of these climate changes would be changes in the length of the growing season and in moisture availability. Assessing changes in moisture availability, particularly in locations in which both temperature and rainfall are projected to increase, requires the analysis of evaporation. We need to know how any increase in rainfall or length of growing season is likely to interact with increased evaporation associated with higher temperatures. Most of the models calculate evaporation and soil moisture internally, but the surface energy and water budgets of the models tend to be rather simple. For example, all but the GISS model represent soil water by using a 15-cm-deep "bucket," which, when full, produces runoff. Moisture is lost through evaporation, at slower rates as the bucket empties (Kellogg and Zhao, 1988). A problem arises in dry regions in which the 1×CO_2 simulation of soil moisture is low. Even if potential evaporation in the model rises with the higher temperatures of 2×CO_2 conditions, there is little or no water to evaporate in the bucket, and, as a result, the estimated change in soil moisture will be very small. In reality, deeper soils can often contain more water, and soil moisture could be reduced significantly as temperatures rise.

Information obtained for the water balance of the GISS and OSU models suggests that potential evaporation will increase in all months because higher air temperatures increase the atmospheric demand for moisture in air that is relatively dry. The GISS model internally estimates a 15% increase in potential evaporation for the Chihuahua grid square and a 2% increase in Puebla. The OSU model projects a 12% decrease in potential evaporation at Chihuahua and a 6% increase in Puebla.

Because of the limitations of modeling soil moisture, we have also used model projections of precipitation, temperature, and solar-radiation changes to estimate exogenous evaporation and moisture availability by the Thornthwaite and Penman methods.[2] Table 18.3 shows the exogenously estimated changes in Penman potential evaporation for the stations of Puebla in central Mexico and Ciudad Obregon in the northwest irrigated region. Current Penman potential evaporation totals ≈1528 mm/year under current (1951–1980) climate in Puebla. With global warming, this could

---

[2] These methods are discussed in Rosenberg (1974).

TABLE 18.3 Changes in Potential Evaporation and Water Availability[a]

	Observed (mm)	Projected				
		GFDL	GISS	NCAR	OSU	UKMO
Ciudad Obregon						
PE	2437	19%	20%	8%	13%	27%
P − PE	−2181	21%	16%	12%	15%	23%
Puebla						
PE	1528	14%	12%	7%	10%	16%
P − PE	−694	45%	33%	23%	18%	30%

[a]PE, potential evaporation (Penman); P − PE, precipitation minus potential evaporation; GFDL, Geophysical Fluid Dynamics Laboratory model; GISS, Goddard Institute for Space Studies model; NCAR, National Center for Atmospheric Research model; OSU, Oregon State University model; UKMO, U.K. Meteorological Office model.

increase from 7% to 16%, depending on the climate model. These results have serious implications for water resources and irrigation in central Mexico, because they imply increased losses from open water surfaces such as reservoirs. In Ciudad Obregon, current potential evaporation in this desertlike environment is 2438 mm/year. The higher temperatures associated with global warming could increase this evaporation 8–27%.

The relationship between increased potential evaporation and changes in rainfall can be approximated over a year by subtracting total annual potential evaporation from annual rainfall to produce an estimate of moisture surplus or deficit. Under current climate, such a calculation for Puebla produces a moisture deficit of 694 mm. That is, potential evaporation is about double precipitation. Global warming could increase this deficit 17–45%, depending on the climate model. A 45% moisture deficit would occur with the GFDL scenario, because this model combines a temperature increase and a large precipitation decrease at Puebla. Although the UKMO model projects the largest potential evaporation increase for Puebla, an increase in precipitation partly offsets this, for a deficit increase of 30%. Similar scenarios occur in the northwest, with moisture deficits increasing in Ciudad Obregon between 15% and 23%. This reduction in water availability would harm both rain-fed and irrigated agriculture in the region. The implications of possible errors in the model projections for precipitation and solar radiation have been investigated through a sensitivity analysis of the Penman calculations. For example, we find that it would require significant decreases in wind speeds (-50%) or solar radiation (-20%) to prevent an increase in potential evaporation with higher temperatures. Moisture deficits could only be reduced by much larger increases in rainfall than currently projected by the models. Only if the climate models are greatly overestimating temperature increases, or if they are very wrong about rainfall and solar radiation, does the future seem optimistic for water availability if global warming occurs in Mexico.

Global Warming and Corn Yields

The possible impacts of global warming on Mexican crop production are being explored with the CERES-MAIZE crop-yield model as part of the EPA study on the international agricultural impacts of global warming. The CERES model uses daily weather data to estimate the effects of climate on yields of corn under water-management regimes ranging from rain-fed to fully irrigated. The model is physiologically based, simulating the influence of solar radiation, degree days, nutrients, and water availability on the development of corn through major phenological stages (Jones and Kiniry, 1986; Adams et al., 1990). It has been suggested that many crops will benefit from higher levels of atmospheric CO_2 and that higher yields may result. The CERES model has been modified to include the direct effects of elevated CO_2 on crop growth and water-use efficiency, but does not account for losses from, and changes in, pests and diseases.

Yields of corn, the major staple in Mexico, have been simulated at several sites in Mexico. The CERES model has been validated for sites near Cuernavaca (in a highland valley south of Mexico City) and Veracruz (on the Gulf Coast) with agricultural experiment data provided by the International Center for Wheat and Maize Improvement (CIMMYT). The model has been run with observed weather data to simulate rain-fed and irrigated yields of corn under typical soil and management conditions. In rural Mexico, corn is typically grown by poorer farmers who cannot afford chemical fertilizers. Thus, we use only 50 kg of nitrogen fertilizer at planting in the model runs, and assume soils of relatively low fertility and moisture content. The planting density is three plants per square meter.

Climate-change scenarios are generated with results of the GFDL, GISS, and UKMO climate models for likely changes in temperature, precipitation, and solar radiation with a doubling of CO_2. Rain-fed and irrigated yields are then simulated under these changed climates with higher concentrations of CO_2 (555 μmol/mol, or 555 ppm) and compared with the yields estimated for the years of observed weather. The results of these experiments, which initially assumed no alterations in planting dates or other input conditions as the climate changed, are shown in Table 18.4.

At the Cuernavaca site, average yields of rain-fed corn for the 1974–1989 period are estimated at 0.40 kg/m² [4.0 tons/hectare (T/ha)]. All rain-fed climate-change scenarios result in drops in corn yields at Cuernavaca, ranging from a 20% drop for the GFDL model to a 60% drop for the UKMO model. Under observed conditions, only one year, 1989, produces yields below 0.1 kg/m² (1 T/ha). The GISS and UKMO models increase this frequency to 7 years, and GFDL, to 2 years. These low yields occur despite overall rainfall increases projected by some models, because higher temperatures create elevated evaporation early in the growing seasons, and in some cases high rainfall leaches nutrients out of the soil.

Irrigated yields for 1974–1989 at Cuernavaca average 0.37 kg/m² (3.7 T/ha), with 207 mm of irrigation water applied during the growing season. Irrigated yields do not drop as much under changed climates, ranging from a drop of 16% in the UKMO case to 24% in the GISS case. Irrigated scenarios for global warming produce less-variable yields, but do not exceed, on average, the current yields. Demand for irrigation water increases by

TABLE 18.4 Changes in Corn Yields[a]

	Observed	GISS	GFDL	UKMO
Cuernavaca (Tlaztizapan)				
Rain-fed yield				
(kg/m²)	0.40	0.30	0.32	0.16
(T/ha)	4.0	3.0	3.2	1.6
Irrigated yield				
(kg/m²)	0.37	0.28	0.30	0.31
(T/ha)	3.7	2.8	3.0	3.1
Irrigation (mm)	207	212	213	191
Veracruz (Poza Rica)				
Rain-fed yield				
(kg/m²)	0.31	0.30	0.27	0.23
(T/ha)	3.1	3.0	2.7	2.3
Irrigated yield				
(kg/m²)	0.31	0.28	0.29	0.28
(T/ha)	3.1	2.8	2.9	2.8
Irrigation (mm)	192	208	215	188

[a]GISS, Goddard Institute for Space Studies model; GFDL, Geophysical Fluid Dynamics Laboratory model; UKMO, U.K. Meteorological Office model.

≈3% compared with current climate, except for the UKMO scenario, in which the demand decreases. Projections for moisture deficits at Puebla, ≈150 km (≈100 miles) from Cuernavaca, indicated by the Penman calculations above, suggest that this extra irrigation water may not be available in a warmer climate.

Recent (1973-1979) rain-fed yields of corn at the site near Veracruz average 0.31 kg/m² (3.1 T/ha) and decrease 13% with the GFDL model, 26% with the UKMO model, and 3% with the GISS model. Irrigated yields currently average 0.31 kg/m² (3.1 T/ha), only decreasing by 10% for the GISS and UKMO models, and 6% for the GFDL model. Irrigation needs increase by 2-12%, depending on the model.

These yield declines and crop failures projected for corn, a crop important to Mexico's food self-sufficiency and for peasant livelihoods, are extremely serious. However, sensitivity analyses indicate the importance of some of the current uncertainties in estimating the impacts of future climatic changes. The results reported above assume no change in the average date of planting, with crops planted on May 20 (Cuernavaca) and June 20 (Veracruz) in both observed and GCM conditions. These are average planting dates under current climate, which allows farmers to balance the risk of late spring rains and early fall frosts. We are currently investigating whether earlier and later sowing dates may offset the negative impacts of a warmer climate. We have also left the crop variety unchanged in these initial experiments and will be seeing if different varieties may adapt better to global warming. It is also possible to design hypothetical genetic coefficients within the CERES model that maintain high yields under global warming, but it is not clear that these coefficients are biologically possible. The sensitivity of the model results to these factors will indicate possibilities for adjusting to global warming through altering the dates of planting and the genetic type or characteristics of the seed.

The amount of fertilizer and irrigation is critically important to the corn yields predicted by the CERES models. The experiments above all assume relatively low fertilizer use because many Mexican producers can only afford to use 50-100 kg of nitrogen fertilizer at planting. If more fertilizer becomes available to more farmers as the climate changes, then some of the yield reductions might be offset. Only with adequate water and fertilization can the direct effects of CO_2 be of significant advantage to C3 plants such as corn. When we run observed and model scenarios with no water or nutrient limitations, we find that the GISS scenario produces 0.39 kg/m² (3.9 T/ha) in Cuernavaca [compared with 0.28 kg/m² (2.8 T/ha) with nutrient-limited, irrigated observed climate], the GFDL model, 0.40 kg/m² (4.0 T/ha), and the UKMO model, 0.39 kg/m² (3.9 T/ha). However, given the environmental and economic constraints and trends in agricultural inputs in Mexico, unlimited water and nutrients are extremely unlikely. The experiment illustrates, however, the sensitivity of corn yields to assumptions about future resource availability.

For export-oriented agriculture and price incentives to producers, it will be Mexico's comparative advantages and disadvantages in agricultural production in a warmer world that may determine the national and regional economic impacts of climate changes. We have not yet undertaken any detailed estimates on other crops or the areas of land that may be threatened by sea-level rises. Many low-lying coastal agricultural areas in Mexico would be threatened by sea-level rises. In some cases, the threat might be of direct flooding of land or of increased storm damages (e.g., along the Gulf Coast), but in others, such as Sonora and Sinaloa, the threat would be to irrigation systems and soil fertility through rising water tables and salinization.

Conclusions

The principal conclusion of our analysis of what climate models can tell us about the regional impacts of global warming is that Mexico is likely to be warmer and drier. Whichever model we use, it seems that potential evaporation will increase, and moisture availability will decrease, even in those cases in which the model projects an increase in precipitation. Sensitivity analyses of our evaporation and moisture-deficit calculations indicate that water availability could increase only if the model results were to produce much higher rainfall and relative humidities, or significantly less solar radiation and wind. Such changes will be possible, of course, as the modeling of clouds and synoptic conditions improves, but at present it seems that a moisture decrease is more likely.

Again, despite differences between the climate-model results, the general direction of changes in Mexican corn yields with global warming is a decrease, whatever the model, and whether irrigation is used or not. Sensitivity analysis indicates that decreases in crop yields will be severe under global warming unless irrigation expands, fertilizer use increases, or new varieties are developed.

References

Adams, R. M., C. Rosenzweig, R. M. Peart et al. 1990. "Global Climate Change and U.S. Agriculture." *Nature* 345: 219–24.

Bach W., H. J. Jung, and H. Knottenburg. 1985. *Modeling the Influence of Carbon Dioxide on the Global and Regional Climate: Methodology and Results.* Paderborn, Germany: Ferdinand Schoeningh.

Cohen, S. J. 1986. "Impacts of CO_2-Induced Climatic Change on Water Resources in the Great Lakes Basin." *Climatic Change* 8: 135–53.

Enge, K. I., and S. Whiteford. 1989. *The Keepers of Water and Earth: Mexican Rural Social Organization and Irrigation.* Austin: University of Texas Press.

Garcia, E. de Miranda. 1965. *Distribucion de Precipitacion en la Republica Mexicana.* Mexico, D.F.: Instituto de Geografia de Universidad Nacional Autonoma Mexicana (UNAM).

Gleick, P. H. 1988. "The Effects of Future Climatic Changes on International Water Resources: The Colorado River, the United States, and Mexico." *Policy Sciences* 21: 23–39.

Gleick, P. H. 1989. "Climate Change and International Politics: Problems Facing Developing Countries." *Ambio* 18(6): 333–9.

Hansen, J., G. Russell, D. Rind et al. 1983. "Efficient Three-Dimensional Global Models for Climatic Studies: Models I and II." *Monthly Weather Review* 3(4): 609–62.

Jenne, R. Unpublished notes. "Data from Climate Models." Boulder, Colo.: National Center for Atmospheric Research.

Jodha, N. S. 1989. "Potential Strategies for Adapting to Greenhouse Warming: Perspectives from the Developing World," in N. Rosenberg, W. E. Easterling, and P. Crosson, eds., *Greenhouse Warming: Abatement or Adaptation.* Pp. 147–58. Washington, D.C.: Resources for the Future.

Jones, C. A., and J. R. Kiniry. 1986. *CERES-Maize: A Simulation Model of Maize Growth and Development.* College Station: Texas A&M Press.

Kellogg, W. W., and Z. Zhao. 1988. "Sensitivity of Soil Moisture to Doubling of Carbon Dioxide in Climate Experiments. Part 1: North America." *Journal of Climate* 1: 348–66.

Liverman, D. M. 1990. "Vulnerability to Drought in Mexico: The Cases of Sonora and Puebla in 1970." *Annals of the Association of American Geographers* 80(1): 49–72.

Manabe, S., and R. T. Wetherald. 1987. "Large Scale Changes in Soil Wetness Induced by an Increase in Atmospheric Carbon Dioxide." *Journal of the Atmospheric Sciences* 44: 1211–35.

Meehl, G. A., and W. M. Washington. 1988. "A Comparison of Soil Moisture Sensitivity in Two Global Climate Models." *Journal of the Atmospheric Sciences* 45(9): 1476–92.

Metcalfe, S. E. 1987. "Historical Data and Climate Change in Mexico: A Review." *The Geographical Journal* 153(2): 211–22.

Mosino, P. A. 1975. "Los Climas de la Republica Mexicana," in Z. Cserna, ed., *El Escenario Geografico.* Pp. 57–171. Mexico, D.F.: Instituto de Geografica de Universidad Nacional Autonoma Mexicana (UNAM).

Parry, M. L., T. R. Carter, and N. T. Konjin, eds. 1988. *The Impact of Climatic Variations on Agriculture.* Volumes 1 and 2. Dordrecht, The Netherlands: Kluwer.

Rosenberg, N. 1974. *Microclimate: The Biological Environment.* New York: John Wiley & Sons.

Schlesinger, M. E., and J. F. B. Mitchell. 1987. "Model Projections of the Equilibrium Response to Increased Carbon Dioxide." *Review of Geophysics* 25: 760–98.

Schlesinger, M. E., and Z. Zhao. 1989. "Seasonal Climatic Changes Induced by Doubled CO_2 as Simulated by the OSU Atmospheric GCM/Mixed-Layer Ocean Model." *Journal of Climate* 2: 459–95.

U.S. Environmental Protection Agency (U.S. EPA). 1989. *The Potential Effects of Global Climate Change on the United States*. Washington, D.C.: U.S. EPA.

Washington, W. M., and G. A. Meehl. 1986. "Climate Sensitivity Due to Increased CO_2: Experiments with a Coupled Atmosphere and Ocean GCM." *Climate Dynamics* 4: 1–38.

Wilson, C. A., and J. F. B. Mitchell. 1987. "A Doubled CO_2 Climate Sensitivity Experiment with a Global Climate Model Including a Simple Ocean." *Journal of Geophysical Research* 92(D11): 13315–43.

World Resources Institute. 1990. *World Resources 1990–91*. Oxford, England: Oxford University Press.

19

The Impact of Expected Climate Changes on Crop Yields: Estimates for Europe, the USSR, and North America Based on Paleoanalogue Scenarios

Gennady V. Menzhulin[1]

Introduction

The weather and climate exert a direct and considerable effect on one of the most important social activities—agriculture, especially farming. For example, American economists estimated mean annual losses in the United States in agricultural production from unfavorable weather conditions to be ≈$10billion, which is almost twice the total loss in all other sectors of the economy. Because territorial and temporal weather conditions in the agricultural areas of other countries, including the USSR, are more variable than in the United States, it can be concluded that in this case, similar estimates for the USSR are even higher. The impact of climatic change and weather variations on food production is greatest in many developing countries, where agriculture, even in the favorable years, cannot meet people's food needs (Pulwarty and Cohen, 1984).

The need to study the multifactor effects of present-day climatic changes on agriculture has been emphasized repeatedly in the reports of international and national organizations, in conferences, and in scientific publications (Clark, 1985; Parry, 1985; Parry et al., 1988a, 1988b). The World Climate Program and the United Nations Environmental Program (UNEP) have included, as one of their major, integral parts, the study of climatic effects on social activities, which emphasizes agricultural-climatology problems (WMO/ICSU/UNEP, 1981). The new program investigating these

---

[1] From the State Hydrological Institute, Leningrad, USSR.

problems, adopted recently as an important part of the Intergovernmental Panel for Climate Change (IPCC), is a continuation of the activity in previous programs.

International scientific programs that address effects of climate on social activities are very timely because consequences of the effect of expected future climate change on agriculture can influence world and regional food situations considerably. Assuming maximum efficiency, estimates of potential world agricultural production look hopeful (Buringh et al., 1975; Jensen, 1978). Asian, African, and South American countries have great possibilities for extending arable lands. However, the development of new crop-cultivation areas is associated with considerable expense and requires the introduction of modern agricultural technology. The economic situation in developing countries does not provide encouragement for the solution of these problems in the near future. Additional land resources in the developed countries of Europe and North America are estimated to be small.

The above addresses the difficulties encountered in trying to solve the present-day world-food problem. The fact that all of this concurs with expected drastic anthropogenic climatic changes is very troubling. To begin to solve this problem, the following questions should be answered: What agricultural effects have current climatic changes had, and what will be the effects of human-induced CO_2-concentration growth? How can possible harmful consequences be decreased? And, how can total economic effects from climatic changes be optimized by transforming crop-cultivation geography and introducing crops that are more tolerant to environmental stress?

These questions can be answered only by joint, coordinated investigations by specialists in climatology, agriculture, plant physiology, and economics (Chen, 1981; Kimball, 1985; Warrick et al., 1986). Because problems related to the agricultural consequences of global climate changes are large-scale and socially important, their global and national significance should be taken into account by those working on their solutions (Callaway et al., 1982; Carbon Dioxide Assessment Committee, 1983; Pulwarty and Cohen, 1984; Terjung et al., 1984; Rosenzweig, 1985; Decker et al., 1985; Land Evaluation Group, 1987). An important part in any combined interdisciplinary investigation of the future world food problem is solving the agrometeorological problem, i.e., estimating the yield effects of changed climate and associated changes in the environment. Such estimates should be incorporated into methods used for the planning of optimum development of agriculture.

Comprehensive investigations of this problem are well-organized and their direction is clear. During recent years, American scientists have been interested in the problem of the impact of anthropogenic climate change on agriculture. (*See*, for example, Sakamoto et al., 1980; Warrick and Riebsame, 1981; Rosenberg, 1982; Ausubel, 1983; Oram, 1985; and Parry and Carter, 1985.)

Among the USSR studies on this problem are the works carried out in the All-Union Research Institute of Agricultural Meteorology (Sirotenko and Boiko, 1980; Pavlova and Sirotenko, 1985). The Climate Changes Department of the State Hydrological Institute in Leningrad, which carries out studies within the framework of the USSR–U.S. Intergovernmental Agreement, pays a great deal of attention to the problems of agricultural consequences of modern climatic changes (Menzhulin, 1976, 1984; 2Menzhulin and Savvateev, 1980, 1981; Koval and Savvateev, 1982; Menzhulin et al., 1983; Koval et al., 1983; Nikolaev et al., 1985; Nikolaev, 1985; Savvateev, 1985; Menzhulin and Nikolaev, 1987; Menzhulin et al., 1987; Budyko and Israel, 1987). In this chapter, we analyze in detail the methodology and basic results of the above-mentioned works.

Updated forecasts of climatic regimes allow the determination of basic components of the regimes' agricultural consequences in the following ways. (1) The forecasts can be accompanied by the analysis of the present-day climate regime and the directions in which it has tended to change during recent years in the main regions of agricultural production. Such an analysis allows the revelation of the social importance of the problem and the estimation of the expected societal reaction to the climatic changes that are possible in the near future. (2) In the estimations of crop-yield response to climatic changes, account should be taken of physiological factors of increased CO_2 impact on photosynthesis and accompanying effects on plants. (3) One of the important characteristic features of future agricultural conditions, along with the potential mean level of crop yields, is the crop-yield year-to-year variability. In addition to possible changes in productivity, the investigations can incorporate estimated changes of variability indices as well. (4) Because of various climatic conditions in different regions, computations of changes in crop-yield indices should be based on detailed territorial scenarios of future climate. Such scenarios will help make forecasts more concrete, which is most important for optimizing agricultural consequences of the modern climatic changes.

Features of Agroclimate Conditions Over the Past Decades

For analyzing the effects of global warming on agriculture, data on crop-yield variations during the past 30 years and in 1910–1930 in the two largest midlatitude agricultural zones of the Northern Hemisphere, which are in the United States and the USSR, can be used to find two considerable global-warming phases during these periods. The first of them, due mainly to natural causes (i.e., increased atmospheric transparency), reached its maximum in the 1930s. The second one, caused mainly by anthropogenic factors (i.e., increases in CO_2 and trace-gas concentrations), has become noticeable since the mid-1970s. In both cases, mean global-temperature growth compared with the preindustrial period was ≈ 0.5 °C and even more in individual years. Data show that during recent years, mean global temperature has continued to climb, which allows us to state that the present-day global warming has no analogue in the preindustrial period of instrumental observations.

Figure 19.1 presents data on mean crop yields of cereals in the USSR and the United States, which are of interest for determining a combined estimate of the global-warming effect on crop productivity. It is clear that year-to-year crop-yield variability is determined by two main factors, technological level and weather variations. The simplest way to distinguish between the effects of these two factors is based on the information on greatest yields. The values in Figure 19.1 (the three upper curves) were calculated from that information. Even in the years of maximum crop yields, weather conditions were not beneficial to agriculture in every USSR and U.S. crop-growing zone. There are reasons, however, to believe that crop yields in the main agricultural regions are limited, to a large extent, by technological factors rather than by meteorological ones. Such a conclusion is confirmed by computations using the model for estimating crop-yield response to technological factors (Menzhulin et al., 1983; Menzhulin and Nikolaev, 1987). Results of these calculations lead us to conclude that crop yields increased during the past decades when weather conditions of some years were favorable.

Returning to the analysis of data represented in Figure 19.1, note the 25–30% decrease in grain yields in the 1930s during several consecutive years in the USSR. Such a drop was caused by meteorological factors, because main agricultural regions of the USSR experienced moisture stress in that period. Note also that this climate change had a global character—several extreme droughts were observed in the United States in the same years, which decreased mean cereal-crop production by approximately the same amount (Warrick, 1984).

In the postwar years in the USSR, atmospheric precipitation conditions that influence crop yields were comparatively favorable for a rather long time. Some decrease in wheat production in the USSR in the early and mid-1960s was apparently caused not by climatic factors, but by the development of vast new areas for growing cereal crops in the West (Virgin Lands), where wheat productivity was and still is noticeably lower than it is in regions with more favorable climatic conditions.

Data on wheat and corn yields in the North American agricultural region show that there was a noticeable tendency for crop yields to drop and for crop-yield variability to increase in the 1930s. Estimated total losses of

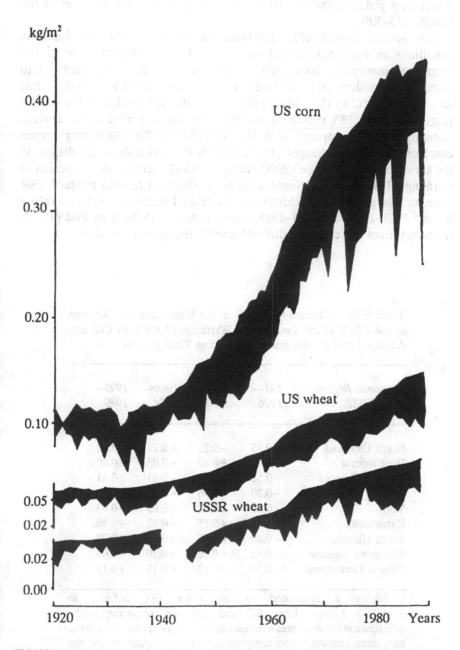

FIGURE 19.1 Cereal-crop yields and productivity trends. (The jagged lines at the bottom of the shaded areas are the average yields; the lines at the top of the shaded areas are the best yields.)

cereal-crop yields in the United States during these years were, as in the USSR, ≈25–30%.

The second considerable deterioration of atmospheric-precipitation conditions in main agricultural regions of the midlatitude zone of the Northern Hemisphere took place in the period after 1975. Taking into account the modern estimates, we see that mean global temperature had grown by ≈0.5 °C by the early 1980s. Then, its growth slowed for several years and, after 1985, increased sharply. The initial period of the current warming can be compared with the warming of the 1930s that caused considerable climatic changes. We can safely say that climatic conditions in the agricultural zone of the USSR changed quickly with the development of warming. The above-mentioned changes in different regions of the USSR were, to a large extent, synchronous. Table 19.1 contains the mean values for the 1961–1974 and 1975–1988 aridity index, as defined by Ped (1975); it characterizes the extent of drought during the growing season.

TABLE 19.1 Changes in Aridity in the Main Economic Regions of the USSR at the Two Global Warmings of the 20th Century, Averaged over Its Beginning and Ending Time Periods[a]

Economic Regions of the USSR	1914–1926	1927–1939	1960–1974	1975–1989
North Caucasus	−0.35	−0.25	+0.21	+0.62
West Siberia	−0.23	+0.30	−0.03	+0.40
Donetsk-Dnieper	−0.58	−0.19	−0.01	+0.41
Ural	−0.20	+0.36	−0.35	+0.46
Volga	−0.59	+0.03	−0.02	+0.03
Kazakhstan	−0.18	+0.15	+0.12	+0.40
South Ukraine	−0.42	+0.03	+0.02	+0.22
Southwest Ukraine	−0.81	+0.06	+0.10	+0.05
Central Chernozem	−0.51	−0.12	+0.18	+0.17

[a] Aridity is calculated as an index (S), defined as $S = \Delta T/\sigma_T - \Delta P/\sigma_P$, where ΔT and ΔP are, respectively, the difference in seasonal mean temperature and precipitation from the long-term climatic mean temperature and precipitation for the season, and σ_T and σ_P are the standard deviations.

The meteorological conditions of seven of the nine USSR regions were noticeably worse during the past year than during the previous 14-year period. This is connected with the anomalies of decreasing precipitation and increasing temperature. The two exceptions were Southwest Ukraine and Central Chernozem, where the amount of precipitation did not decrease, and, as a result, meteorological conditions did not change much.

The general determination of meteorological conditions in the USSR is further supported by the data in Table 19.2. It shows the change in aridity from the beginning to the end of two periods of global warming, 1914–1939 and 1961–1988. Aridity increased over nearly the entire agricultural zone of the European USSR during the global-warming period of 1914–1939; exceptions were some northern regions in the European part of the USSR as well as the south of West Siberia.

Predictably, some specific features of the global-warming effect on agriculture that manifested themselves in the 1930s were also observed during the warming of recent years. The tendencies of climatic-condition changes in the northern and southern agricultural regions of the European USSR and West Siberia were in opposite directions. However, during the current period of global warming in the western agricultural regions of the USSR, climatic conditions improved compared with those in 1914–1939.

Aridity distribution patterns for the principal agricultural states of the United States during the two periods of global warming in this century are given in Table 19.3. During the 1930s, as in the USSR, there was a striking tendency for climatic conditions to deteriorate. Exceptions were some of the southern and eastern states, including Arkansas, Kentucky, Illinois, and Virginia. The aridity distribution pattern for the United States during 1961–1988 shows that the deterioration of meteorological conditions was not as severe as it was during the 1930s or as in the recent period in the principal agricultural regions of the USSR. Nevertheless, the geographical pattern of relatively greater aridity in the northwestern states is similar for both the current and past periods.

The year-to-year variability in aridity changed from 1961–1974 to 1975–1988. Figures in the third and fourth columns of Tables 19.2 and 19.3 show the coefficient of variation calculated as a deviation of the aridity index (S) for the two above-mentioned periods. This coefficient has grown considerably during the latter period, both in the United States and in the USSR's grain belt. For example, in the central region of the European part of the USSR, this coefficient has increased by > 50%. Here, as in the case of the mean value of the trends in aridity, the definite likelihood of changes of the coefficient of variation during the global warming of the 1930s and nowadays is noteworthy, and is especially apparent in the USSR. Changes in aridity and meteorological variability have considerably affected trends and variability of crop yields in both the United States and the USSR,

TABLE 19.2   Changes in Aridity and the Yearly Climatic Variability in the USSR, from Global-Cooling to Global-Warming Time Periods in the 20th Century [a]

	Aridity		Yearly Climatic Variability	
			Between 1914–1926 and 1927–1939 (%)	Between 1961–1974 and 1975–1988 (%)
	1914–1939	1961–1988		
Leningrad	−0.07	−0.16	−20	−19
Estonia	+0.86	−0.86	−7	−18
Latvia	+0.49	−0.55	+3	−24
Lithuania	−0.08	−0.97	+7	−16
Minsk	−0.44	−0.36	+22	+18
Rovno	+0.44	−0.51	+31	+43
Karpaty	+1.06	−0.50	+15	+15
Vinnitsa	+1.66	−0.78	+13	+22
Moldova	+1.90	−1.35	+9	+15
Kherson	+0.60	+0.12	−26	+6
Poltava	+0.67	−0.37	+15	+75
Bryansk	−0.02	+0.14	+22	+57
Tula	−0.37	+0.42	+19	+45
Lipetsk	−0.15	+0.16	+3	+74
Lugansk	−0.99	+0.10	−24	+47
Krasnodar	−1.45	+0.23	−35	+11
Volgogard	−0.84	+0.54	−28	+46
Nish. Novgorod	−0.13	−0.09	−13	+19
Tatarstan	−0.69	−0.18	−21	+59
Orenburgh	+0.70	+0.76	−3	+11
Chelyabinsk	+0.47	+0.20	−15	−21
Kokchetav	−0.09	+0.19	−9	−25
Karaganda	+1.13	+0.58	+9	−16
Altay	−0.20	−1.60	+6	−22

[a] Aridity is calculated as given in Table 19.1 Changes in climatic variability are calculated as the percent change in the coefficient of variation calculated for the yearly aridity index, S.

TABLE 19.3 Changes in Aridity and the Yearly Climatic Variability in the Principal Agricultural Areas of the United States, from Global-Cooling to Global-Warming Time Periods in the 20th Century [a]

| | Aridity | | Yearly Climatic Variability | |
			Between 1914–1926 and 1927–1939 (%)	Between 1961–1974 and 1975–1988 (%)
	1914–1939	1961–1988		
Washington	+1.31	+0.47	−1	+36
Idaho	+1.35	+0.54	−5	−12
Montana	+1.53	+0.43	−15	0
Wyoming	+0.81	+1.50	+17	−3
Utah	+1.31	+1.51	+14	−26
Colorado	+1.23	−0.84	+37	+33
North Dakota	+1.30	+0.84	+28	+47
South Dakota	+1.78	+0.49	+8	−7
Nebraska	+1.05	+0.11	+22	+23
Kansas	+1.12	−0.51	+34	+21
Oklahoma	+0.48	−1.19	−3	+4
Minnesota	+0.63	+1.02	+15	+48
Iowa	+1.90	+0.07	+49	+52
Missouri	+0.09	+0.10	+27	−1
Arkansas	−0.02	−0.67	+6	+4
Wisconsin	+1.09	0	0	+20
Illinois	−0.05	−0.91	+28	+5
Kentucky	−0.21	−0.40	+8	−18
Tennessee	+0.30	−0.31	−11	+46
Indiana	+0.27	−0.45	+19	+2
Michigan	+0.60	−0.53	−6	+19
Ohio	+0.52	−0.75	+16	−5
West Virginia	+0.40	−0.35	+20	−12
Virginia	−0.05	+0.30	+3	+20
Pennsylvania	+0.15	−0.48	+4	−2

[a] Aridity is calculated as given in Table 19.1 Changes in climatic variability are calculated as the percent change in the coefficient of variation calculated for the yearly aridity index, S.

especially in those regions where the meteorological regime during the growing season is the main factor influencing crop productivity.

This analysis has revealed that, on the whole, the current stage of global climate changes has had unfavorable consequences for agriculture in midlatitudes of the Northern Hemisphere, where the main agricultural areas of the United States and the USSR are situated. This raises two urgent questions: How will further development of global warming affect agricultural production, and will unfavorable tendencies that have already manifested themselves continue or are they temporary? The answer to these questions can be obtained by analyzing the effect on crop productivity of expected future changes in temperature, precipitation, and CO_2 concentration.

Physiological Effects of Increased CO_2 Concentration on Plant Productivity

Along with photosynthetically active radiation, CO_2 is an input to photosynthesis and a plant morphogenesis factor; therefore, human-induced increases in CO_2 will directly influence the plant's growth and productivity along with the influence of the changes in temperature and moisture. Results of experimental investigations carried out during recent years yielded important information about the direct effects of increased CO_2 concentration on physiological processes that determine crop productivity under various conditions.

Plants have several different types of photosynthetic metabolisms. The main feature of plants with the C3 photosynthesis type, which includes the majority of widely distributed plant species (e.g., wheat, rice, potatoes, and cotton) is that their photosynthetic rate increases considerably with increases in CO_2 concentration. In plants with the C4 type of photosynthesis (e.g., corn, sugar cane, and sorghum), photosynthesis is slightly increased when CO_2 concentrations exceed current levels. These kinetic effects are explained by the different chemical affinity of the enzymes that catalyze the reaction of CO_2 fixation (Rose, 1989; Laysk, 1977). According to many authors' data, Michaelis' constant (numerically equal to the CO_2 concentration at which the photosynthetic rate is half the physiological maximum) is near 400 μmol/mol (400 ppm) for the enzyme that catalyzes the primary CO_2 fixation in C3 plants (ribulozobisphosphate carboxylase, or RUBISCO), whereas for the corresponding enzyme in C4 plants [phosphoenolpyruvate (PHEP) carboxylase], it is considerably smaller (50 μmol/mol). Despite the fact that the interaction among internal photosynthetic factors can disguise the kinetic effects of initial CO_2 binding to a certain extent, a small value of Michaelis' constant for PHEP-carboxylase is responsible for the more rapid leveling off of the photosynthesis-rate-vs.-CO_2-concentration curves plotted for C4 plants. Carbon dioxide exchange in leaves of corn reaches its plateau when the CO_2 concentration in the air is equal to (or even lower than) its present level. For wheat leaves, the plateau of photosynthesis is reached when CO_2 concentrations in the air are almost three times present concentrations (Cure, 1985).

Photorespiration phenomena play an important role in the productivity of C3 plants (Laysk, 1977). Photorespiration is one of the possible ways of synthesizing glycine and serine in vegetable cells and stems from the oxygenation of the initial CO_2-accepter ribulosebisphosphate (RuBPh) in chloroplasts. Expenditure of RuBPh for photorespiration leads to loss of dry matter, which decreases productivity. Rates of carboxylation and oxygenation of RuBPh depend, first of all, on the ratio of the concentrations of CO_2 and O_2. Increased content of CO_2 in chloroplasts caused by increased external CO_2 levels should, in the end, stimulate carboxylase activity of RUBISCO and decrease the portion of RuBPh being oxidized, which will favorably affect productivity. In this case, alternative biosynthetic pathways of glycolate and serine are activated, though their concentration in the tissue decreases slightly (Keerberg and Viil, 1982). Investigations of kinetic properties of the photosynthetic apparatus of plants grown under modern gas-exchange regimes are thorough and undoubtedly important. However, they cannot give final answers to questions of productivity changes caused by long-term exposure of plants to high CO_2 concentration. Short-term experiments with plants in elevated CO_2 concentration characterize, mainly, initial photosynthetic reactions and cannot reveal the physiological processes responsible for biosynthetic processes. In recent years, several important studies have been carried out that allow us to determine the effects of the morphogenetic impact of increased CO_2 content on the growth, metabolic activity, and productivity of plants.

One of the important problems in determining the influence of increased CO_2 concentration on plants is understanding how the photosynthesis rate changes with CO_2 concentration. Experiments show that photosynthesis per unit area of leaves of plants raised under conditions of elevated CO_2 concentration is partially depressed. This depression is caused, mainly, by a decrease of RUBISCO enzyme content in chloroplasts (Keerberg and Viil, 1982); variations in other parameters of photosynthetic metabolism of such plants are insignificant.

As for the influence of increased CO_2 concentration on plant respiration, the results of experiments are contradictory. For example, for wheat that has been raised over a long period with a CO_2 concentration twice the normal, the intensity of respiration decreased twofold. Results of experiments with leguminous plants were just the reverse. In general, depressed respiration favors increased plant productivity.

Studies of plant transpiration and mineral nutrition under increased CO_2 concentrations are of great interest. Such plants are characterized by lower stomatal conductance, which is usually associated with stomatal closure. An important consequence of this effect is the decrease in evapotranspiration. Generally, plants raised with increased CO_2 concentrations are characterized by a higher productivity per unit amount of water transpired (Mauney et al., 1978). Experiments have shown that both cotton plants (C3) and corn (C4) grown with a twice-normal CO_2 concentration are characterized by a transpiration efficiency coefficient that is also twice the normal (Rose, 1989; Acock and Allen, 1985). This is observed for different mineral nutrition levels as well. Plants grown under different water conditions show different responses to increased CO_2 concentrations. For C3 plants, the relative rise of CO_2 assimilation is larger in conditions of low soil-moisture content, though the absolute increase in photosynthesis rate under conditions of optimum moisture is larger. The sensitivity to moisture deficiency of C3 and C4 plants grown with increased CO_2 concentrations is decreased (Gifford, 1977, 1979; Gifford and Morrison, 1985).

Experiments where the dynamics of plant growth are studied are very important for understanding CO_2-induced effects on plants. Results of practically all such experiments show that under optimum moisture conditions, the amount of dry matter from leaves of 1-year-old C3 plants is considerably increased by the end of the period of intensive growth. The leaf area of C4 plants, whose photosynthesis responds slightly to higher CO_2 concentration, also increases. Such an effect is observed both under favorable and unfavorable moisture conditions (Rose, 1989). Accelerated growth of C3 plants cultivated under increased CO_2 is observed during the active vegetation phase. For example, bushing out of wheat begins earlier and is more intensive when ambient CO_2 is increased over normal (Neales and Nicholls, 1978).

Some experiments have found that the phase of flowering and ripening of cereals is prolonged under CO_2 enrichment. For example, large leaf area and a long ripening phase lead to larger ears of corn. A considerable increase in seed mass is peculiar to cotton plants, whose crop productivity under doubled CO_2 concentration exceeds that of reference plants by 80% (Mauney et al., 1978).

Valuable results were obtained in experiments on growing cereals under various CO_2 regimes. For C3 plants other than cereals, doubling the CO_2 concentration results, on average, in a 26% increase in crop yield and a 40% increase in dry-matter weight. For cereals, an increase in the yield of seeds is almost twice the increase in total biomass (Goudriaan and de Ruiter, 1983). This effect could be especially important under the conditions of the expected rise of atmospheric CO_2 concentration because of the role cereals play in present-day agriculture.

The rising CO_2 concentration of the atmosphere is, undoubtedly, a powerful physiological and climatic factor in increasing crop productivity.

However, some experiments have shown that it can induce adverse effects on productivity, including the senescence of some plant species (Omer and Horvath, 1983) and the lengthening of ripening time for cereals. However, this harmful effect is likely to be compensated for by the expected lengthening of the growing season with global warming.

In experiments in which plants were cultivated under different moisture conditions, it was demonstrated that the relative increase of CO_2 assimilation is greater under moisture-deficit conditions for practically all species. The moisture-deficit sensitivity of wheat plants grown under increased CO_2 concentrations decreases (Gifford, 1977, 1979). So, a conclusion can be made that an increase in atmospheric CO_2 concentrations leads to an increase in mean crop productivity, and along with this, a decrease in year-to-year variability in the regions subject to droughts.

The results of the majority of experiments to date show that increased atmospheric CO_2 concentration favorably affects plant growth and productivity. Analyses of experimental data have supplied the valuable information needed to develop models of crop productivity under conditions of increased CO_2 concentration of the atmosphere. Beginning in the mid-1970s, several models were developed in the State Hydrological Institute in Leningrad through the USSR-U.S. collaboration on present-day climate changes and their consequences. These models include experimental data on physiological reaction of plants to atmospheric CO_2 growth. The next section of this paper contains estimates of the climate-induced changes in productivity obtained with the aid of these models.

Estimates of Future Crop-Productivity Conditions

To understand the impacts of expected climatic change on crop cultivation in some geographical zones and countries, the possible types of changes in agroclimatic conditions should first be described. The estimates of such important indices as the climatic productivity norm of various crops, the crop's year-to-year yield variability, the duration of the warmer period, etc., are connected with this problem. To conduct an agroclimatic study, it is necessary first to decide what scenarios of future climatic changes on the different time scales should be considered. In this connection, the present-day climatology offers two possibilities.

The first consists of the use of scenarios of future conditions developed using global-climate models (Wilks, 1988); the second one is the use of paleoclimatic reconstruction (Budyko, 1980; Budyko et al., 1987). Climate models have two drawbacks. The first, that results obtained using different climatic models do not always agree, has already been mentioned by specialists (Kalkstein, 1991). The second drawback is that practically all the computations and scenarios necessary for developing such models are related to factors affecting large-scale global warming (e.g., a doubling of atmospheric CO_2).

It is clear that studying the impacts of such warming on agriculture, though of great interest, cannot elucidate all of their specific changes in the next few decades. The character of changes in atmospheric precipitation over main agricultural regions in temperate latitudes of the Northern Hemisphere is nonmonotonic, as mentioned in several well-known climatological investigations (e.g., Budyko, 1980; Budyko et al., 1987). This phenomenon should manifest itself at the initial and, possibly, the intermediate stages of global warming ($\approx$1.0–1.4 °C average annual temperature of surface air). It is clear that there is great interest in estimating agroclimatic consequences of global warming, especially during the intermediate stage of global warming, and in comparing them with conditions in the past decades and the distant future. Recently, nonequilibrium climatic models have been used for simulating scenarios of climate-change dynamics, which are useful in studies of agricultural consequences of climatic changes (Hansen et al., 1988).

It seems reasonable to use paleoclimatic reconstructions to estimate the future evolution of agroclimatic regimes. At present, the most complete data with a solid scientific foundation are those on temperature and precipitation variations in the Northern Hemisphere for the Holocene climatic optimum (8000 years ago), Eemian (Sangamon, Mikulino) interglacial ($\approx$125,000 years ago), and the Pliocene climatic optimum ($\approx$4 million years ago). According to the paleoclimatic data for these three periods, surface temperature exceeded the normal climate by 1.3, 2.2, and 3.4 °C, respectively.

To obtain a reliable forecast of changes in agroclimatic conditions during the next decade, special models and methods that allow the estimation of the two most important agroclimatological parameters must be used: the level of crop productivity and the year-to-year variability of yields. To assess crop productivity, a parameterized model was developed in the State Hydrological Institute in Leningrad (Koval et al., 1983). This model takes into account temperature and precipitation changes during the growing season and the increase in CO_2 concentration. To estimate year-to-year variability in crop yield, an empirical method based on statistical analysis of long-term crop yield and regression dependencies of crop-yield variations on the changes of temperature and annual or seasonal precipitation was used (Menzhulin and Nikolaev, 1987). The main feature of both the model and the empirical method is that they are adjusted for the possibilities of climate forecasting by paleoclimatological methods.

Patterns presented in Figure 19.2 are the results of calculations of possible productivity-level changes (in percent change from normal) in C3-cereal crops (wheat, barley, oats, rye, and others) that are predicted to occur by the year 2000 in Europe, North America, and the USSR. These calculations assume that the atmospheric CO_2 concentration will reach 380 μmol/mol by then.

In the European agricultural regions, changes in C3-cereal-crop productivity are not expected to be significant. The increase in productivity in northern Europe can be explained by the fact that the present normal climatic productivity values in these regions are lower than they are in the southern regions. That is why the values expressed by percent are high.

In contrast to the homogeneous pattern of the change in productivity in Europe, the pattern in the grain belt of the USSR shows considerable differences in the estimates across the region. Here, the emphasis should be primarily on the zone with the relatively small productivity increases in the vast area at 55° N. This result appeared to be due to the paleoclimatic scenario for the Holocene optimum used in these calculations. In this scenario, there is a slight decrease in annual precipitation.

By contrast, the change in C3-cereal productivity for the North American grain belt (Figure 19.2) is forecasted to be negative. This region includes the following states, important for U.S. agriculture: Minnesota, Iowa, Illinois, Wisconsin, Indiana, and Missouri. In the paleoclimatic scenario of the Holocene optimum, the negative crop-yield effects due to precipitation decreases for these areas are not totally compensated for by the positive effects of increased CO_2 concentrations.

Besides obtaining a detailed pattern of crop productivity changes over these areas, it is important to get an estimate of the possible change in productivity for the entire Northern Hemisphere in accordance with the Holocene optimum for the year 2000. To obtain this estimate, it was assumed that changing hydrothermal conditions can influence directly both crop productivity and the shifting of agricultural zones; i.e., the areas that were barren until now could be cultivated.

Figure 19.3 shows the pattern of the changing coefficient of variation for wheat yields in grain zones of Europe, the USSR, and North America, taking into account the climatic scenario of the Holocene optimum. As expected, the pattern of precipitation increase in Europe will be homogeneous, because by the year 2000, in most countries, year-to-year variability of wheat yields will change only slightly. It is estimated to decrease in southern regions more than in other ones, although even there, the estimates of such decre2ases do not exceed 10%. From the scenario for the Holocene climatic optimum by the year 2000, it follows that over a part of the USSR-European territory north of ≈53° N. and west of ≈60° E., one can expect the variability in wheat yields to increase (Figure 19.3). As in the case of the above estimates for the crop-productivity level, this is related to a possible decrease in annual precipitation. Deterioration of moisture

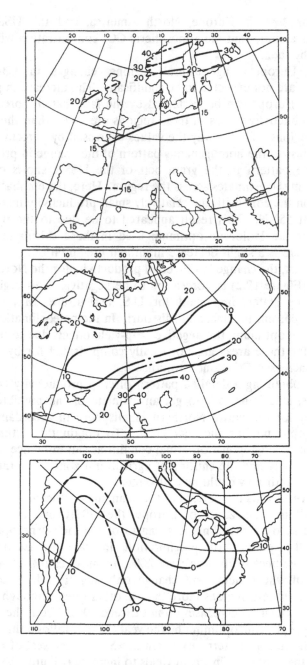

FIGURE 19.2 Changes in the productivity index (%) of C3 cereal crops for the Holocene-optimum scenario as an analog of global warming by 1 °C for Europe (top panel), the USSR (middle panel), and North America (bottom panel).

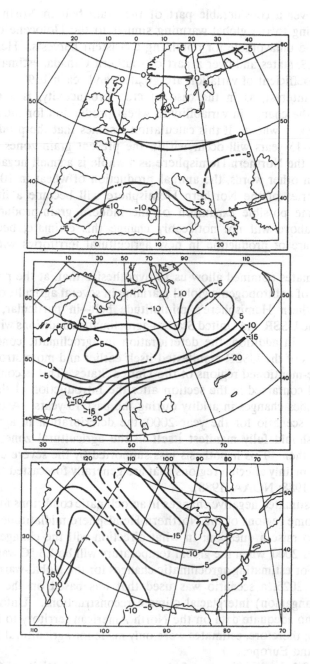

FIGURE 19.3 Changes in the year-to-year coefficient of variation (%) of wheat yield for the Holocene-optimum scenario as an analog of global warming by 1 °C for Europe (top panel), the USSR (middle panel), and North America (bottom panel).

conditions over a considerable part of the grain belt in North America (corresponding to the global warming similar to the Holocene optimum) appears to be unfavorable for obtaining consistent harvests. Here, in the indicated U.S. states and over a part of southern Canada, estimates of the increasing coefficient of variation are greatest and reach 15%.

Also of interest, as in the case of crop productivity, is a combined estimate of the change in variability in wheat production for the Northern Hemisphere as a whole. If this calculation assumes that the production of wheat in 10–15 years will occur within the present grain zones, then the estimate for the Northern Hemisphere as a whole is a small negative value ($\approx -10\%$); in other words, the annual production of wheat in 10–15 years by all countries of the Northern Hemisphere will become a little more stable. Some possible expansion of the modern grain-producing zone mentioned above will not noticeably change this estimate, because the probable share of production in new agricultural territories will be very small.

The estimates obtained allow us to hypothesize that, at the present-day initial stage of anthropogenic global warming in several agricultural regions of the Northern Hemisphere's midlatitude belt—in particular, in some regions of the USSR and United States—agroclimatic conditions will develop unfavorably. In addition, the deterioration in agroclimatic conditions in those regions of the USSR manifested itself earlier and more strongly than in the above-mentioned regions of the United States. This is confirmed by the material contained in the section after the Introduction of this report, which describes changes in aridity during the past 15 years. According to the climatic scenario for the year 2000, the deterioration in agroclimatic conditions should fully manifest itself in the agricultural zone of North America by then. This hypothesis was supported by the severe drought of 1988, which mainly affected regions that are generally forecasted to be drier (Schneider, 1988; NOAA, 1988).

Having estimated negative changes in agroclimatic conditions for the near future for some regions of the Northern Hemisphere midlatitude belt, it is important to raise a question: In what direction will the changes develop after the year 2000, when the global temperature will be 1.3 °C warmer than it is now? For estimating agroclimatic changes for the global-warming level expected in 2025, a scenario was used that was based on the Mikulino (Eemian, Sangamon) interglacial climate reconstructions. Unfortunately, there were no adequate data on the North American territory to include in the scenario; therefore, estimates were only made for agricultural regions in the USSR and Europe.

Figure 19.4 shows estimates of changes in the productivity of C3 cereals up to the year 2025 made as before, accounting for changes in temperature

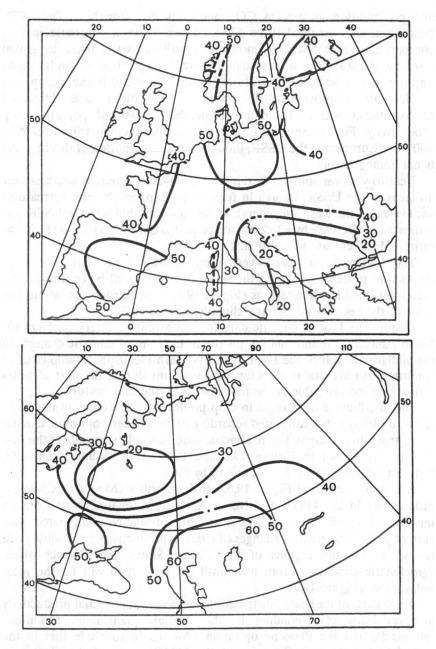

FIGURE 19.4 Changes in the productivity index (%) of C3 cereal crops for the Eemian-interglacial scenario as an analog of global warming by 2 °C for Europe (top panel) and the USSR (bottom panel).

and precipitation as well as CO_2-concentration growth. The CO_2 concentration is assumed to be 420 μmol/mol by then. Increase in CO_2 concentration and favorable moisture conditions correspond to global warming of a scale that will cause a tendency for agroclimatic conditions to improve in practically all agricultural regions of the USSR and Europe. A considerable improvement in moisture conditions and increased photosynthesis will result in a considerable growth of potential crop productivity. For example, for C3 cereals, crop productivity will increase by ≈40% for Europe and the USSR, though spatial distribution of this increase is not homogeneous.

Year-to-year variability over the years 2000–2025 generally decreases for all areas in the USSR, though in regions north of 55° N., the decrease of such values is the least. In general, wheat-yield variability in the USSR was estimated to decrease by ≈30% up to the year 2025; in the year 2000, it was estimated to decrease by up to 20%.

The tendency for year-to-year crop variability of European grain production to decrease during the years 2000–2025 will be continued and even intensified in some regions (Figure 19.5). For Europe as a whole, the value of the decrease of this agroclimatic index is 20%.

Despite the lack of climatic data for the Mikulino interglacial era for North America, the information on the mechanism of climatic changes in the midlatitude belt of the Eastern and Western Hemispheres allows us to confirm that unfavorable agroclimatic conditions during the next 15 years should become favorable in the first quarter of the 21st century.

The distribution of changes in crop productivity as well as in the crop-yield-variability index, calculated according to the Pliocene optimum climate reconstruction corresponding to climatic conditions of the middle of the 21st century, are shown in Figures 19.6 and 19.7. It is assumed that CO_2 concentration will reach 560 μmol/mol by then.

As can be seen from Figure 19.6, which describes changes of C3-cereal productivity in the USSR and North America, estimates range from an increase of <35% to nearly 100% in crop productivity compared with current yield. The greatest changes of climatic productivity (>50–60%) will be typical for the regions of the United States and Canada where agroclimatic conditions from now until the year 2000 will be the most unfavorable (Figure 19.7).

An analysis of the spatial distribution of changes of potential productivity and variability corresponding to the Eemian (Sangamon, Mikulino) interglacial and the Pliocene optimum allow us to conclude that in the agricultural regions of Europe and the USSR, most of the change in potential crop productivity will have occurred by ≈2025–2030. This can be seen from patterns for these areas, where changes in the wheat-yield coefficients of variation are similar. The only exceptions are the agricultural

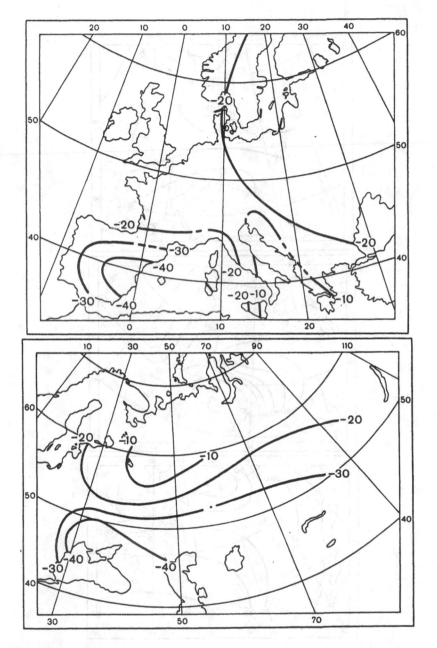

FIGURE 19.5 Changes in the year-to-year coefficient of variation (%) for the Eemian-interglacial scenario as an analog of global warming by 2 °C for Europe (top panel) and the USSR (bottom panel).

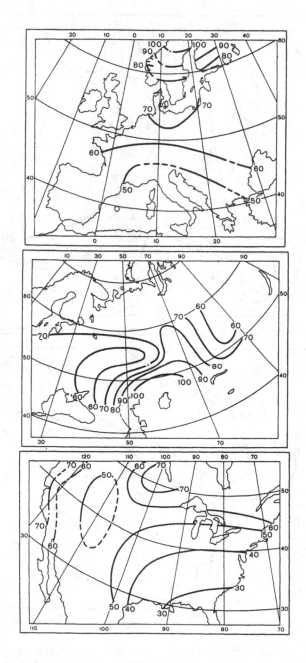

FIGURE 19.6 Changes in the productivity index (%) of C3 cereal crops for the Pliocene-optimum scenario as an analog of global warming by 3–4 °C for Europe (top panel), the USSR (middle panel), and North America (bottom panel).

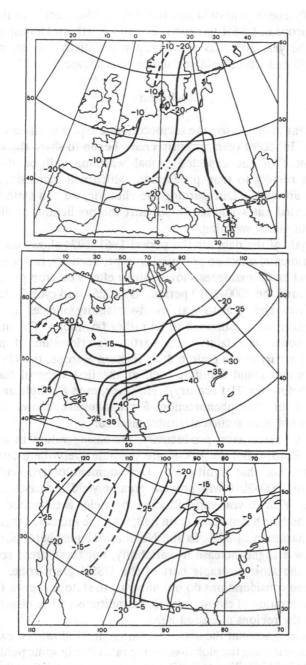

FIGURE 19.7 Changes in year-to-year coefficient of variation (%) for the Pliocene-optimum scenario as an analog of global warming by 3–4 °C for Europe (top panel), the USSR (middle panel), and North America (bottom panel).

region of the Pyrenean peninsula and the very southern areas of the cereal zone of the European USSR territory. For the Northern Hemisphere as a whole up to the year 2050, C3-cereal-crop productivity was estimated to increase by 55% and wheat-yield variability, to decrease by 20%.

Conclusion

It is important to emphasize the socioeconomic aspect of climate impacts on agriculture. In recent years, scientists have begun to share the view that the consequences of the expected global warming will be disastrous, especially with respect to crop production, which will increasingly suffer from frequent droughts. Disregarding the rhetoric, we can state that the analyses we carried out would hardly support the predictions of disastrous consequences of global warming.

While analysis of the data for the period 1961–1974 shows warming to have had a detrimental effect on potential crop productivity, future scenarios based on paleoclimatic analogues show positive effects on crop productivity, particularly beyond the 2000–2010 period. By 2025 and beyond, changes in crop productivity and year-to-year yield-variability indices in all the agricultural regions were estimated to be rather favorable. The only thing that presents some uncertainty is a partial deterioration of moisture conditions over a part of the grain belt in Eastern Europe, the USSR, and North America that could manifest itself at the first stage of the global warming (i.e., before the 21st century). However, even a simple analysis of the significance of this phenomenon for agriculture shows that this uncertainty has not been sufficiently substantiated.

As for the C3-cereal-crop productivity, increasing values in the years 2000–2010 for Europe and the USSR are possible everywhere but, in the major belts, are not that significant. Development of crop-cultivation technology until 2000–2010 will provide greater yield increases than will climate change. Year-to-year variability in crop yields is estimated as likely to increase in the North American grain belt, with the increase ranging from 10–15% to a maximum of 30%. In the Northern Hemisphere, portions of the USSR show a slight increase in variability, but these areas constitute only ≈15% of the most favorable part of the USSR grain zone. All the above-mentioned considerations do not allow us to state that even the most noticeable deterioration of crop-production conditions will be disastrous for agriculture for the regions discussed here.

If one takes into account the above scenarios, the following views of the consequences of retarding the global-warming rate cited in some publications can be presented. According to available estimates, restricting the use of carbon fuel is unlikely to stop the global-warming rate. Such measures, if adopted on an international scale, may retard global warming at a level corresponding to that of the onset of the 21st century, i.e., to climate conditions comparable to the Holocene optimum. However, as the above analysis showed, such a climate regime is characterized by unfavorable effects on crop productivity and variability in the major agricultural regions of the USSR and North America. Yet, warming beyond Holocene optimum conditions was shown to be beneficial for all agricultural regions in the Northern Hemisphere.

This paper warns that to fight the negative consequences of climate changes by controlling CO_2 emissions will bring about the stabilization of global warming at a level that is very unfavorable for agriculture. Instead, it may be desirable to counter the expected negative climatic effects on agriculture over the next one to two decades with better management, more intensive agricultural production technologies, and similar measures. Conclusions about suggested projects to retard global warming due to decreasing CO_2-concentration growth should be drawn only after a thorough and comprehensive analysis of all the economic and ecological consequences of expected climatic changes.

References

Acock, B., and L. H. Allen. 1985. "Crop Responses to Elevated Carbon Dioxide Concentrations," in B. R. Strain and J. D. Cure, eds., *Direct Effects of Increasing Carbon Dioxide on Vegetation.* Pp. 33–97. Washington, D.C.: U.S. Department of Energy. (DOE/ER-0238.)

Ausubel, J. H. 1983. "Can We Assess the Impacts of Climatic Changes?" *Climatic Change* 5(1): 7–14.

Buring, P., H. Van Heenst, and G. Staring. 1975. *Computation of the Absolute Maximum Food Production of the World.* Wageningen, The Netherlands: Wageningen Agricultural University.

Budyko, M. I. 1980. *Climate in the Past and Future.* Leningrad: Hydrometeoizdat (in Russian).

Budyko, M. I., and Y. A. Israel, eds. 1987. *Anthropogenic climate changes.* Leningrad: Hydrometeoizdat (in Russian).

Callaway, J. M., F. J. Cronin, J. W. Currie, J. Tawil. 1982. "An Analysis of Methods and Models for Assessing the Direct and Indirect Impacts of CO_2-Induced Environmental Changes in the Agricultural Sector of the U.S. Economy." Richland, Wash.: Pacific Northwest Laboratory, Battelle Memorial Institute. (PNL-4384.) Carbon Dioxide Assessment Committee. 1983. "Changing Climate." Washington, D.C.: National Academy of Sciences.

Chen, R. S. 1981. "Interdisciplinary Research and Integration: The Case of CO_2 and Climate." *Climatic Change* 3(4): 429–47.

Clark, W. C. 1985. "Scales of Climatic Impacts." *Climatic Change* 7: 5–27.

Cure, J. D. 1985. "Carbon Dioxide Doubling Responses: A Crop Survey," in B. R. Strain and J. D. Cure, eds., *Direct Effects of Increasing Carbon Dioxide on*

Vegetation. Pp. 99–116. Washington, D.C.: U.S. Department of Energy. (DOE/ER-0238.)

Decker, W. L., V. Jones, and R. Achutuni. 1985. "The Impact of CO_2-Induced Climate Change on U.S. Agriculture," in M. R. White, ed. *Characterization of Information Requirements for Studies of CO_2 Effects: Water Resources, Agriculture, Fisheries, Forests and Human Health*. Pp. 69–93. Washington, D.C.: U.S. Department of Energy. (DOE/ER-0236.)

Gifford, R. M. 1977. "Growth Pattern, Carbon Dioxide Exchange and Dry Weight Distribution in Wheat Growing under Different Photosynthetic Environments." *Australian Journal of Plant Physiology* 4: 99–110.

Gifford, R. M. 1979. "Growth and Yield of CO_2-Enriched Wheat under Water-Limited Conditions." *Australian Journal of Plant Physiology* 6: 367–78.

Gifford, R. M., and J. I. L. Morrison. 1985. "Photosynthesis, Water Use and Growth of a C4 Grass Stand at High CO_2 Concentration." *Journal of Photosynthetic Research* 7: 77–90.

Goudriaan, J., and H. E. de Ruiter. 1983. "Plant Response to CO_2 Enrichment at Two Levels of Nitrogen and Phosphorus Supply. 1. Dry Matter, Leaf Area and Development." *Netherlands Journal of Agricultural Sciences* 31: 157–69.

Hansen, J., I. Fung, A. Lacis et al. 1988. "Global Climate Changes as Forecast by Goddard Institute for Space Studies Three-Dimensional Model." *Journal of Geophysical Research* 93(D8): 9341–64.

Jensen, W. F. 1978. "Limits to Growth in World Food Production." *Science* 201: 317–20.

Kalkstein, L. S., ed. 1991. "Global Comparison of Selected GCM: Runs and Observed Climate Data." Prepared for the Office of Policy, Planning and Evaluation, Climate Change Division, U.S. Environmental Protection Agency. Newark, Del.: Center for Climatic Research, University of Delaware.

Keerberg, O. F., and A. Viil. 1982. "Systems of Regulation and Energetic Supply of Reductional Pentoso-phosphate Cycle," in *Physiology of Photosynthesis*. Pp. 104–18. Moscow: Nauka Publishing House (in Russian).

Kimball, B. A. 1985. "Adaptation of Vegetation and Management Practices to a Higher Carbon Dioxide World," in B. R. Strain and J. D. Cure, eds., *Direct Effects of Increasing Carbon Dioxide on Vegetation*. Pp. 185–204. Washington, D.C.: U.S. Department of Energy. (DOE/ER-0238.)

Koval, L. A., G. V. Menzhulin, and S. P. Savvateev. 1983. "On the Principles of Constructing Parameterized Models of Crop Productivity." *Transactions of the State Hydrological Institute* 280: 119–29 (in Russian).

Koval, L. A., and S. P. Savvateev. 1982. "Using Crop Productivity Models for Calculating the Impact of Climatic Changes on Crop Yields," in I. V. Popov and S. A. Kondrateyev, eds., *The Problems of Land Hydrology*. Pp. 219–23. Leningrad: Gidrometeoizdat.

Land Evaluation Group. 1987. "Implications of Climatic Warming for Canada's Comparative Position in Agricultural Production and Trade." Guelph, Ontario: University of Guelph.

Laysk, A. H. 1977. *Photosynthesis and Respiration Kinetics of C3 Species*. Moscow: Nauka Publishing House (in Russian).

Mauney, J. R., K. E. Fry, and G. Guinn. 1978. "Relationship of Photosynthetic Rate to Growth and Fruiting of Cotton, Soybean, Sorghum and Sunflower." *Crop Science* 18: 259–63.

Menzhulin, G. V. 1976. "The Effect of Climate Change on Crop Productivity." *Transactions of the Main Geophysical Observatory* 365: 41–8 (in Russian).

Menzhulin, G. V., and S. P. Savvateev. 1980. "The Effect of Modern Climate Changes on Agricultural Plant Productivity," in M. I. Budyko, ed., *The Problem of Atmospheric CO_2: Transactions of a Soviet-American Symposium.* Pp. 186–97. Leningrad, Hydrometeoizdat (in Russian).

Menzhulin, G. V., and S. P. Savvateev. 1981. Present-Day Climate Change and Agricultural Productivity." *Transactions of the State Hydrological Institute* 281: 90–103 (in Russian).

Menzhulin, G. V., M. V. Nikolaev, and S.P. Savvateev. 1983. "Estimates of Technological and Climate Components of Cereal Yields." *Transactions of the State Hydrological Institute* 280: 111–9 (in Russian).

Menzhulin, G. V. 1984. "The Effect of Modern Climate Changes and CO_2 Concentration Increase on Crop Productivity." *Soviet Meteorology and Hydrology Journal* 4: 95–101.

Menzhulin, G. V., and M. V. Nikolaev. 1987. "Methods for Studying Year-to-Year Variability of Cereal Crop Yields. *Transactions of the State Hydrological Institute* 327: 113–31 (in Russian).

Menzhulin, G. V., L. A. Koval, M. V. Nikolaev, and S. P. Savvateev. 1987. "On the Estimates of the Effect of Modern Climate Changes on Agriculture: Scenario for North American Wheat Production. *Transactions of the State Hydrological Institute* 327: 132–46 (in Russian).

National Oceanic and Atmospheric Administration. 1988. "U.S. Drought 1988: A Climate Assessment." Washington, D.C.: NOAA, Climate Office, U.S. Department of Commerce.

Neales, T. F., and A. O. Nicholls. 1978. "Growth Responses of Ambient CO_2 Levels." *Australian Journal of Plant Physiology* 5: 45–59.

Nikolaev, M. V., G. V. Menzhulin, and S. P. Savvateev. 1985. "Some Regularities of Cereal Crop Yield Variability in the U.S.S.R. and the U.S.A." *Transactions of the State Hydrological Institute* 339: 61–81 (in Russian).

Nikolaev, M. V. 1985. "On the Technological and Weather-Climate Variability Constituents of Cereal Crop Yields in the U.S.A." *Transactions of the State Hydrological Institute* 339: 48–60 (in Russian).

Omer, St. L., and S. M. Horvath. 1983. "Elevated Carbon Dioxide Concentration and Whole Plant Senescence." *Ecology* 64: 1311–4.

Oram, P. A. 1985. "Sensitivity of Agricultural Production to Climatic Change." *Climatic Change* 7(1): 129–52.

Parry, M. L. 1985. "Estimating the Sensitivity of Natural Ecosystems and Agriculture to Climatic Change." *Climatic Change* 7(1): 1–4.

Parry, M. L., and T. R. Carter. 1985. "The Effect of Climatic Variations on Agricultural Risk." *Climatic Change* 7(1): 95–110.

Parry, M. L., T. R. Carter, and N. T. Konijn, eds. 1988a. *The Impact of Climatic Variations on Agriculture. Vol. 1. Assessments in Cool Temperate and Cold Regions.* Dordrecht, Germany: Kluwer Academic Publishers.

Parry, M. L., T. R. Carter, and N. T. Konijn, eds. 1988b. *The Impact of Climatic Variations on Agriculture. Vol. 2. Assessments in Semi-Arid Regions.* Dordrecht, Germany: Kluwer Academic Publishers.

Pavlova, V. N., and O. G. Sirotenko. 1985. "On the Use of Dynamic Models for Estimating the Effect of Possible Changes and Fluctuations of Climate on Crop Capacity." *Transactions of the All-Union Institute of Agricultural Meteorology* 10: 81–90 (in Russian).

Ped, D. A. 1975. "On the Aridity and Humidity Index." *Transactions of the Hydrometeorological Center of the USSR* 156: 19–39.

Pulwarty, R. S., and S. J. Cohen. 1984. "Possible Effects of CO_2 Induced Climate Change on the World Food System: A Review." *Climatology Bulletin* 18(2): 33–48.

Rose, E. 1989. "Direct (Physiological) Effects of Increasing CO_2 on Crop Plants and Their Interactions with Indirect (Climatic) Effects," in J. B. Smith and D. A. Tirpak, eds., *The Potential Effects of Global Climate Change on the United States.* Appendix C: Agriculture, vol.2. Washington, D.C.: Office of Policy, Planning and Evaluation, U.S. Environmental Protection Agency.

Rosenberg, N. J. 1982. "The Increasing CO_2 Concentration in the Atmosphere and Its Implication for Agricultural Productivity Effects Through CO_2-Induced Climate Change." *Climatic Change* 4: 239–54.

Rosenzweig, C. 1985. "Potential CO_2-Induced Climate Effects on North American Wheat-Producing Regions." *Climatic Change* 7: 367–89.

Sakamoto, C., S. Leduc, N. Strommen, and L. Steyaert. 1980. "Climate and Global Grain Yield Variability." *Climatic Change* 2: 349–61.

Savvateev, S. P. 1985. "On the Problem of Calculations of Agricultural Plant Productivity by Parameterized Models," in I. V. Popov and S. A. Kondratyev, *The Problems of Land Hydrology.* Pp.213–6. Leningrad: Hydrometeoizdat (in Russian).

Schneider, K. 1988. "Drought Cutting U.S. Grain Crop 31% this Year." *New York Times:* Aug. 12: A1.

Sirotenko, O. D., and A. P. Boiko. 1980. "The Use of the Imitation Model of the Soil-Plant-Atmosphere System for Estimating the Influence of CO_2 Concentration Fluctuations on Agrocenoses Productivity," in M. I. Budyko, ed., *The Problems of Atmospheric CO_2: Transactions of a Soviet-American Symposium.* Pp. 243–51. Leningrad: Hydrometeoizdat (in Russian).

Smith, J. B., and D. A. Tirpak, eds. 1989. "The Potential Effects of Global Climate Change on the United States." Report to Congress (draft). Appendix C, vols. 1 and 2. Washington, D.C.: Office of Policy, Planning and Evaluation, U.S. Environmental Protection Agency.

Terjung, W. H., D. Liverman, J. T. Hayes et al. 1984. "Climatic Change and Water Requirements for Grain Corn in the North American Great Plains." *Climatic Change* 6(2): 193–220.

Warrick, R. A. 1984. "The Possible Impacts on Wheat Production of a Recurrence of the 1930s Drought in the U.S. Great Plains. *Climatic Change* 6(1): 5–26.

Warrick, R. A., and W. E. Riebsame. 1981. "Societal Response to CO_2-Induced Climate Change: Opportunities for Research." *Climatic Change* 3(4): 387–428.

Warrick, R. A., R. M. Gifford, and M. L. Parry. 1986. "CO_2 Climatic Change and Agriculture. Assessing the Response of Food Crops to the Direct Effects of Increased CO_2 and Climatic Change," in B. Bolin, B. R. Doos, J. Jager, and R. A. Warrick, eds., *The Greenhouse Effect, Climatic Change and Ecosystems. A Synthesis of the Present Knowledge.* Pp.393–473. New York: John Wiley and Sons.

Wilks, D. S. 1988. "Estimating the Consequences of CO_2-Induced Climate Change on North American Grain Agriculture Using General Circulation Model Information." *Climatic Change* 13(1): 19–42.

World Meteorological Organization/International Council of Scientific Union/United Nations Environmental Program (WMO/ICSU/UNEP). 1981. "On the Assessment of the Role of CO_2 in Climate Variations and Their Impact." Geneva: WMO.

20

Perspectives on Potential Agricultural and Resource Effects of Climate Change in Japan

Ryohei Kada[1]

Introduction

The term sustainability is quite new to Japanese agriculture. The main reason for the absence of the concept of sustainable development is that rice-based Japanese agriculture has developed and continued for nearly 2000 years under the temperate, monsoon climate. Needless to say, rice is produced in paddy fields, which necessarily contain a certain depth and amount of water during the growing period. Rice production has been an ecologically and biologically sustainable system, which has enabled farmers to produce rice in paddy fields year after year for a very long time.

However, such a sustainable system has been gradually changing since the early 1960s, when Japan entered into a highly industrial economic-development stage. The rice-production system has changed from one with relatively low input and low output to one with high input and high output. In other words, Japanese agriculture has rapidly increased the intensity of land use per unit land area, which has caused such problems as deterioration of soil fertility, soil and ground-water pollution, and other conservation problems. Recently, as global environmental problems such as global warming and desertification have become serious global issues, Japanese people have been more and more concerned about these issues, too.

The main purpose of this paper is threefold. First, we outline the historical development and current status of the cropping patterns and land-use systems in Japanese agriculture. Second, we examine the current problems and policy issues of Japan's land-use systems. Finally, we discuss

---

[1] From the Faculty of Agriculture, Kyoto University, Japan.

the long-term development and research needs for keeping Japanese agriculture and its land-use systems sustainable.

Locational and Climatic Conditions of Japanese Agriculture

Japan consists of four major islands with many smaller ones, which stretch from north to south between 45°5' and 27° north latitude over a distance of 3000 km. The country can be divided into two climatic areas: the North, which experiences even subfrigid conditions, and the South, which is temperate to subtropical. Because Japan is mostly in the temperate-monsoon area, it experiences hot, humid summers and cold, dry winters.

These regional climatic variations can be illustrated by the contrasting winter weather of the Pacific Ocean and the Japan Sea sides of the country. Areas facing the Japan Sea receive extremely heavy snowfall, with depths of up to 3.77 m, one of the highest annual snowfalls in the world. In contrast, the Pacific side experiences monsoon conditions, which are not only cold and windy, but are also very dry and can cause serious water shortages and danger of fire.

Wide differences in climate also provide Japan with a great variety and abundance of vegetation. Heavy rainfall is an important factor in governing the agriculture of Japan. Annual rainfall figures in Japan average ≈1800 mm, ranging from 1500 to 2500 mm, in contrast with the United States, where rainfall averages ≈700 mm.

The Economic Situation of Japanese Agriculture

One of the major factors restricting the development of Japanese agriculture is that there is so little land for agricultural purposes. About 66% of the total land area in Japan is mountainous. Of the remaining 34%, 20% is given over to urban use, which leaves the agricultural land area in Japan at only ≈14%.

In contrast to the major agricultural countries, the percentage of arable land of Japan is one of the lowest. In particular, the land given over to meadows and pastures in Japan is only a tiny percentage of the total land area, which means that the dairy and livestock production in this country is very restricted. Japan's land-use figures exemplify the nature of Japanese agriculture—the large percentage of forested area and the limited arable and pasture land make for a highly intensive farming structure.

The total farming population in Japan fell from ≈16% (26.6 million) in 1970 to ≈7% (19.2 million) in 1987. However, it is important to point out that the agricultural labor force is governed by the seasonality of agriculture in Japan. Those who find it profitable to work in agriculture over the summer months may find it equally profitable to turn to industry or other occupations during the winter. People over 60 years of age make up 39% of the agricultural work force; farming is becoming increasingly popular among people retired from city jobs.

Japan's agricultural situation is made clearer by the role of Japanese agriculture in the gross domestic product (GDP). The agricultural percentage of the GDP fell from 8% in 1970 to 3% in 1985, a trend similar to that in the major European countries. This figure is relatively high for a country with so little agricultural land, in contrast to the United Kingdom, in which it is 2% of the GDP, and the United States, where it is 2%. These values indicate, again, the highly intensive nature of Japanese agriculture and illustrate its importance when a major European Community (EC) producer such as France—with an extremely high percentage of farming land—achieves a similar percentage of GDP from agriculture.

While the agricultural population was decreasing dramatically between 1970 and 1988, more and more farmers tended to take off-farm jobs and to become part-time farmers.

Agricultural Production

Since the mid-1960s, the nation's dietary habits have diversified, resulting in sharp declines in the demand for and the annual production of rice. Japan is currently facing oversupplies of several major farm products, including rice and mandarin oranges, while its low rates of production of other commodities such as wheat, soybeans, and feed grains emphasize the need for diversification. The decline in the growth of products such as rice is countered by the growth of livestock products, from 15% of total agricultural output in 1960 to 27% in 1987. Japan's area for cultivation is small in comparison with her competitors'; thus, the increase in livestock is connected with the need for a highly intensive farming system and increases the need to import feed.

Commercial production of livestock in Japan started expanding around the mid-1960s and is still in a developing stage, which makes the situation totally different from those countries with a long history of raising livestock. Obviously, at the moment, it is difficult for Japanese producers to compete with producers from Australia and the United States. However, in recent years, the scale of production has been enlarged steadily, and efforts are being made to offer competitive prices. This growth in the Japanese livestock industry has brought about an expansion in poultry and dairy products, milk cows in particular. Looking further at the compositional changes in agricultural output, rice decreased from 47.4% in 1960 to 30.9% in 1987; wheat and barley decreased from 5.5% to 1.8%; vegetables increased from 9.1% to 19.9%; fruits increased from 6.0% to 7.7%; livestock products have undergone a large increase, from 15.2% to 27.2%; and sericulture has decreased from 3.0% to 0.5%.

Agricultural Land Use

The pattern of land use in Japan is changing constantly. The most noticeable change is the overall reduction in Japan's agricultural land. The total agricultural land use has been reduced by 8%, from 5.796×10^{10} m^2 (5796 thousand ha) in 1970 to 5.317×10^{10} m^2 in 1988. This reduction is basically calculated by subtracting the amount of agricultural land added, from the land ruined. The total area of added cultivated land was 2.37×10^8 m^2 in 1988, and the total area of cultivated land ruined in the same year was 4.66×10^8 m^2, which includes 4.04×10^8 m^2 for artificially transformed land. In Japan, there has been a strong outside demand for agricultural land, and although the Japanese government has endeavored to retain the superior farmland, it has sold some of it. It is remarkable that the upland-field area increased by 2%, while the paddy-field area has decreased 15% during the same period (1970–1988). The increase of the upland-field area is basically due to the expansion of the meadows developed in the mountainous areas. Meadow land has increased by 122% since 1970. The total meadow land in 1980 was $\approx 6.36 \times 10^9$ m^2.

In 1988, paddy fields constituted 54% of the total agricultural land. Of the 46% of upland-field area, 24% was normal upland field, 12% was meadows, and 10% was under permanent field. Although the importance of meadow land to the Japanese economy is growing, it still constitutes only 12% of Japan's total agricultural land area.

In recent years, the Japanese government has endeavored to combat the surplus rice production through diversification. The Rice Crop Diversion Policy was initiated in 1971, and the Paddy Field Reorganization Policy was introduced in 1978. The total area converted from rice production to production of some other crop in 1988 was 8.17×10^9 m^2, which was 15.4% of the total cultivated area.

The diversion of paddies from rice production is concentrated on production of vegetables, soybeans, forage crops, wheat, and barley. For example, in 1988, 8.17×10^9 m^2 diverted from rice included 1.20×10^9 m^2 for vegetable production, 1.35×10^9 m^2 for forage crops, 1.34×10^9 m^2 for wheat and barley, and 0.96×10^9 m^2 for soybeans. The fact that forage crops constitute the largest area of diversification from rice helps the shift toward livestock production. Diversification is also being promoted by the increase of the multipurpose paddy field, which can produce a variety of other crops as well as rice.

It is not surprising that the actual planted area was reduced by 12% during the period from 1970 to 1987. The major crop reduced was rice, of which the area decreased from 2.923×10^{10} m^2 in 1970 to 2.146×10^{10} m^2 in 1987, a 26.6% reduction. Wheat and barley were also reduced on a large scale, with a total reduction of ≈20%.

The main increase in the aggregate planted area has occurred in forage and manure crops. This area rose from 7.36×10^9 m^2 in 1970 to 10.89×10^9 m^2 in 1987, a total increase of 48%. This increase is basically a result of the increase in cultivated land that was diverted from rice. It is important to add that Japan has limited options in diversifying to other crops from rice, basically because of the high cost in comparison with the cost of imported products. Of course, this is a situation that the EC also has to face. Consequently, the Japanese are particularly interested in recent policy developments, especially in relation to the Common Agricultural Policy, introduced in 1957.

An international comparison of farmland areas reveals Japan's great disadvantage in land area. The United States has 77 times more farmland than Japan. The United Kingdom, which is about two-thirds the size of Japan, has three times as much farmland. Farming must become increasingly intensive to compete with these countries. The figures illustrating farmland area per farm household emphasize the small size of Japan's farming units. The United States, with 1.75×10^6 m^2 (175.2 ha) per farm household, has 146 times more farmland per household than Japan. More significantly, the United Kingdom, with 0.77×10^6 m^2 in comparison with Japan's 1.2×10^4 m^2, has 64 times more farmland per household, and France has 25 times more. Farmland prices are high in Japan. At 343 Yen/m^2 (1372 thousand Yen/10 acres) it is 33 times more expensive than farmland in the United States, 16 times more expensive than in France, 9 times greater than in the United Kingdom, and 3 times greater than in Germany. These very high land prices provide the Japanese farmer with a great problem. Difficulties in expansion mean that production must constantly be made more intensive if Japanese agriculture is going to be able to compete abroad. It could be argued, however, that the value of the Japanese farmer's land is almost the same as that of most European farmers.

Farm Size and Productivity

Farm size and productivity are obviously dictated by this intensive agricultural system. The size of the paddy field reached its peak in 1970, at 6.22×10^4 m^2 (62.2 ares) per farm household, and in 1983, it had been reduced to 6.02×10^4 m^2. On the other hand, the growth of livestock and greenhouse production continues at a steady pace. Dairy cattle have increased by 8.4 times, from 3.4 head per farm household in 1965 to 28.6 head in 1988; beef cattle have increased by 7.8 times, from 1.3 head in 1965 to 10.2 head in 1988; hogs have increased by 35.8 times, from 5.7 head in

1965 to 203.9 head in 1988; laying chickens have increased from 27 head in 1965 to 1356 head in 1988, which is 50.2 times the number in 1965. Chickens other than layers have increased by 28.6 times, from 892 in 1965 to 25,500 in 1988. The number of greenhouses per farm household has also undergone a tremendous increase.

The reduction in the total cultivated area of rice and the changes in paddy-field size were brought about by the Rice Crop Diversion Policy. Livestock production is still in a period of growth that will probably continue for some time yet.

Let us explain the long-term change in the area of Japanese farmland. From the year 700 until the late 16th century, yield of rice remained at a constant level of ≈ 0.16 kg/m^2 (≈ 1.6 tons/ha). From that period until the mid-1870s, a gradual increase occurred in the yield of rice. However, it was during the 110 years after the Meiji Restoration in 1872 that the most dramatic increase occurred, basically because of strenuous efforts made by both the government and the people of Japan. Before World War II, the higher productivity was achieved through several methods, including improved rice varieties, use of animal power, and improved fertilization. The small-farm management system was also improved through a heavier output of both fertilizer and labor. After the war, however, the higher productivity was brought about by labor-saving techniques, mainly through mechanized farming.

The difference in productivity on farms with different cultivation areas basically shows that the smaller the area of agricultural production, the longer the hours worked per unit area, and the higher the secondary production costs. For example, farmers with agricultural areas of $<3 \times 10^3$ m^2 (0.3 ha) worked 82 hours/10^3 m^2 in 1982. Farmers with $>5 \times 10^4$ m^2 (5 ha) worked only 32 hours/10^2 m^2.

Several interesting facts about current agricultural production and productivity trends should be considered. First, the agricultural productivity of Japan has improved at a rate similar to that in other countries under the high-level economic growth of the country during the 1960s. However, due in part to the sharp increase of wages, the increase in the prices of farm products in Japan has proved steeper than in the West. The rate of increase in prices of agricultural products slowed down considerably during the period of stabilized economic growth from 1976 to 1983. The reason for this is that farm-product supply and demand eased and wages and commodity prices in general stabilized. Note, however, that with the reduced rate of farming-population decline and slackening growth of agricultural production, the circumstances surrounding the improvement of productivity have changed. The annual rate of agricultural-production increase in the years 1960–1970 was 2.3%, whereas in 1976–1983, it was only 0.9%. On the other hand, the annual rate of farming-population decline fell from 3.7% in the years 1960–1970 to 3.0% in the period 1976–1983.

There are several approaches to promoting improvement in agricultural productivity in Japan. Upgrading the productivity of the extensive farming system is one approach. This would necessitate a substantial expansion in scale, both through an increase in the mobility of farmland-use rights and the systemization of agricultural production. Second, Japanese agriculture would benefit from an increased use of farming technology as a way to increase efficiency and production. Third, the maintenance of agricultural production and the increased promotion of higher-value-added cash crops are considered essential for furthering productivity improvement. To achieve this, it is important that Japanese agricultural producers strive toward the following: undertaking domestic production of crops that can be raised in Japan, promoting development of new demands within the country, expediting an increase in agricultural-product exports; and growing more higher-value-added cash crops.

Changing Cropping Patterns

Under the monsoon conditions of high summer temperatures, and high humidity, Japan's vegetation is highly productive. The average production of organic matter in Japanese forests, for example, is, on a dry-matter basis, 1.3–2.3 kg/m^2 (13–23 tons/ha), which is as much as tropical forests producing 1.0–3.5 kg/m^2 and much more than in Mediterranean Europe, where the average forest production is between 0.25 and 1.5 kg/m^2. Such a high natural production capacity has enabled Japan to grow rice as the most productive, population-supportive crop for nearly 2000 years.

In the paddy field, where rice grows continuously for a long time, soil tends to possess the following characteristics. First, because the paddy field is covered by water during summer time, decomposition of organic matter in the soil is very slow, even under high temperatures. During winter time, soil becomes dry, but decomposition is minimal due to cold temperatures. As a whole, the decomposition of organic matter can be minimized in the paddy field. In addition, rice straws and roots remain, which keeps soil fertility relatively constant. As a result, rice production tends to be relatively easily stabilized over long periods.

Second, ferrous ions are susceptible to deoxidation to ferric ions in a watered paddy field. Then, the pH of soil changes from acidic to neutral, which makes rice plants grow well.

Third, phosphorus ions separated from ferrous compounds become available for rice plants when decomposition occurs. Furthermore, many minerals dissolved in water are supplied to paddy fields during the irrigation period. These minerals can be conserved and used as nutrients for growing plants, because it is difficult for water to leak out through a field's basic soil. Finally, soil erosion caused by heavy rainfall is minimized when terraced paddy fields are located stepwise on the slope of a mountain.

For all these reasons, Japanese farmers have endeavored to create paddy fields as much as possible out of the total agricultural land (i.e., paddy plus upland fields). In areas where winter is not very cold, paddy fields have been used to produce barley and other upland crops in winter. This system of crop rotation has been long established in various parts of Japan. Because the average farm size in Japan is very small, farmers have tried to produce as much as possible per unit of land, and have adopted crop-rotation systems. Livestock and human manure as well as decaying leaves have been widely used to fertilize soil, which established a self-contained, recycling system of paddy-land use.

However, when Japan entered into a high economic-growth period, Japanese agriculture started to change very dramatically. More and more farm-household members took up nonfarm employment, and the agricultural population started to decline very rapidly. As a result, many rice farmers became part-time farmers, who produced rice as a monoculture. The remaining farmers tended to be specialized, for example, in vegetable or livestock production. Rice farmers produced only rice, with heavy use of machinery, chemicals, and chemical fertilizer. The intensity of land use became very much lowered. In livestock farming, feeds were mostly purchased to increase the number of livestock.

These changes have not only broken down the crop-rotation system, but have also created agricultural-pollution problems such as an accumulation of livestock manure. On the other hand, soil fertility among crop farmers has gradually deteriorated due to the heavy dependence on chemical fertilizer.

Demand-side factors and policy factors have also influenced the nature of agricultural-land-use systems in Japan. The most important factor may be the reduction of rice consumption in Japan.

Although rice is still a stable crop in the Japanese diet, per capita consumption of rice decreased from the peak of 115 kg in 1962 to 72 kg in 1988. Alternatively, Japanese people are eating more and more livestock and dairy products. This demand-side change has necessarily reduced the area of rice paddies. The Japanese government has encouraged rice farmers to divert rice production to other crops in the paddy fields with government subsidies. However, crops diverted from rice are not as profitable as rice, and some paddy fields have been abandoned without any production. Naturally, the self-sufficiency ratio of food production to the total supply in Japan has rapidly decreased.

Perspectives on the Future and Research Needs

To increase self-sufficiency in domestic food production and to use paddy fields more efficiently, one of the most important agricultural-policy measures in Japan is reorganizing cropping patterns of paddy fields under these changing circumstances. For this, the Ministry of Agriculture has adopted and implemented a long-term project to remodel the system of land use toward a more ecologically adaptive and economically viable one. One of the core plans is to reorganize the use of paddy fields so that it combines rice as a main crop with other upland crops, such as wheat, barley, soybeans, or forage, on a regional basis. To this end, the federal and local governments in Japan encourage, for example, the establishment of a so-called manure-bank system in each locality for using organic manure (substituted for chemical fertilizer) and maintaining soil fertility in the region. The basic idea behind such efforts is to create a system of regionally based cycles in crop production by cooperation between livestock farmers and cash grain farmers. Several research projects have also been started to examine the sustainability of alternative land-use systems by researchers of the government agricultural research institutes and universities.

Concluding Remarks: A Time of Change

Since the high-economic-growth period, Japanese agriculture has become more and more energy-intensive and resource-exploitative. As a result of increased use of chemical fertilizers and agricultural chemicals, yield has been increased and stabilized, and quality has been improved. At the same time, however, several problems have emerged in this transformation process, such as overproduction of rice and some other crops, deterioration of soil fertility, and the dangers of an unsafe food supply. These problems are seen not only by the agricultural side, but also by consumer groups and the general public of Japan. There is increasing concern among Japanese consumers about the safety of domestic and imported agricultural products.

It is a time of change in the system of Japanese agricultural production and land use. More and more effort should be devoted to revitalizing Japan's centuries-old rice-based sustainable agriculture.

Bibliography

Bojo, J., et al. 1990. *Environment and Development: An Economic Approach*. The Netherlands: Kluwer Academic Publishers.

Kada, R. 1980. *Part-Time Family Farming*. Tokyo: Center for Academic Publications.

Kada, R. 1990. *Environmental Conservation and Sustainable Agriculture*. Tokyo: Ienohikari Publishers (in Japanese).

Morrison, R. E. 1990. *Global Change and the Greenhouse Effect*. Washington, D.C.: Congressional Research Service, Library of Congress.
Trudgill, S. 1990. *Barriers to a Better Environment*. Belhaven Press.

PART SIX

Review Chapters

21

Global Climate Change: Effects on Agriculture

Timothy Mount[1]

Lessons from the Energy Crisis

When the Organization of Petroleum Exporting Countries (OPEC) organized an embargo and raised the price of oil sold to industrial nations in 1973, the "energy crisis" was born. An immediate task faced by economists in the United States was to determine what effect higher prices for fuels would have on the economy. However, most models of the U.S. economy at that time did not incorporate the price or quantity of imported oil as explicit variables. Consequently, the scope of initial analyses was limited to indirect approaches, such as evaluating the effects of inflation associated with higher oil prices.

A similar situation exists now with respect to the "environmental crisis." Studying the effects of climate change on agriculture is handicapped by the fact that most models of agricultural production do not incorporate climate variables explicitly into the analytical framework. Typically, weather is treated as a random factor that is site-specific and that can be captured by a random residual. As a result, there is no direct way to determine the effects of changing patterns of temperature and rainfall on production by using these models. Indirect methods, such as shifting intercepts to reflect yield changes, are used by default to make existing models workable for analyzing the new environmental issues.

After the energy crisis, economists responded by developing new analytical techniques that were eventually incorporated into models of the economy. For example, theoretical work by Christensen et al. (1971) combined the theory of substitutability among factors and the flexible form

---

[1] From the Cornell Institute for Social and Economic Research in the Department of Agricultural Economics, Cornell University, Ithaca, N.Y.

of the translog cost function. Berndt and Wood (1975) used this model to add energy and materials as input factors to the traditional factors of labor and capital. Given these new methods and new data, questions such as whether energy was a substitute or a complement to labor and the effects of higher oil prices on employment could then be addressed directly.

In addition to developing new analytical techniques, new models were constructed that treated energy as the central issue. The global model developed by Edmonds and Reilly (1985) is a good example of such a model. The overall result of these developments was that the scope of analyses of energy problems expanded dramatically from the time immediately after the OPEC embargo. It seems reasonable to expect that a similar sequence of steps will occur in the analysis of how climate change affects agricultural production. This sequence can be summarized as follows:

Step 1. Introduce exogenous shocks or shifts to existing models.
Step 2. Develop new analytical techniques for incorporating climate variables into models of agricultural production.
Step 3. Develop new aggregate models of agricultural production that include climate variables explicitly.

Existing Models

Four papers on agricultural production are reviewed here. The papers by Kane et al. (1992) and by Adams et al. (1992) illustrate step 1. Kane et al. use a model of world grain trade [Static World Policy Simulation (SWOPSIM)] to investigate the effects of yield changes that are attributed to a doubling of CO_2 in the atmosphere. The yield changes are incorporated into the model as exogenous shifts in the supply curves. Because the yields increase for some countries (e.g., the USSR) and decrease for others (e.g., the United States), the net effects are found to be relatively small for world production. Nevertheless, the effects for individual countries and regions could be large, which illustrates the importance of regional differences and distributional effects among countries. In addition, Kane et al. emphasize that there is considerable scientific uncertainty about the future magnitudes of climate change and, therefore, that their results should be treated as predictions that are conditional on the input assumptions.

The analysis by Adams et al. uses a mathematical programming model of U.S. agriculture that is regionally disaggregated [the U.S. Mathematical Programming Model (USMP)]. This type of model represents the main characteristics of soil and climate more accurately than does a model based on national averages. The objectives of the analysis are to determine the cost of expanding the area directed to forestry in the United States as a way to sequester carbon, and of adopting practices to reduce emissions of carbon and other trace gases from agricultural production. The analytical framework emphasizes the importance of physical resources, such as land and regional diversity, within a country in assessing the effects of climate change. The same model has been used in earlier research (Adams et al., 1990) to assess the effects of climate change on U.S. agriculture. This study was a forerunner to the global analysis by Kane et al.

The paper by Lewandrowski and Brazee (1992) discusses the effects of farm programs in the United States on the incentives for farmers to adapt to climate change. Government intervention in the market could hinder or help this adaptive process, but current programs are not designed to encourage adaptation. The importance of policies that affect irrigation is expected by Lewandrowski and Brazee to increase if climatic change makes rainfall less reliable for crop production. Lewandrowski and Brazee conclude that government support programs could be much more expensive if climate change occurs unless greater flexibility in the choice of cropping patterns by producers is encouraged through new policies.

The paper by Kaiser et al. (1992) represents an example of step 2 in the sequence listed in the first section. The Kaiser et al. model contains a climate component, a soil–crop-yield component, and a management-decision component that are linked together. The objective of the model is to develop a generic framework that is not site-specific and that can be applied in different regions, given information about climate, soil, the array of cultivars, economic cost, and farm technology. The main characteristics of the model are (1) it is dynamic, and, therefore, can address the adaptive processes of how production patterns and production levels change over time; (2) the riskiness of production is reflected in the management-decision criterion, and this riskiness can also change over time as climate changes; and (3) major features of agricultural production are explicit (e.g., the choice among different cultivars, the number of available field days, and the moisture content of grain at harvest). The main limitation of the current model is that the links from production in the selected region (southern Minnesota) to the market for grains is relatively simple, and, consequently, the different adaptive processes that are due to economic responses to changing production patterns are not yet explicit. However, these types of links will probably be developed in the future.

The main advantage of the Kaiser et al. framework is that it provides new insights into how and why production patterns change. First, yield changes depend on the mix of cultivars selected (e.g., early vs. late varieties). In fact, new cultivars could be introduced into the array of available options to show how farmers might, for example, replace existing varieties with new drought-resistant varieties in response to drier conditions. Second, the choice of cultivar is not determined exclusively by its yield potential. The availability of field days for planting and the cost of drying grain are also important factors. Finally, the framework allows one to evaluate the effects of changes in the pattern of rainfall and temperature through the growth cycle from planting to harvest. This is important because of the vulnerability of cultivars to moisture deficits at specific periods in the life cycle (i.e., at germination, when flowering, and just prior to harvest).

In summary, the two step 1 models (Kane et al., 1992; Adams et al., 1992) use comparative statics to evaluate the effects of climate change, and exogenous shifts in yields to distinguish among different scenarios. The step 2 model (Kaiser et al., 1992) provides more insight into why yields change and into the dynamics of adaptation. Nevertheless, there is still a long way to go before a fully integrated global model of agricultural production and climate change can be developed. On the basis of experience with energy models, step 3 models of agricultural production and climate change will appear in the future.

The Need for Long-Run Analysis

The time scales for the analysis of climate change are long (i.e., >50 years), and the changes are expected to be gradual. For this type of analysis, the implications of other factors will be just as important for agricultural production as will climate change. Probably the most important factors are the growth in world population and the increasing inequity of incomes between industrialized and developing countries.

The changes in population, income, and energy use that occurred during the first half of the 1980s can be seen in Table 21.1. Developing nations accounted for 76% of world population but generated only 14% of world income in 1986. The rate of population growth was over twice as high in developing countries as in industrialized countries, but the two rates for real income growth were very similar. Consequently, the rate of real income growth per capita is much smaller (0.18%/year) in developing countries than in industrialized countries (1.36%/year). Because food is a necessity, the increasing inequality of the distribution of income will have a negative effect on the demand for food worldwide.

Population growth and the size of the income gap between developing countries and industrialized countries will have major effects on the demand for food. It is likely that the productive potential of U.S. agriculture will be needed in the future if all countries are to attain adequate levels of nutrition. To a large extent, the demand for food exports from the United States will depend on how successful policies are in reversing current demographic and economic trends, particularly with respect to the slow growth of income per capita in developing countries.

Greater use of energy is associated with higher levels of income but also with higher emissions of greenhouse gases. Hence, climate changes caused

Table 21.1 A Comparison of Industrialized and Developing Countries: Population, Income, and Energy Use

	Industrialized Countries	Developing Countries	All Countries
Population			
Quantity (×10⁹)[a]	1.192 [24.45%]	3.684 [75.55%]	4.876
Growth (%/year)[b]	0.76	1.98	1.67
Income			
Quantity ($×10¹²)[ac]	13.10 [86.15%]	2.10 [13.84%]	15.20
Growth (%/year)[b]	2.13	2.16	2.16
Energy use			
Quantity[a]			
(J×10²⁰)	2.128 [74.51%]	0.728 [25.49%]	2.856
(Btu×10¹⁵)	201.7 [74.51%]	69 [25.49%]	270.7
Growth (%/year)[b]	0.25	3.79	1.07
Income per capita			
Quantity ($×10³)[ac]	10.988	0.571	3.118
Growth (%/year)[b]	1.36	0.18	0.46
Energy per capita			
Quantity[a]			
(J×10²⁰)	1.785	0.197	0.586
(Btu×10¹⁵)	169.2	18.7	55.5
Growth (%/year)[b]	−0.51	1.7	−0.59

[a]For 1986.
[b]Average for 1980–1986.
[a]1986 U.S. dollars.

Source: Derived from D. Chapman and T. Drennen (1990).

by global warming will be affected by the growth of the use of energy. One of the most interesting results in Table 21.1 is the contrast in growth rates of energy use per capita between the industrialized countries (−0.51%/year) and the developing countries (1.70%/year). The disparities in the size of the population and level of income per capita between industrialized and developing countries will make it difficult to stabilize global energy use and the emissions of greenhouse gases. This is the main issue addressed by Chapman and Drennen (1990).

The overall conclusion for modeling the effects of global change on agricultural production is that models must integrate components for the size and composition of population, the distribution of income, and the use of fossil fuels. The common practice in economic modeling of treating demographic variables as exogenous and ignoring distributional effects is no longer warranted, given the long time horizons needed for the analysis. One positive note is that the interrelationships that exist between energy use and climate will make it sensible to build on the analytical developments for modeling energy use. A major limitation of standard energy models is that the spatial representation of economic activity is insufficiently developed to model the effects of climate on agriculture.

Conclusions

Several conclusions emerge from reviewing the four models of agricultural production. (1) The scientific uncertainty about the nature of climate change implies that current analyses of the effects on agricultural production are conditional on the input assumptions. Consequently, sensitivity analyses are justified to identify which parameters in a model matter. (2) The scope of analyses of the effects of climate change should be global, because of the interdependencies among countries through international grain markets, and long-run, because of the gradual nature of climate change. (3) Information about physical resources, such as soil characteristics and irrigation, is more important for analyzing climate change than it is in traditional economic models of agriculture. (4) Interdisciplinary approaches to modeling agricultural production are needed to understand biological as well as economic adaptations to climate change. (5) Existing types of government interventions in world markets may hinder the ability of farmers to adapt to climate change. (6) For long-run analyses, population growth and the equity of income between industrialized and developing countries will be important factors in U.S. agricultural production. (7) Energy use is an important determinant of both income levels and environmental quality.

The overall recommendation that emerges from this review is that the scientific uncertainty about climate change should not be used as an excuse for delaying study of the effects on agriculture. New methods need to be developed to address agriculture, population, energy use, and the distribution of income in a long-run framework. Even if climate does not change appreciably, the growing inequity of incomes between industrialized and developing countries and the continued growth of world population justify this type of modeling effort.

References

Adams, R. M., C-C. Chang, B. A. McCarl, and J. M. Callaway. 1992. "The Role of Agriculture in Climate Change: A Preliminary Evaluation of Emission-Control Strategies," in J. M. Reilly and M. Anderson, eds., *Economic Issues in Global Climate Change: Agriculture, Forestry, and Natural Resources*. Pp. 273–87. Boulder, Colo.: Westview Press.

Adams, R. M., C. Rosenzweig, R. M. Peart et al. 1990. "Global Climate Change and U.S. Agriculture." *Nature* 345: 219–24.

Berndt, E. R., and D. O. Wood. 1975. "Technology, Prices, and the Derived Demand for Energy." *The Review of Economics and Statistics* LVII: 259–68.

Chapman, D., and T. Drennen. 1990. "Equity and Effectiveness of Possible CO_2 Treaty Proposals." *Contemporary Policy Issues* VIII(3): 16–28.

Christensen, L. R., D. W. Jorgenson, and L. J. Lau. 1971. "Conjugate Duality and the Transcendental Logarithmic Production Function." *Econometrica* 39: 255–6.

Edmonds, J., and J. Reilly. 1985. *Global Energy: Assessing the Future*. Oxford: Oxford University Press.

Kaiser, H. M., S. J. Riha, D. G. Rossiter, and D. S. Wilks. 1992. "Agronomic and Economic Impacts of Gradual Global Warming: A Preliminary Analysis of Midwestern Crop Farming," in J. M. Reilly and M. Anderson, eds., *Economic Issues in Global Climate Change: Agriculture, Forestry, and Natural Resources*. Pp. 91–116. Boulder, Colo.: Westview Press.

Kane, S. M., J. M. Reilly, and J. Tobey. 1992. "A Sensitivity Analysis of the Implications of Climate Change for World Agriculture," in J. M. Reilly and M. Anderson, eds., *Economic Issues in Global Climate Change: Agriculture, Forestry, and Natural Resources*. Pp. 117–31. Boulder, Colo.: Westview Press.

Lewandrowski, J. K., and R. Brazee. 1992. "Farm Programs and the Climate Change: A First Look," in J. M. Reilly and M. Anderson, eds., *Economic Issues in Global Climate Change: Agriculture, Forestry, and Natural Resources*. Pp. 132–47. Boulder, Colo.: Westview Press.

22

Evaluating Socioeconomic Assessments of the Effect of Climate Change on Agriculture

Steven T. Sonka[1]

Introduction

This paper will suggest high-priority research directions that should be undertaken as we strive to develop an improved understanding of the potential effects of global change on agricultural systems. Besides papers in this volume, this discussion will draw on the results of an extensive analysis recently completed for the Organization for Economic Cooperation and Development (OECD) (Sonka, 1991). That review focused on methodological enhancements needed to provide better information to decision makers about the potential agricultural impacts of climate change.

This paper focuses on those studies that provide quantitative estimates of future implications of climate change. Nonempirical analyses and those that assess the effectiveness of alternative mitigation strategies [such as Adams et al. (1992)] are not the main focus of this discussion.

This paper is organized into four sections. First, relevant decision makers and their interests in climate change are defined. Then, the framework used previously (Sonka, 1991) to review methodologies will be presented along with key results of that analysis. Next, two key deficiencies of current approaches to assessing impacts of climate change are identified. The paper concludes with a brief summary and a list of high-priority directions for socioeconomic research efforts.

---

[1] From the Department of Agricultural Economics, University of Illinois at Urbana-Champaign.

Methodology and a Decision-Making Perspective

Specification of a particular set of procedures as preferred implies general agreement on the purpose driving the underlying analysis. It appears that the underlying purpose is still evolving for agricultural-impact assessments of climate change (AIACCs). As will be documented later, most previous studies served a very narrow purpose: to detail that a changed climate could have profound consequences for the agricultural-production system.

This paper uses a decision-making perspective. Doing so reflects the assumption that future AIACCs will be justified by the need for societal decision makers to make many difficult choices. To make such choices, decision makers need information that will aid in problem identification, evaluation of alternative actions, and/or selection of preferred responses. From a decision-making perspective, demonstrating potential effects (the purpose of previous and current empirical studies) at best contributes information to problem identification, but adds little to the other two needs.

Clearly, more powerful and focused methodologies are required. To suggest the types of improvements needed, one must identify the relevant decision makers and specify their interests.

Who Are the Decision Makers?

To understand the methods most appropriate for decision making, it is useful to clearly define the types of decision makers who can be expected to find the research results relevant. For AIACCs, the following classification of actors is appropriate: public sector—political leaders, bureaucrats and institutional managers, and the general populace; private sector—agricultural producers, agribusiness managers, and consumers.

For political leaders, the desirability of political action, including legislation, is a natural concern. Bureaucrats and institutional managers have many nonpolitical factors to consider. (Examples include research initiatives or changes to water-management systems.) In addition, bureaucrats are likely to have latitude in implementing legislative imperatives. Depending on each nation's political system, the concerns of the general populace, at a minimum, influence the decisions of political leaders.

Agricultural producers and agribusiness managers make long-term investments that are based, usually implicitly, on some expectation about the future characteristics of climate. Further, agricultural producers will make short-term adaptation decisions in response to a changing climate. Both the nature and degree of these adaptations may be extremely important in determining the ultimate effect of climate change. And, finally, consumers are affected by the price and quantity of climate change's eventual effect on agriculture.

Key Issues Facing Decision Makers

Identifying decision makers allows us to directly consider the types of decision questions of interest. Deriving information that will allow decision makers to better understand the effects of alternative choices is, of course, the real reason that society should use scarce resources to analyze AIACCs.

In broad terms, three key decision issues can be expected to dominate society's interest in AIACC: food security (both domestically and internationally), the economics of food consumption and production, and investment in agricultural enterprises and institutions.

Food Security. One of the basic concerns of humankind is food security, an issue with local, national, and global dimensions. Climate change could enhance or diminish a local area's comparative advantage in agriculture. In those societies involved in the international trade of agricultural products, however, local production capabilities must be considered in the context of changing international supply-and-demand balances. Although food security can be a chronic or long-term issue, short-term scarcity (because of heightened year-to-year variability of climatic conditions) is also an important social concern.

Food in the Economy. In all societies, food production and/or consumption are major economic activities. Changing local and international supply factors can markedly affect revenues from production and costs to consumers. In both cases, profound implications for the general economy are possible. For example, an increased cost for food would mean fewer funds available for the purchase of other goods, but may mean higher wealth in food-producing areas. Interregional shifts in comparative advantage and associated employment and welfare changes are particularly relevant here.

Investment in Agricultural Enterprises and Institutions. Both agricultural producers and managers of agricultural institutions will make investment decisions that can greatly affect the economic and social impacts of climate change. Methods to assess the impacts of climate change, therefore, must reflect all of the forces—both within and outside the decision maker's organization—that these managers could face, as well as the range of current and future choices they would consider.

Providing information relevant to decisions about the future impacts of climate change on these three broad issues should drive the evaluation of methodological alternatives. It is important to note that these issues, although obviously linked, are clearly not of equal social priority. Here, the three issues have been listed in declining order of societal priority, and any methodological trade-offs that must eventually be made should recognize that ordering.

Previous Approaches to AIACC

Because of space limitations, only a summary of the OECD review (Sonka, 1991) will be presented in this paper. The framework developed for that analysis will be applied to some papers in this volume as well.

That framework consisted of two parts: an analysis of assumptions, data sources, and modeling approaches used to conduct the analysis; and a consideration of the output of the analyses and description of the types of effects measured. For the former component, 10 climatic factors and modeling considerations were detailed: source of climate-change-variable estimates, climatic factors considered, timing of climate change, geographic unit of analysis, modeling unit, economic model(s), motivation of decision maker, assumptions about changes in nonclimatic factors, incorporation of uncertainty about the extent of climate change, and agricultural activities considered. For each study, an assessment was made of the extent to which each of the following six effects were measured: agricultural production, shifts in regional comparative advantage, long-term environmental effects, farm profitability, national or international food stocks, and secondary economic effects.

Nineteen empirical analyses were reviewed.[2] A comprehensive discussion of the findings is available in Sonka (1991). The following remarks will emphasize critical aspects of the perspectives used in framing the key underlying assumptions and in organizing the output estimated from these studies. Of the assumptions underlying these analyses, three are especially significant: the geographic unit of analysis, the timing of the specified climate change, and the manner in which nonclimatic factors are incorporated in the analysis.

Geographic Unit of Analysis

Modeling agricultural impact assessments can and is done at a variety of levels of aggregation. These include yields of the plant or field, profit of the farm firm, and productivity and/or social welfare regionally or nationally.

---

[2] Adams, 1989; Adams et al., 1988; Arthur et al., 1986; Bergthorsson et al., 1988; Dudek, 1988; Kane et al., 1989; Kettunen et al., 1988; Kokoski and Smith, 1987; McKeon et al., 1988; Pitovranov et al., 1988; Rosenzweig, 1985, and unpublished observations, 1989; Salinger, 1988; Santer, 1985; Smit, 1987; Smith et al., 1989; Smith and Tirpack, 1988; Williams et al., 1988; Yoshino et al., 1988.

Most studies reviewed for the OECD study focused their analyses on agricultural systems in developed nations of the Northern Hemisphere. Although this region is certainly important for analysis, evaluation of implications of climate change in the developing nations is certainly needed. Liverman (1992) estimated productivity effects on Mexican agriculture and demonstrated the critical importance of evaluation in developing nations.

Timing of Climate Change

In all but one of the studies, the climate change hypothesized was assumed to be instantaneous. Except for Pitovranov et al. (1987), analysis of the adjustment from the current climate to the specified alternative was not conducted. Extreme events, usually drought, were considered in several of the analyses reviewed. However, year-to-year variability and the role of extreme events within those dynamics were not evaluated in these cases.

Assumptions about Changes in Nonclimatic Factors

Evaluation of the interaction between climate change and key variables such as population growth and changes in resource availability were strikingly absent. In a few cases, the potential for technological change of some type was considered.

The net effect of these two perspectives is to define a world in which climate has undergone a major shift but all other factors are like those of the early 1980s. Such a world seems implausible at best. Further, analyses can be conducted that are not so rigidly constrained. For example, the paper by Kaiser et al. (1992) is a notable exception to this approach. Although the process linking the farm-level model to current market forces is a significant weakness, the overall approach does suggest the promise inherent in more dynamic characterizations of climate change. Also Kane et al. (1989, 1992) have enhanced their investigation of the interplay between potential climate change and other changes likely in society in studies conducted since the OECD review.

Effects Measured

The studies reviewed all explicitly measure the physical-production effects of a change in climate. However, no study evaluates all six of the potential effects identified. Three studies included measurements on four of the effects.

Comparative advantage, environmental effects, and profitability are relatively popular factors to evaluate. It should be noted, however, that even though a single study may have considered many factors, the analysis may not have been accomplished in a completely integrated fashion. For example, the environmental effects may be derived from detailed modeling at the field or watershed area. Profitability, however, may be assessed with an aggregate model of a regional economy.

A surprising finding was the paucity of studies that directly addressed impacts on food stocks. One theory is that food security and food-sector economics are key driving forces fueling societal interest in this topic. If that perspective is correct, the methods currently used for AIACC do not lead to results that pertain to the basic decision issues that society must address.

The works of Kane and of Adams and their colleagues illustrate the importance of considering the consumption aspects of the food economy. Although it is necessary to evaluate direct effects on food production, incorporating the demand for food is also necessary to comprehensively estimate the welfare effects of climate change. Changes in food prices can distribute the impacts of climate change throughout the global economy and, in some cases, can offset the economic effects of food-production changes within a region or nation.

Future Directions

Early in this paper, I suggested that a decision-making perspective could be productively used as the driving force for defining an appropriate methodology. Types of decision makers and issues of likely relevance were identified. The three decision issues expected to dominate society's interest in AIACC were noted as food security (with both national and international dimensions), the economics of food consumption and production, and investment alternatives in agricultural enterprises and institutions. Providing relevant information about the future impacts of climate change on these three broad issues, therefore, should shape the evaluation of the appropriateness of any methodological alternatives.

Developing methodology must be recognized as evolutionary in nature. Therefore, rather than suggesting that previous work was methodologically deficient, the discussion in the preceding section should be viewed as indicating needed future directions. The remainder of this section will discuss two major thrusts that should underlie future analyses: explicitly addressing changes in nonclimatic factors, and investigating the dynamic aspects of a changing climate.

Explicitly Addressing Changes in Nonclimatic Factors

If concerns about food security are a major justification for conducting future AIACCs, then future population sizes and resource availability are just as important as changes in temperature or precipitation. In addition to population, key factors include the agricultural land base, economic growth, water availability, input costs, and socially imposed restrictions on agricultural practices. For example, the Adams et al. (1992), Fosberg et al. (1992), Richards (1992), and Bowes and Sedjo (1992) studies in this book all indicate the importance of considering trade-offs in land use between forestry and agriculture. Accurately predicting all these factors, of course, is an unreachable goal. Considering potential changes in these factors as they interrelate with the AIACC is a reasonable expectation, however.

Further, using current values for nonclimatic factors does make an implicit assumption about the level of all the nonclimatic factors. The assumption is that such factors would stay at their current levels forever. If there is one assumption that is clearly implausible, it is exactly the one that is used most often. Significant changes in factors such as climate, population, and resource availability will not go unnoticed. Therefore, failure to explicitly consider decision-maker responses is just as serious as ignoring the changes in the underlying factors themselves. Indeed, failure to explicitly consider such responses significantly weakens the credibility of the AIACCs reviewed previously. The types of behavioral responses that warrant consideration are adoption of alternative management practices that producers can implement with existing technologies, changes in food-consumption patterns as consumers respond to differing supply regimes for food, and efforts to create and implement technological improvements that mitigate or exploit alternative climatic regimes.

The process of considering nonclimatic factors is related to incorporation of behavioral adaptations. For example, greater use of irrigation is one currently available management practice that might offset yield reductions from adverse temperatures or declines in natural rainfall. This response assumes that adequate irrigation water is economically available, a critical assumption. Failure to address uncertainties such as the availability of water for irrigation may send false signals about the resilience of the agricultural system.

Tracing the Path of a Changing Climate

Decision makers live in a challenging, complex, and constantly changing present. Analyses that abstract too far from that setting lessen their usefulness, no matter how technically well done. The year 2030 (or any other year 40 to 50 years in the future) may be an excellent reference point for technical and climate reasons. From a decision perspective, particularly when the choice involves action to reduce the emission of greenhouse gases today, economic-impact data tied to such a reference point lose meaning.

Instead, the dynamics of changing conditions as society moves to that future are of extreme relevance.

The technical feasibility of meaningfully tracing the impacts of a changing climate is a serious issue. Two major deficiencies in our analytical capabilities must be addressed. First, atmospheric scientists' understanding of climate change is uncertain. Even tracing alternative paths of change may overwhelm the credibility of existing knowledge bases. Second, accurately modeling economic systems over time has proven challenging to economists, and forecast credibility is a significant concern. However, the task needed here does not require the same precision that forecasting quarterly gross national product for each of the next 5 years does. Instead, broader implications of the impacts of climate change and societal adaptations are the goal. This perspective may make the process achievable.

Addressing climate change as an evolving process opens tremendous opportunities. One is the explicit analysis of the potential for greater frequency of extreme climate events. Some analysts feel that greater variability may be a feature of a changing climate. If so, that variability may have profound, relatively immediate implications for decision makers. When food security and the economics of the global food system are key issues, variability of supply can be as economically relevant as the average levels of these variables. As noted previously, Kaiser et al. (1992) illustrates both the complexity of modeling necessary and the potential richness of results available from modeling a changing climate and its effect on agriculture. The Lewandrowski and Brazee (1992) paper, focusing on farm-program implications conceptually demonstrates the important role that institutional restraints can have.

Summary and Research Priorities

Results of previous studies have fallen short of the goal of providing the information that decision makers need to respond economically, effectively, and equitably to the climate-change problem. Previous studies have almost exclusively assumed an instantaneous shift in climate conditions with no change in nonclimate variables. By doing so, the analyses have had relatively little to say about the potential effect of climate change on food security and on economic activity in the food-production and -consumption sectors. This lack is the key deficiency of work conducted so far.

It is a complex, resource-intensive task to conduct empirical studies that consider changes in nonclimate factors and strive to incorporate the dynamic implications of a changing climate. Clearly, those researchers who conducted the analyses reviewed earlier recognized the need for considering these perspectives. They faced a serious constraint in doing so, however, because of resource limitations and the lack of tested protocols and procedures that could be adapted to meet the needs of specific study situations.

Several socially important and intellectually exciting research priorities are available to economists and other social scientists willing to devote sustained and creative efforts to the analysis of the impacts of climate change on agriculture. Three seem urgent. First, the scope of analysis must be broadened to include change in key supply-and-demand factors in addition to climate. Doing so allows for the analysis of potential food security as well as the investigation of welfare effects on producers and consumers. Second, the decision aspects of uncertainty must be investigated more fully—in particular, the difficulties decision makers will face as they respond to a changing climate. Although comparative static analysis of equilibrium circumstances is often useful, many of the troubling aspects of climate change for decision makers are in managing an uncertain set of transitions. Third, the locational aspects of analysis must be expanded to include climate-change implications for developing nations. Semiarid nations with large populations (e.g., Mexico and India) are especially attractive candidates for analysis.

Responding to the priorities just noted, especially the first two, has several important implications. First, the focus may well shift from hypothetical direct effects occurring decades in the future to policy uncertainties to be faced over the next decade. Second, economists will have to incorporate paradigms from other areas of study to successfully pursue these goals. For example, modeling a world in which all key factors can change is likely to require a more complex form of sensitivity analysis. Scenario analysis, as practiced in the business-policy field, would seem to have value in this effort. Further, the uncertainty implied in a changing climate eclipses the type of variability modeled by postulating a known probability function. The ambiguity paradigm of decision theory, which says that one of several probability distributions may be in force, is intriguing for the changing-climate issue.

Although it is challenging to employ paradigms from other areas of study, such mergers often have handsome payoffs. It is quite likely that economists will be able to make substantial contributions to society in evaluating climate change—if the risky steps needed to aggressively conceptualize the problem are undertaken. A major factor needed to encourage such risk taking is the presence of sustained funding initiatives for work investigating the interrelations of climate and agriculture.

References

Adams, R. M. 1989. "Global Climate Change and Agriculture: An Economic Perspective," paper presented at the American Agricultural Economics Association (AAEA) annual meeting, Baton Rouge, La., July 30–Aug.25, 1989.

Adams, R. M., C. C. Chang, B. A. McCarl, and J. M. Callaway. 1992. "The Role of Agriculture in Climate Change: A Preliminary Evaluation of Emission-Control Strategies," in J. M. Reilly and M. Anderson, eds., *Economic Issues in Global Climate Change: Agriculture, Forestry, and Natural Resources.* Pp. 273–87. Boulder, Colo.: Westview Press.

Adams, R. M., B. A. McCarl, D. J. Dudek, and J. D. Glyer. 1988. "Implications of Global Climate Change for Western Agriculture." *Western Journal of Agricultural Economics* 13: 348–56.

Arthur, L. M., V. J. Fields, and D. F. Kraft. 1986. "Towards a Socio-Economic Assessment of the Implication of Climate Change for Agriculture in Manitoba and the Prairie Provinces," report submitted to Atmospheric Environment Services, Department of Agricultural Economics, University of Manitoba, Winnipeg, March 1986.

Bergthorsson, P., H. Bjornsson, O. Dyrmundsson, B. Gudmundsson, A. Helgadottir, and J. V. Jonmundsson. 1988. "The Effect of Climatic Variations on Agriculture in Iceland," in M. L. Parry, T. R. Carter, and N. T. Konijn, eds., *The Impact of Climate Variations on Agriculture, Vol. 1: Assessments in Cool Temperate and Cold Regions.* Dordrecht, The Netherlands: Kluwer.

Bowes, M. D., and R. A. Sedjo. 1992. "Climate Change and Forestry in the U.S. Midwest," in J. M. Reilly and M. Anderson, eds., *Economic Issues in Global Climate Change: Agriculture, Forestry, and Natural Resources.* Pp. 252–72. Boulder, Colo.: Westview Press.

Dudek, D. J. 1988. "Assessing the Implications of Changes in Carbon Dioxide Concentrations and Climate for Agriculture in the United States," in M. L. Parry, T. R. Carter, and N. T. Konijn, eds., *The Impact of Climate Variations on Agriculture, Vol. 1: Assessments in Cool Temperate and Cold Regions.* Dordrecht, The Netherlands: Kluwer.

Fosberg, M. A., R. A. Birdsey, and L. A. Joyce, and M. A. Fosberg. 1992. "Global Change and Forest Resources: Modeling Multiple Forest Resources and Human Interactions," in J. M. Reilly and M. Anderson, eds., *Economic Issues in Global Climate Change: Agriculture, Forestry, and Natural Resources.* Pp. 235–51. Boulder, Colo.: Westview Press.

Kaiser, H. M., S. J. Riha, D. G. Rossiter, and D. S. Wilks. 1991. "Agronomic and Economic Impacts of Gradual Global Warming: A Preliminary Analysis of Midwestern Crop Farming," in J. M. Reilly and M. Anderson, eds., *Economic Issues in Global Climate Change: Agriculture, Forestry, and Natural Resources.* Pp. 91–116. Boulder, Colo.: Westview Press.

Kane, S., J. Reilly, and R. Bucklin. 1989. "Implications of the Greenhouse Effect for World Agricultural Commodity Markets," paper presented at the Western Economic Association Conference, Lake Tahoe, Calif., June 1989.

Kane, S., J. Reilly, and J. Tobey. 1992. "A Sensitivity Analysis of the Implications of Climate Change for World Agriculture," in J. M. Reilly and M. Anderson, eds., *Economic Issues in Global Climate: Agriculture, Forestry, and Natural Resources.* Pp. 117–31. Boulder, Colo.: Westview Press.

Kettunen, L., J. Mukula, V. Pohjonen, O. Rantanen, and U. Varjo. 1988. "The Effects of Climatic Variations on Agriculture in Finland," in M. L. Parry, T. R. Carter, and N. T. Konijn, eds. *The Impact of Climate Variations on Agriculture, Vol. 1: Assessments in Cool Temperate and Cold Regions.* Dordrecht, The Netherlands: Kluwer.

Kokoski, M. F., and V. K. Smith. 1987. "A General Equilibrium Analysis of Partial Equilibrium Welfare Measures: The Case of Climate Change." *American Economic Review* 71: 331–41.

Lewandrowski, J. R., and R. Brazee. 1992. "Government Farm Programs and Climate Change: A First Look," in J. M. Reilly and M. Anderson, eds., *Economic Issues in Global Climate Change: Agriculture, Forestry, and Natural Resources.* Pp. 132–48. Boulder, Colo.: Westview Press.

Liverman, D. M. 1992. "Global Warming and Mexican Agriculture: Some Preliminary Results," in J. M. Reilly and M. Anderson, eds., *Economic Issues in Global Climate Change: Agriculture, Forestry, and Natural Resources.* Pp. 332–52. Boulder, Colo.: Westview Press.

McKeon, G. M., S. M. Howden, D. M. Silburn, J. O. Carter, J. F. Clewett, G. L. Hammer, P. W. Johnston, P. L. Lloyd, J. J. Mott, B. Walker, E. J. Weston, and J. R. Willcocks. 1988. "The Effect of Climate on Crop and Pastoral Production in Queensland," in G.I. Pearman, ed., *Greenhouse: Planning for Climate Change.* Pp. 546–63. Melbourne, Australia: University of Melbourne.

Pitovranov, S., V. Iakimets, V. Kiselev, and O. Sirotenko. 1988. "Effects of Climatic Variation on Agriculture in the Subarctic Zone of the USSR," in M. L. Parry, T. R. Carter, and N. T. Konijn, eds., *The Impact of Climate Variations on Agriculture, Vol. 1: Assessments in Cool Temperate and Cold Regions.* Dordrecht, The Netherlands: Kluwer.

Richards, K. 1992. "Policy and Research Implications of a Recent Carbon-Sequestering Analysis," in J. M. Reilly and M. Anderson, eds., *Economic Issues in Global Climate Change: Agriculture, Forestry, and Natural Resources.* Pp. 288–308. Boulder, Colo.: Westview Press.

Rosenzweig, C. 1985. "Potential CO_2-Induced Climate Effects on North American Wheat-Producing Regions." *Climatic Change* 7: 367–89.

Salinger, M. J. 1988. "Climatic Warming: Impact on the New Zealand Growing Season and Implications for Temperate Australia," in G. I. Pearman, ed., *Greenhouse: Planning for Climate Change.* Pp. 546–75. Melbourne, Australia: University of Melbourne.

Santer, B. 1985. "The Use of General Circulation Models in Climate Impact Analysis —A Preliminary Study of the Impacts of a CO_2-Induced Climatic Change on West European Agriculture." *Climatic Change* 7: 71–93.

Smit, B. 1987. "Implications of Climate Change for Agriculture in Ontario." *Climate Change Digest*.

Smit, B., M. Brklacich, R. B. Stewart, R. McBride, M. Brown, and D. Bond. 1989. "Sensitivity of Crop Yields and Land Resource Potential to Climate Change in Ontario, Canada." *Climatic Change* 14: 153–74.

Smith, J. B., and D. A. Tirpak, eds. 1988. *The Potential Effects of Global Climate Change on the United States*. Washington, D.C.: U.S. Environmental Protection Agency.

Sonka, S. T. 1991. "Methodological Guidelines for Assessing the Socio-Economic Impacts of Climate Change on Agriculture," in *Climate Change*. Pp. 21–44. Paris: Organization for Economic Cooperation and Development.

Williams, G. D. V., R. A. Fautley, K. H. Jones, R. B. Stewart, and E. E. Wheaton. 1988. "Estimating Effects of Climatic Change on Agriculture in Saskatchewan, Canada," in M. L. Parry, T. R. Carter, and N. T. Konijn, eds., *The Impact of Climate Variations on Agriculture, Vol. 1: Assessments in Cool Temperate and Cold Regions*. Dordrecht, The Netherlands: Kluwer.

Yoshino, M., T. Horie, H. Seino, H. Tsujii, T. Uchijima, and Z. Uchijima. 1988. "The Effects of Climatic Variations on Agriculture in Japan," in M. L. Parry, T. R. Carter, and N. T. Konijn, eds., *The Impact of Climate Variations on Agriculture, Vol. 1: Assessments in Cool Temperate and Cold Regions*. Dordrecht, The Netherlands: Kluwer.

23

Implications of Global-Change Uncertainties: Agricultural and Natural Resource Policies

W. Kip Viscusi[1]

Introduction

Although global-change research is a comparatively new area of economics, the papers presented in this volume reflect a substantial degree of sophistication, in terms of both the economic modeling and the empirical analysis that underlies them. Perhaps the main difference between these papers and my own is the greater weight I place on the substantial uncertainties involved. Note, however, that the papers by Reilly (1992) and Yohe (1992) do address the role of uncertainty in a fundamental manner.

The Efficiency Objective

Let me begin with an indication of the disciplinary biases that I bring to the task of commenting on these papers. My overall approach is that of economic analysis, and, more specifically, I advocate an economic-efficiency perspective. This view contrasts with that of Norgaard and Howarth (1992), who question the legitimacy of benefit-cost analysis and economic analysis more generally. Their thoughtful paper challenges economics in a variety of ways. Most fundamentally, they question the pertinence of conventional economic analysis to the fundamental issues posed by global change.

I view these attacks as useful in highlighting important areas of economic inquiry. However, as a critique of the economics of global change, such criticisms appear premature. Because economic analysis is still being developed to address many of these issues, it is too early to assess the

---

[1] From the Department of Economics, Duke University, Durham, N.C.

adequacy of these tools in addressing this class of problems. Moreover, I believe that we can obtain profitable insights by applying the existing economic tools to many of the difficult problems that arise in this area.

One issue raised by Norgaard and Howarth (1992) is sustainability. In particular, they ask, to what extent should we view sustainability as a constraint that we will impose on acceptable policies? Must our policies result in a path of future outcomes that lead to sustainable development for future generations?

I rephrase this issue to ask whether it is efficient from an economic standpoint to impose sustainability as a constraint. In general, I think that it is not. Typically, we are not faced with extreme outcomes such as those involving sustainable and nonsustainable future-development paths. The range of outcomes is a continuum, not a pair of discrete alternatives. Moreover, even if we are subjected to choices involving sustainable and nonsustainable outcomes, there may be some probability of nonsustainability that we will be willing to tolerate as we trade off this concern against other objectives. It is unlikely that we will adopt a policy that reflects an absolute unwillingness to bear any risk whatsoever. Thus, our objective should not be to maximize the probability of sustainable development when we cannot ensure a probability of sustainability of 1.0. Rather, there will be other concerns that we must take into account.

I believe that the source of our differences is the following. Being an able economist, Richard Norgaard has correctly formulated the policy-choice problem, and he believes that after we solve this problem, we will arrive at a policy that ensures sustainable future development as the optimal solution. Because he is so confident of this answer, he is willing to impose the constraint of reaching this policy outcome and then to manipulate variables such as the discount rate to ensure that we achieve this objective. Although it may be that this will be the optimal policy, my preference is to establish the justification for this or any policy from the ground up rather than from the top down. In particular, I would like to see sustainable paths of development emerge as the optimal policy choice after we have given careful and complete consideration to the relative merits of all policy alternatives, rather than assuming *ex ante* that a particular policy course is correct. Developing the optimal policy from the ground up will not only better enable us to ensure that we have selected the best policy, but will also serve to motivate people to adopt it.

Communicating the Uncertainties

The papers included in this volume indicate that global-warming problems are often difficult to grapple with conceptually. The society that must take action several decades from now and the informational base at that time may be quite different from today's. The optimal-control model of Reilly (1992) and the farmer's-choice problem structured by Yohe (1992) illustrate that the structure of these problems is often not trivial. However, as the papers by Drennen and Chapman (1992) and by Gollehon et al. (1992) indicate, it is possible to make considerable progress in assessing some of the specific empirical elements of these models. Moreover, as Reilly (1992) suggested, many of the applicable policy principles, such as the need to promote international cost-effectiveness in our climate-change policies, can serve to enhance the efficacy of current policies considerably.

Uncertainty is a major component in these policy analyses. Reilly (1992) noted the role of uncertain damage functions that prevail, and all the authors indicate a general awareness of the overall uncertainties pertaining to the entire area of global change. Drennen and Chapman (1992), for example, pointed out that there is substantial imprecision in our knowledge of future temperature ranges. Similar statements are made in most treatments of climate change. However, what is not clear is what these temperature-change ranges mean. Are they point estimates, best guesses, or an impressionistic confidence interval?

Generally, the authors are not acting as classical statisticians establishing 95% confidence intervals based on carefully executed experiments. Indeed, the standard models of classical statistics appear wholly inappropriate to this subjective forecasting exercise. These ranges seem to be more akin to the ranges promulgated by economic forecasters. Forecasts of gross national product (GNP) by one forecaster may have one particular range, but the span of estimates for all major econometric forecasters may be considerably greater.

A fundamental need in the climate-change literature is to develop some common approach to assessing the uncertainties that are present and to indicate what we mean by these uncertainties. Can we assess a subjective probability distribution for the temperatures that may prevail? Knowledge of the entire range of outcomes and the probabilities attached to them is more meaningful than knowing ranges but not the context. This approach will necessarily be Bayesian (Raiffa, 1968), rather than adhering to classical statistics.

It is important to obtain some sense of the extent of our uncertainty, because we cannot operate on the basis of mean estimates alone when there are substantial nonlinearities in the payoffs that might result from the policy choices we make. The extreme cases of nonlinearities are the sustainability issues raised by Norgaard and Howarth (1992), but even at a more conventional level, such as the shape of the damages function in the Reilly (1992) model, there may also be nonlinear effects.

Perhaps the best starting point for addressing uncertainties is to develop a more uniform vocabulary for indicating the uncertainties that are present. Several authors who attended the conference from which these papers emerged alluded to these uncertainties. William Easterling, for example, indicated that he doesn't "take the numbers all that seriously." Duane Chapman's slides included typed versions of estimates as well as crossed-out versions of these numbers with handwritten replacements, which presumably reflect refinements that capture the uncertainty of his forecasting exercise. Similarly, Richard Norgaard declared that "we can't predict the future."

There appears to be a general appreciation of the fundamental uncertainties that affect this area, but there is less agreement about how we should approach these uncertainties and incorporate them into the analyses. Perhaps the main message is that there is considerable uncertainty and, in the parlance of Bayesian decision theory, we should spread the tails of our probability distribution appropriately to reflect this uncertainty. In addition, rather than simply indicating ranges of estimates or expected temperature ranges without stating confidence levels, it would be preferable for authors to indicate their subjective assessment of the distribution of outcomes and to attach explicit probabilities to the most important estimates in their analyses. Perhaps then we could not only appreciate the extent of the uncertainty, but we could also modify the analysis appropriately in situations where there are nonlinear payoffs. It is perhaps most important at this stage of global-warming research to establish a uniform basis for communicating our understanding of the problem and for sharing the information that we do have. Establishing a uniform vocabulary for addressing these uncertainties should head the priority list.

One form of uncertainty that is recognized by the authors is the uncertainty of future damages. Perhaps the most fundamental aspect of this uncertainty pertains to the values that future generations will place on these outcomes. The main complication that arises in this instance is the endogeneity of the income that future generations will have, which in turn will affect their valuation of climate change. The decisions we make today will affect the well-being of future generations and the values they will place on future policy outcomes. Perhaps the best reference point to use in assessing the values of future generations is to ask what valuations would prevail if we adopted an efficient course of policies from now until some future point in time.

The role of uncertainty also introduces the role of regret. Reilly (1992) raises the topic of a no-regrets policy with respect to climate change by indicating that some individuals have advocated that we should at least take the measures that are optimal from the standpoint of reducing environmental damage —such as using gasoline efficiently—wholly apart from any consideration of climate change. In my work with Wesley Magat for the Environmental Protection Agency (EPA), I am developing an estimate of the optimal energy tax rate based on such considerations.

The role of no-regrets policies is even more complex than the concerns raised by Reilly (1992) once we introduce the complication of uncertainty. It is one thing to suggest that we pursue a policy option that we would choose not to regret in the future, given the certain damages that we know will prevail, but if we also introduce the possibility of lotteries in terms of outcomes, regret will arise if the lottery outcome is unfavorable. We may express regret in the future if we do not take the course that turns out to be the correct one. The desire to avoid future regret over current policy decisions may introduce an additional component into our decisions. Whether we recognize future regret as a legitimate concern or dismiss it as a psychological anomaly that arises when we make choices under conditions of uncertainty may affect whether these phenomena should influence policy decisions.

Predicting Behavioral Responses to Risk

Predicting the effect of many of these uncertainties on future outcomes depends in large part on the behavior of economic actors and not simply on the models built by academics and the empirical insights we obtain from existing data. As the paper by Yohe (1992) indicates, a central factor that will drive future outcomes is how individuals process information. Yohe discusses the difference in results between the "dumb" farmer and the "clairvoyant" farmer. In reality, we have a farmer who has cognitive limitations, as do we all. Moreover, when we give these farmers choices that involve probabilistic components, we create additional complications in the decision process. A fundamental concern is how farmers and other decision makers will respond to information provided to them about climate change, particularly when this information may be highly uncertain.

My recent research with Wesley Magat suggests that these responses can be quite complex (Viscusi et al., in press). The focus of our efforts has been on providing information about uncertain environmental risks, with the objective of providing research input that will be valuable to EPA in situations where there is scientific uncertainty, i.e., where there is uncertainty about the extent of the risk as opposed to uncertainty regarding our valuation of risk. The precise risk-policy context of our research—cancer and nerve disease—does not limit the applicability of our results for other situations in which information is uncertain. We focused on issues in which the government did not necessarily know the risk but was attempting to convey the range of uncertainty to individuals to enable them to undertake actions to avert the risk in some manner. For example, the government might indicate that the risk of cancer in one study was 150 per million, whereas the risk in a second study was 200 per million. The question then becomes, how do individuals react to different risk information? Ambiguous and possibly conflicting information is a key component of the entire climate-change area.

Our results do not indicate that people are irrational, but they do suggest that certain specific aspects of the structure of the information presented play an important part in influencing the response. The samples for this study consisted of >600 respondents in Greensboro, N.C., an area whose residents are not too dissimilar from the rural decision makers considered in Yohe's paper (1992). Three of our principal findings follow. First, subjects place a greater weight on the second study mentioned to them, even in situations in which no temporal order of the studies is indicated. Second, in situations where an explicit temporal order is indicated, subjects almost invariably place a much greater weight on the second study than on the first. This is a plausible response if the second study builds on the first and is more credible because it uses more recent scientific knowledge and, presumably, is more pertinent to current risk-exposure levels. Finally, if subjects are given a range of possible outcomes, they display what we term "risk-ambiguity aversion" for any given mean risk indicated. Consider two pairs of studies with a median risk of 150 per million. If the first pair of studies indicates risks of 140 and 160 per million, respectively, and a second pair indicates risks of 100 and 200 per million, then the perceived risk for the second pair of studies will be greater than the first because of the greater risk spread. Respondents are fearful of worst-case outcomes and the chance that the more alarmist study may be right.

These types of results certainly do not exhaust the classes of anomalies and other considerations that must be taken into account when assessing how we can communicate the uncertain classes of risks that arise in the context of global warming. However, we must confront such considerations if we are to move from our current policy-analysis situation to one in which farmers, consumers, firms, and others take decentralized actions. Ideally, the government will not be a passive observer in the adaptation process, but will stimulate it by providing information that these decision makers need. What information we present and how we present it will be important concerns in trying to induce a rational response.

Moreover, even if we present risk information effectively, systematic biases will continue to prevail. For example, individuals tend to overreact to low-probability events, underreact to larger risks, and overreact to risks given frequent media coverage. There is a wide range of conflicting biases that will affect responses to information about global change, making it difficult to predict likely responses and to design policies that will be most effective. These issues are not intractable, but they remain open questions that must be addressed if we are to convey information that will lead to sound decisions.

Uncertain Government Actions

The uncertainties that are present also include political uncertainties. The paper by Gollehon et al. (1992) highlights the potential role of the government in altering the effects of climate change on our well-being. An important class of governmental policies consists of public-works projects. The U.S. Department of the Interior embarked on a massive effort in this century to provide water throughout the western United States through policies that most economists would judge inefficient. If climate change significantly alters the distribution of water throughout the United States, it is likely that farmers and other groups in areas where water is now plentiful will seek expensive public-works projects to bail them out of the future losses that will be inflicted. In effect, from the standpoint of equity, if there are significant losses, these groups will be able to hold us hostage and to obtain public-works projects. These potential costs can be diminished by enabling farmers and municipal users to undertake adaptive measures now, rather than assuming that the government has an obligation to ensure that there are no adjustment costs.

The extent of these adjustment costs may be massive and difficult to quantify. Economists are very good at examining the costs and benefits within a particular line of activity, but policies that lead to massive upheavals and the need for large-scale adjustments require a specific policy analysis of the impacts, which has not yet been forthcoming.

One of the major uncertainties affecting policy decisions pertains to regulations that might be issued by the government. These, in effect, are policy-control variables that can be manipulated to foster the control of global warming. However, many of these regulations will change independently of any global-warming concerns. For example, our continuing efforts to promote air-pollution control and compliance with the provisions of the revised Clean Air Act will have an effect on pollution emissions. Both the enforcement of current regulations and the promulgation of potentially different regulations in the future must be taken into account when assessing the additional steps that must be taken to address the problems of global warming. The greatest policy uncertainty in need of study is the effect of regulatory actions on pollution in a world of incomplete compliance.

Changes in other nations are even less subject to policy control than are changes in the United States. Deforestation policies and energy-conservation efforts in other countries are among the most notable examples of such changes. Because the sources of the problems being addressed are global, any activities that are outside the control of the United States will also influence the base level of pollution that will affect global warming. If there is a nonlinear relationship between the level of pollution and the economic implications of global warming, then concerns about the effect of other countries' pollution on the base level of global warming will clearly be important.

The Limits of Insurance

The natural response in situations of uncertainty is to look to insurance as the solution. Can private or government-provided insurance address these risks? Unfortunately, the insurance industry will be substantially affected by the shift in the character of the uncertainties that it faces. Perhaps the chief line of insurance likely to be influenced is that for property damage, particularly in coastal regions and for other property susceptible to floods. In addition, government insurance programs such as farm price supports are also likely to be notably affected.

Private insurance efforts generally rely on actuarial tables that are based on long periods of experience with insuring a class of risks. By introducing a major shock to this system, global warming will substantially affect the workings of the insurance market.

Although conceptual analysis of these issues is important, substantial lessons may be learned by examining the performance of insurance markets in response to other major shocks in the past. For example, the liability-insurance crisis of the 1980s provides information about how the markets are likely to respond to other shocks. Will insurance rationing result as it did in the early 1980s? Will firms choose to withdraw from particular lines of insurance? If so, what will be the effect on individual risk-taking behavior and on the rationale for government intervention in these markets?

The Role of Discounting

One of the recurring issues raised by the papers in this volume is the choice of the appropriate rate of discount. Although matters become complex when there are multiple risks, for most long-term problems a higher discount rate will typically lead to a greater weight on current generations compared with future ones. Perhaps the main difficulty is to select a specific rate of discount for use in evaluating climate-change policies. A very high rate of discount will make policy discussions, including all those discussed in this book, largely irrelevant, because the saliency of present costs will dominate distant-future concerns.

We can provide some limited economic guidance. Zero discounting is certainly not appropriate, because abandonment of discounting altogether would necessarily make any permanent damage to the environment a trump card that would dominate all other concerns. It is equally clear that the 10% rate of discount that is frequently suggested by the Office of Management and Budget (OMB) is too high as an inflation-adjusted rate of interest to be used in evaluating government policies.

It has often been suggested that OMB must resort to such a high rate of discount as a form of discipline to control the biases in federal agencies' analyses of different policy options. However, there is no reason to assume that these biases are greater for projects with distant-future implications than for those that are focused on current impacts. It would be preferable to distinguish the agency-discipline issue from the appropriate discounting issue so that we do not distort our long-term policy choices in an effort to promote more honest policy assessments.

Our current economic approach and our political process more generally are excessively oriented toward immediate rewards. Far too little attention is devoted to crises whose consequences are not yet apparent or to the possibility, however small, that there will not be a crisis. Long time lags and the existence of some scientific uncertainty tend to foster complacency. Policy analysts must do a better job in assessing the discounted expected value of various policy alternatives and taking those actions that advance society's long-term interests.

It has been suggested that the rate of discount that should be used for health effects and other environmental impacts should be different from those used for monetary impacts because these outcomes are not traded on financial markets. Typically, however, benefit assessments put all the outcomes in terms of a common measure, such as society's willingness to pay for the outcomes, so this issue appears to be more of a distraction than a fundamental concern.

Moreover, in a series of papers written with Michael Moore, I estimated the implicit rate of discount for workers with respect to long-term occupational health risks, with the result that these discount rates are significantly different from zero, but they are not significantly different from prevailing financial rates of return (Moore and Viscusi, 1990a, 1990b; Viscusi and Moore, 1989). No convincing empirical rationale has been provided for discounting global-change policies differently from other policies with long-term implications. Perhaps the major need is for a consistent discount-rate policy across all policy areas, such as that now being developed by the U.S. General Accounting Office.

Conclusion

At this stage we are better at posing questions than we are at providing answers. The series of papers dealing with global-change uncertainties and the impact of climate change more generally have closed some of the important gaps in our knowledge. In particular, they formulated some of the general structure of the policy problems that arise, from the macro context of societal choice to the narrower decision problems faced by farmers and other individuals. In addition, they began the task of developing an empirical base for implementing models of global-change uncertainties, which is essential if we are ever to move from hypothetical policy discussions to actual policy decisions.

The need for empirical economic analysis within the context of the global-warming-policy debate is as great as it is in other economic contexts. Climatological theories concerning the prospects of global warming—and climate change more generally—may be of substantial interest to participants in the climate-change debate. Moreover, discussions of the competing theories and the alternative mechanisms at work are often fruitful. However, not until we quantify the extent of the likely climate change and the time when this change is likely to occur can we begin the task of assessing whether society can or should take concrete actions to address these issues.

Similarly, within the context of economic analysis, we need to know much more than we do now in order to develop a firm policy prescription. Economists are latecomers to the climate-change debate; the early stages of this debate were dominated largely by scientific concerns. However, any ultimate policy decisions to take action to reduce global warming or to do nothing will necessarily hinge on the relative merits of each of these alternatives. Providing a firmer assessment of the empirical effects associated with the global-warming-policy options is one of the key remaining tasks that economists involved in climate-change research must address.

References

Drennen, T., and D. Chapman. 1992. "Biological Emissions and North-South Politics," in J. M. Reilly and M. Anderson, eds., *Economic Issues in Global Climate Change: Agriculture, Forestry, and Natural Resources.* Pp. 215–31. Boulder, Colo.: Westview Press.

Gollehon, N. R., M. R. Moore, M. Aillery et al. 1992. "Modeling Western Irrigated Agriculture and Water Policy: Climate–Change Considerations," in J. M. Reilly and M. Anderson, eds., *Economic Issues in Global Climate Change:*

Agriculture, Forestry, and Natural Resources. Pp. 148–67. Boulder, Colo.: Westview Press.

Moore, M. J., and W. K. Viscusi. 1990a. "Discounting Environmental Health Risks: New Evidence and Policy Implications." *Journal of Environmental Economics and Management* 18: S-51–62.

Moore, M. J., and W. K. Viscusi. 1990b. "Models for Estimating Rates for Long-Term Health Risks Using Labor Market Data." *Journal of Risk Uncertainty* 3: 381–401.

Norgaard, R. B., and R. B. Howarth. 1992. "Sustainability and Intergenerational Environmental Rights: Implications for Benefit-Cost Analysis," in J. M. Reilly and M. Anderson, eds., *Economic Issues in Global Climate Change: Agriculture, Forestry, and Natural Resources.* Pp. 41–55. Boulder, Colo.: Westview Press.

Raiffa, H. 1968. *Decision Analysis.* Reading, Mass.: Addison-Wesley.

Reilly, J. M. 1992. "Climate-Change Damage and the Trace-Gas-Index Issue," in J. M. Reilly and M. Anderson, eds., *Economic Issues in Global Climate Change: Agriculture, Forestry, and Natural Resources.* Pp. 72–88. Boulder, Colo.: Westview Press.

Viscusi, W. K., W. A. Magat, and J. Huber. In press. "Communication of Ambiguous Risk Information." *Theory and Decision.*

Viscusi, W. K., and M. J. Moore. 1989. "Rates of Time Preference and Valuations of the Duration of Life." *Journal of Public Economics* 38: 297–317.

Yohe, G. W. 1992. "Imbedding Dynamic Responses with Imperfect Information into Static Portraits of the Regional Impact of Climate Change," in J. M. Reilly and M. Anderson, eds., *Economic Issues in Global Climate Change: Agriculture, Forestry, and Natural Resources,* Pp. 200–14. Boulder, Colo.: Westview Press.

Data and Research Priorities

24

Data Centers and Data Needs:
Summary of a Panel Discussion

Timothy Mount[1]

Introduction

The recognition that human activities can create environmental problems on a global scale has created a need for new research on the causes and effects of these problems and on the development of solutions. The nature of these problems has implications for the types of data needed to support research. In general, data requirements will be very demanding, because answers to the questions posed must be based on sound empirical evidence. Theory can provide structure for doing research, but many of the questions, such as how well agriculture will adjust to climate change, are fundamentally empirical.

Given the global scope and complexity of many environmental problems, data resources should be enhanced in the following areas: (1) global coverage, putting a greater emphasis on completeness and multinational sources; (2) commodity and product flows rather than financial aggregates; (3) physical and biological links to social and economic activity; and (4) spatial information about social and economic activity.

The data requirements for any one of these four items are daunting. The overall conclusion must be that no single organization or computing system can do it all. Cooperation and coordination among different organizations will be essential. Given the magnitude of these data problems, it is highly desirable to get started on establishing an infrastructure for providing data for research as soon as possible.

---

[1] From the Cornell Institute for Social and Economic Research and the Department of Agricultural Economics, Cornell University, Ithaca, N.Y.

This report summarizes the topics discussed by a group of professionals who provide data for public use or who use data for research. The primary objective of the session was to evaluate where we are in the development of an infrastructure for meeting the data requirements of research on the human dimensions of environmental changes, and what the next steps in this development should be. The discussion was organized around two panels. The first panel was composed of people involved with providing data or supporting research in other ways. The second panel was composed of people who are primarily users of data for research. A summary of the issues raised by each panelist is given below.

Panel I: Providing Data For Research

Ann Gray, Cornell Institute for Social and Economic Research, Ithaca, N.Y.: "The Documentation of Data Files for Interdisciplinary Research"
- The data files needed for research on climate change are large and have complex structures.
- The formats used in different data files and the quality of documentation vary widely.
- It is unlikely that any single data-management system can provide the power and flexibility to handle all sources of data.
- The lack of standardization of documentation, rather than computing hardware and software, is the main limitation in providing data electronically for interdisciplinary research.
- The procedures adopted by the U.S. Bureau of the Census for documentation should be considered as a standard for other agencies.

Barbara Aldrich, U.S. Bureau of the Census, Washington D.C.: "A Basis for Standardizing the Documentation of Data for Public Use"
- Copies of the "1990 Census of Population and Housing: 1988 Dress Rehearsal Summary Tape File 1A Technical Documentation" were distributed.
- The major components of the standard structure for documentation were discussed.
- The abstract gives an overview of the nature and contents of the file.
- The citation for a data file is included for referencing the file in publications.
- A "how-to-use" section describes the documentation and the structure of the file.
- Other components include a subject locator, a list of tables, table outlines, a summary-level sequence chart, and a data dictionary.
- Appendices to the documentation include information on classifications, definitions, data accuracy, collections and processing

procedures, and user notes about additional information that may be available after the documentation is released.

Ron Konkel, National Aeronautic and Space Administration, Washington D.C.:
"Merging Satellite Data with Social and Economic Data: The Role of CIESIN"

* The objectives of the Consortium for International Earth Science Information Network (CIESIN), the Earth Observing System (EOS), and the Data Information System (DIS) were discussed.
* EOS is designed to collect data from satellites. These data are transmitted to DIS, which is a ground-based system. Both systems have high priority in the U.S. Congress.
* The mission of CIESIN is to make it possible for people working in scientific disciplines other than earth science to use the EOS data.
* The CIESIN Data Base Policy Committee has established policies for sharing data internationally, preserving data, establishing data archives, evaluating data quality, and setting international standards for quality.
* The goals of data distribution are to share techniques of processing and to provide data at a low cost.

Paul Baxter, Oak Ridge National Laboratory, Oak Ridge, Tenn.: "Information Technology and Global-Change Science"

* A description of major existing and near-term information technologies that could benefit research on global change was given.
* There is a need for new software to take advantage of the higher speeds of new computing hardware.
* New database capabilities include fourth-generation programming languages and relational database-management systems. The latter require time to set up and care in devising the structure of data files and in maintaining the accuracy of data.
* In the near future, attention should be directed to expert systems, information-discovery tools, truth maintenance, neural networks, visualization, uncertainty representation, and automated reasoning.

Roberta Balstad Miller, National Science Foundation, Washington, D.C.:
"Building an Infrastructure for Research"

* Research on global change implies analyzing changes within a changing system and requires that broadly based databases exist. Federal agencies need to address specific policy issues, which limits their ability to develop this type of database.
* It is not practical to wait until all data have been collected before starting to do research.

- The focus of data collection should be for long time scales on a global basis. This implies that the cost will be large and, therefore, that the cost should be shared among agencies.
- Getting data from many countries will necessitate the establishment of protocols and standards for data collection.
- General agreement that the collection of data is important is needed, because data collection will compete for funds with research activities.

Panel II: Data Needs of Researchers

Richard Stuby, U.S. Department of Agriculture, Washington, D.C.: "The Importance of Demographic Processes for Environmental Research"
- In most studies of global change, environmental factors are usually treated as independent or causal. Demographic variables are treated as dependent or as constants. In reality, population dynamics may be causal or, at a minimum, interactive with environmental (physical, biological) factors.
- Focus on the causes of climatic change tends to ignore human adaptation in consumption patterns and the effects of reasonable mitigation policies.
- The real competition for research funding is between the macro view of systems scientists, who want to investigate interactions within complex systems, and the micro view of reductionists or laboratory scientists, who want to work on increasingly smaller, often more arcane, elements of the larger systems and bear no responsibility for system functioning.

Otto Doering, Purdue University, West Lafayette, Ind.: "The Level of Disaggregation Needed for Research on Trade and Agricultural Policy"
- Adaptation to climate change is different from trying to stop climate change. Predicting changes is not enough. It is essential to have a true picture of reality, but existing data are inadequate to do this.
- Existing data on weather in the United States are often collected at airports and in metropolitan areas. Consequently, it is difficult to establish baseline data on climate for agricultural regions.
- The changes in climate that we are trying to identify are small relative to year-to-year variability at any typical location.
- Weather characteristics at critical times over the growth cycle for different crops are needed to assess the effects of climate change on agricultural production.
- Despite the long time horizons of research on climate change, it is essential to have more detailed regional data on weather to evaluate the effects of weather on agriculture.

John Reilly, U. S. Department of Agriculture, Washington, D.C.: "The Quality of Existing Data Sources on Agriculture and Energy Use"

- Traditional research relates the causes of climate change to energy supply and the effects to agriculture. However, there are important links between agricultural causes and feedbacks that should be considered, such as the use of biomass for fuel and the effects of higher concentrations of CO_2 on crop production.
- There is a need for better international data, but the ability to collect data will be influenced by the way the data are used. For example, if the data are used to monitor adherence to international standards on emissions, it will be much harder to get additional data for research.
- The benefits from collecting environmental data should be identified more clearly. Broad, general conclusions may be inaccurate if localized data sets are unavailable. In addition, there is a need for resource indicators to measure the changing value of environmental assets.
- The tradition within the Economic Research Service (ERS) at the U.S. Department of Agriculture is to collect data for commercially significant commodities and for relatively large farms. However, this is not necessarily appropriate for understanding issues like food security, water quality, and soil erosion.
- More effort should be given to sampling land area to understand and monitor the effects of climate change. If accurate historical series are developed, these data can be used to calculate the resource indicators needed for policy analysis.

Conclusions

My summary of the panel discussion leads to the following general conclusions: (1) Research on climate change will require data that are both longitudinal and spatially disaggregated. (2) Better information is needed on commodity flows and resource inventories. (3) No single agency has a mission that spans all the types of data needed. (4) Because the costs of collecting and assembling the required data are large, agencies will have to share the costs. (5) New computing hardware and software offer exciting possibilities for managing large datasets effectively. (6) International standards for data quality and coverage should be established. (7) A major limitation in working with many datasets electronically is the poor quality of documentation. (8) The standards for documentation developed by the U.S. Bureau of the Census could be used by other agencies to solve the problem of inadequate documentation.

25

Setting Priorities for
Global-Change Research in Agriculture

John M. Antle[1]

Introduction

Global change presents unprecedented challenges to researchers. The global scale and diversity of scientific and policy questions pose daunting methodological and organizational problems. The potential consequences of either action or inaction on the policy front raise the stakes to unprecedented levels. How are agricultural researchers to respond to these challenges?

I attempt to address this question by first reviewing some lessons from past agricultural research, and then presenting a simple framework for *ex ante* assessment of research priorities. Combined with what we have learned from the papers presented in this volume, I make a preliminary attempt to assess research priorities for agricultural economics.

Past Lessons

We have an extensive literature on agricultural research and many decades of experience in both the high-income countries such as the United States and in the developing countries. For the purposes of addressing global-change issues, it is significant to note that the high-income countries, where much of the most successful research originated, are in the temperate zone of the Northern Hemisphere, whereas most developing countries are in the tropics, both humid and arid. One of the most important results from studies of agricultural research is that rates of return on investments have been high in "favorable" environments—in temperate and certain

---

[1] From the Department of Agricultural Economics and Economics, Montana State University, Bozeman.

well—endowed humid tropical areas. But agricultural research has been much less sucessful in "unfavorable" environments, particularly in certain parts of the humid tropics and generally in the semiarid and arid tropics. The unfavorable humid environments often suffer from intractable pest problems and poor soils, whereas the arid environments are the most resource-poor from the standpoint of moisture and soils.

A second lesson is that the adoption of new technologies is a function of profitability. Profitability in turn depends on several factors, including suitable infrastructure—both physical, such as roads, and institutional, such as a well-functioning legal system—and on various aspects of economic policy, including agricultural, macroeconomic, and trade policy.

A third lesson is that technology transfer is generally much simpler than policy transfer. That is, disseminating a new seed variety and associated management skills and inputs in a favorable natural and economic environment is much easier than making the institutional and political changes that may be necessary to create a favorable policy environment for adoption of the new technology.

A fourth lesson, taken from the risk literature, is that those who are at the greatest risk, be it from natural or man-made catastrophies, are the poor. Rich individuals as well as rich nations are the ones who can best afford to insure against risk. Insurance can be acquired in a variety of forms, and wealth is the tried and true form of "self insurance."

Finally, from the policy and political-economy literature it must be learned that a key factor motivating policy decisions is income distribution. Here we have something to learn from the French. In their language, *la politique* means both politics and policy. Economists need to understand that policy and politics are inseparable.

The Certainty Case

In a certain world, independent investment opportunities should be ordered according to their net present value. When the benefits of one investment depends on other investments, the problem is more complex. In that case, investment portfolios must be constructed, and the portfolios should be ordered according to net present value. Applying this principle to global-change research, it is important to emphasize that the benefits of various lines of disciplinary research are interdependent. Consequently, it is not methodologically correct to assess research priorities for, say, climate change separately from those for ozone depletion. Similarly, it would be incorrect to set priorities for agricultural technology development separately from those for agricultural economics, and policy research. Adopting alternative agricultural technologies to solve greenhouse-gas-emissions problems, for example, may require a corresponding set of economic policies. The expected benefits of a proposed technology cannot be assessed without evaluating the likely rate of adoption under expected economic conditions.

A proposed typology of the constraints on technology adoption can be summarized using the simple Venn diagram presented in Figure 25.1. An innovation can be characterized in three dimensions: technological, economic, and political. *Ceteris paribus*, research that is feasible in all three dimensions is the most likely to be successful and should receive the highest priority. Projects that satisfy two dimensions should receive lower priority, and those that satisfy only one should receive the lowest priority.

The Uncertainty Case

The world of possible global change is very uncertain. In this regard, it is useful to categorize research into two types: technology- and policy-development (TP) research, which aims to increase expected social welfare, given probabilities of uncertain events; and uncertainty-reduction (UR)

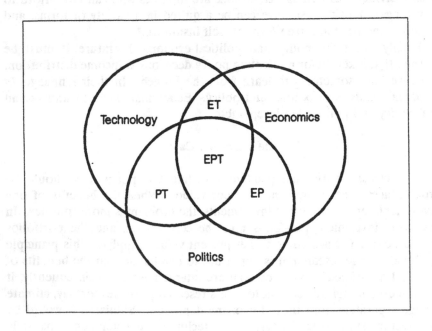

FIGURE 25.1 Setting research priorities according to economic (E), technological (T), and political (P) feasibility.

research, which attempts to reduce the uncertainty associated with global change. Given that global change affects agriculture and vice versa, TP and UR research can be represented schematically as in Figure 25.2. UR refers to the understanding of the interactions between agriculture on the one hand and global climate and other environmental characteristics on the other. TP research for adaptation and mitigation affects the impact of global change on agriculture; TP research on greenhouse-gas-emissions reduction, through policy intervention or technological innovation, affects the impacts that agriculture has on global change.

A variety of decision models can be used to analyze research-priority setting under uncertainty. For example, a conventional model of risk aversion would take the form

Max {[Expected net returns (TP) | UR] − [Risk premium (TP) | UR]}
UR, TP

Thus, the decision problem is solved sequentially. First, TP research priorities are set to maximize expected returns net of the risk premium,

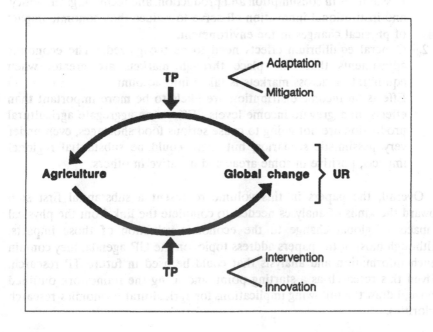

FIGURE 25.2 Technology- and policy-development (TP) and uncertainty-reduction (UR) research in relation to agriculture and global change.

conditional on the level of uncertainty. Second, uncertainty research priorities are set, given the decision rule for TP research as a function of UR. Over time, as uncertainty is reduced through UR, the decisions can be reassessed.

Setting Priorities

Using this framework, what have we learned from papers in this volume, and what can we say about setting research priorities for global-change research in agriculture? A review of the papers shows that most fall into the UR category: they are concerned with reducing uncertainty about the interactions between global change and agriculture. Perhaps the most significant theme that emerges from these papers that needs to be communicated to the general scientific community and the general public is that there is a critical distinction between the physical impacts of global change and the economic consequences of those physical impacts. In this connection, several very important subthemes emerge:

1. Adaptation has great potential to offset physical changes. Subsitution possibilities in consumption and production, and technological, policy, and institutional innovation all serve to reduce the economic impact of physical changes in the environment.
2. General equilibrium effects need to be recognized. The economic adjustments that take place through markets are greater when equilibration across markets is taken into account.
3. Effects on income distribution are likely to be more important than effects on aggregate income levels. Effects on aggregate agricultural production are not going to cause serious food shortages, even under very pessimistic scenarios, but there could be substantial regional impacts, positive in some areas and negative in others.

Overall, the papers in this volume represent a substantial first step toward the kinds of analyses needed to complete the link from the physical impacts of global change to the economic valuation of those impacts. Although most of the papers address topics on the UR agenda, they contain much information and analysis that could be used in future TP research. Given this research as a starting point, and using the framework outlined above, I draw the following implications for agricultural-economics-research priorities.

A Conceptual Framework for Interdisciplinary
Analysis of Global-Change Issues

Returning to the discussion of Figure 25.1, I argued that priority should be given to strategies that are technologically, economically, and politically feasible. But how is that assessment to be made? We have already seen physical scientists drawing policy conclusions—for example, that CO_2 emissions should be reduced by an arbitrary amount by an arbitrary date—without adequate economic or policy analysis to justify such a conclusion. Clearly, economists need to communicate to physical scientists the data needs for appropriate policy analysis. Economists must play a central role because they have the conceptual understanding and analytical tools needed to assess the economic and policy feasibility of alternative strategies of addressing global-change issues. A first priority, therefore, must be to extend the existing conceptual framework and analytical tools so they can be used to address global-change issues. Economists, in collaboration with scientists from other disciplines, need to develop a conceptual and analytical framework in which to integrate disciplinary data to assess the technical, economic, and policy feasiblity of alternative strategies to address global-change issues.

Intervention vs. Innovation and Adaptation

The feasibility of various proposed interventions needs to be investigated. How feasible are the various regulatory and institutional arrangements that are being proposed, such as tradable permits and debt-for-nature swaps? How do they compare to possible technological solutions? To what degree will adaptation mitigate the physical and economic impacts of global change?

Distributional Impacts of Global Change

Everything we know about political economy and the economic modeling conducted thus far suggests that the political and economic consequences of global change depend critically on distributional impacts. Unfortunately, the greatest physical uncertainties are in the regional impacts. Nevertheless, both TP- and UR-economics research can pay much more attention to distributional impacts, so that further analyses can be conducted when better physical data become available.

Impacts of Global Change on Third World Countries

Most of the physical and economic modeling and analysis have focused on the northern latitudes and high-income countries. Yet, most of the poor people who are at highest risk from any kind of adverse physical or economic change live in the tropics. The data currently available also suggest that the third world will become an increasingly important actor in causing global change—by emitting large amounts of greenhouse gases, for example. Thus, any technological or policy solutions must be compatible with economic and political conditions in the developing countries. Here, our recent knowledge of and experience with innovation and diffusion of agricultural and industrial technologies, and the policy environment in developing countries, is critically important. This area needs to be high on the TP-research agenda.

Predicting Technology

As Tom Schelling rightly pointed out at the "Global Change: Economic Issues in Agriculture, Forestry, and Natural Resources" conference, held Nov. 19–21, 1990, in Washington, D.C., the greatest uncertainty in global change may not be in predicting physical changes but rather in predicting the path of future technology. There is a wealth of knowledge in economics and agricultural economics about research-and-development processes, induced innovation, technology adoption, and impacts of technological change on productivity. Most research, however, has focused on explaining observed innovation, not on forecasting future innovations. To appreciate how difficult this is, consider progress that has occurred in computer technology in the past 10 years. Who would have believed 20 or even 10 years ago that this manuscript would be written using a small desktop computer that cost $4500 and that rivals the speed and power of a mainframe of 1980 vintage that cost millions of dollars? Who yet would hazard to forecast technology 30, 50, or 70 years into the future, as is required if we are to assess the impacts of global change?

26

Research Priorities Related to the Economics of Global Warming

Michel Potier and Tom Jones[1]

Introduction

Because we represent an international organization associated with the ongoing process of negotiations toward a climate-change convention, you will not be surprised to hear that our list of global-warming research priorities is dictated by the kind of information that will assist the policy makers of our member governments. Basically, the information of an economic nature that is needed in international fora falls into one of the following four categories:

1. We need a better understanding of the actual size of the potential costs of responding to climate change, and of their distribution over time, that key economic sectors such as agriculture, transport, and energy will have to face if particular climate-change adaptation or mitigation strategies are adopted. We need to assess the implications of such costs for economic growth and for international trade.

2. Although it is a very complicated task, we need to shed more light on demonstrated economic benefits that could be derived from climate-change policies.

3. In case there is eventually an agreement on a particular response strategy, we need proper advice on the kind of policy instruments most likely to lead to efficient results. Economic instruments offer

---

[1] From the Economics Division, Environment Directorate, Organization for Economic Cooperation and Development (OECD), Paris. The opinions expressed in this paper are those of the authors and do not necessarily represent those of the OECD.

some potential advantages in this regard, but these tools have not yet been tested in an international context.

4. It will be important to design an economically efficient international agreement. Then, even if such an accord can be reached, there will also be a need to devise rules to ensure that it can be sustained.

We in the OECD see one key research requirement within each of the four categories, and each is described below. We conclude by offering some suggestions for research in the agriculture and forestry sectors specifically.

Cost Issues

There is a need to develop a multisector, multicountry, dynamic applied general equilibrium (AGE) model to quantify the medium- and long-term economic costs of reducing emissions of greenhouse gases on a global basis. A survey undertaken by the Organization for Economic Cooperation and Development (OECD, 1991) shows that there are several applied general equilibrium models that analyze costs in a single-country context. But, the climate-change problem is a global one, and there are few consistent global models that can tackle this issue on the world scale.

With a multisector, multicountry, dynamic AGE model, there will be the possibility of simulating economy-wide and global costs of alternative policies, including different types of international agreements to limit emissions. With such a model, we could address many important questions about policy issues that are currently being debated in international fora, such as the following:

- What would be the effects on economic growth, international trade patterns, and welfare if some countries implemented agreed targets for emissions reduction, while others did nothing or implemented less stringent ones?
- What would be the economic effects of specifying targets in different ways, e.g., in terms of absolute levels of current emissions or on an equal per capita or a gross domestic product (GDP) per capita basis?
- What are the costs of achieving agreed targets by alternative policies, e.g., command-and-control strategies vs. economic instruments such as a "carbon tax"?
- What kind of reimbursement schedules of the carbon-tax revenues to individual countries should be sought to minimize the risk of the agreement breaking down?
- What are the fiscal measures needed to achieve revenue neutrality with respect to the reimbursed tax revenues? Whether these funds give rise

to indirect tax cuts, social contribution cuts, or investment subsidies may have different implications for economic growth.

• What are the economic implications of instituting an international vs. a national tax-administration system?

The OECD has started developing such a model, which has been named GREEN (General Equilibrium Environmental model). At this stage, GREEN covers only CO_2 and ignores other greenhouse gases such as chlorofluorocarbons (CFCs) and methane. However, we consider this modeling exercise an important step in the right direction.

GREEN is currently built to cover eight producing sectors (agriculture; coal mining; crude oil; gas; refined oil and coal products; electricity, gas, and water distribution; energy-intensive industries; and non-energy-intensive industries and services) and four consumer sectors (food, beverages, and tobacco; fuel and power; personal transport equipment; and other goods and services). Furthermore, it is disaggregated into eight main regions: North America, Europe [the European Community (EC) and the European Free Trade Area (EFTA)], the Pacific area, the oil-exporting developing countries, the other developing countries, the new industrial countries, China, and Eastern Europe and the USSR. Initial results from GREEN are expected to be available in 1991. Additional research outside the OECD should also be encouraged in order to test the strength of the results being obtained through the OECD effort.

Benefit Issues

This is an area where there seems to be substantial disagreement among researchers. Nordhaus (1990), on the basis of a doubling of CO_2, has come up with a rough calculation that potential damage from global warming is worth only ≈ 0.25–2% of the U.S gross national product (GNP). But this estimate is challenged by those who believe that the existing estimates of economic damage potentially associated with global warming are understated for at least three reasons.

First, it is argued that these estimates do not include certain categories of damage. For example, the Nordhaus estimate does not include damage from increased frequency and severity of storm events, such as hurricanes. Second, they do not value environmental losses (e.g., the loss of species or forest). Third, they make no allowance for consumers' willingness to pay a price—for example, for the amenity of avoiding excessive temperature peaks.

There is merit in developing a solid conceptual analytical framework for analyzing benefits, indicating which kinds of damage or loss of amenity should be assessed, and which are measurable in market terms. Unfortunately, most of the work done on benefits so far, as reflected in the final report of the Intergovernmental Panel on Climate Change (IPCC) Working Group 2 (IPCC, 1990), has restricted the concept of benefits to the measurement of short-term physical (rather than socioeconomic) impacts. Such a framework should explicitly incorporate temporal issues, because the time path of benefit streams is an important element of any response policy. It should also explicitly consider distributional issues in terms of the countries or economic sectors that will be affected most by global warming.

It is also increasingly clear that, because of the uncertainty associated with valuing the economic benefits of abating global warming, particular importance has to be attached to the benefits that might accrue in other environmental arenas. In other words, it is critical to get a clear idea worldwide of the nature and magnitude of the spinoff benefits in other areas of environmental policy, if any, that might arise from any action taken to limit greenhouse-gas emissions. For example, actions taken to improve energy efficiency, to protect tropical rain forests, or to promote biological diversity would also help to reduce the atmospheric concentrations of greenhouse gases.

Potential Contribution of Economic Instruments

A lot of work has already been done in academic circles on the theoretical advantages of economic instruments, such as taxes and tradable emission permits. It has been demonstrated theoretically that, in terms of economic efficiency, there is not much of a choice between emission charges and tradable emission permits. Some OECD countries have already decided to use such instruments as part of their policy response for controlling greenhouse gases. CFCs are taxed in Denmark; CO_2 emissions are taxed in Finland, and they will soon be taxed in Sweden. The United States has developed a program for international trading in CFC-production and -consumption allowances in accordance with the articles of the Montreal Protocol.

So far, the preference of European countries seems to be in favor of emission charges (taxes), whereas the United States seems to favor a tradable-permits approach. However, emission charges and tradable-permit systems both raise several practical questions that will need to be addressed before significant progress can be made in the use of these instruments.

With respect to taxes, there is little doubt that a significant reduction in emissions will require a large tax. But there are a lot of uncertainties about the impacts of such a tax, due to uncertainties on both the demand and supply sides. This is an area where more research is clearly needed. Similarly, the effects of such a large tax on economic growth is unclear. It is likely that a substantial carbon tax would slow economic growth, but this slowdown could conceivably be offset by efficiency gains realized by reducing distorting taxes elsewhere in the economy. These questions need to be addressed, and this is an area where AGE models of the type discussed above can usefully contribute.

A related issue to be investigated with respect to taxes is that of equity. There is a real possibility that a carbon tax would have a regressive impact on poorer households, which spend a higher proportion of their budget on heating and lighting than do richer households. Conceptually, the impact of the carbon tax on income distribution could be blunted by making compensatory adjustments within existing taxes. The challenge for governments is to develop tax policy changes that could be made revenue neutral on average, if not for all affected groups. The whole issue of equity in climate-change-response policies involving taxes has often been discussed in very general terms, but it has not been investigated systematically.

For tradable emission permits (particularly for an international system), a major issue to be resolved will be an agreement on the initial allocation of rights. In this regard, several options exist. (1) Permits could be allocated on the basis of historical emission levels, but in an international context, such a system would reward countries that are already developed while penalizing the developing countries. (2) Permits could be allocated by auction to the highest bidders, but such a system would favor the polluters with the greatest ability to pay. (3) GNP could be used as a basis for allocating permits. This would favor countries with low fossil-fuel consumption per unit of GNP (e.g., France and Japan) and harm countries that burn a lot of fossil fuels inefficiently (e.g., China and India). (4) Permits could be allocated on the basis of population, but this rule would favor countries with a low CO_2 emissions per capita (e.g., China and India).

Each of these allocation rules has its drawbacks, but the main problem is that they all favor one group of countries and harm another. Consequently, it is unlikely that any worldwide agreement can be reached through the adoption of any one of these allocation rules individually. An urgent need, therefore, exists to look for some form of combination allocation formula. This is clearly a priority research area.

A third economic instrument that we should look at more closely is subsidies. We have to deal with subsidies, particularly in relation to possibilities for "sink enhancement" (i.e., increasing the capacity of the environment to absorb greenhouse gases). One example is afforestation. Indeed, afforestation or a reduction in deforestation has an effect similar to that of emission abatement. In theory, if emissions are taxed, afforestation should be subsidized ("negatively taxed"). But the preservation and restoration of forests would provide the international community with benefits other than those associated with climate change, particularly with respect to biological diversity. Therefore, it is of paramount importance to get some feeling about the whole range of benefits to be derived from a large subsidy program directed at replanting trees, or to stopping deforestation. It is also important to understand how existing subsidies could be restructured to enhance prospects for reduced global warming, or for a better adaptation to it.

Negotiating an International Convention

We also need to develop practical policy guidance relevant to ongoing negotiation processes, particularly for coping with the "free-rider" issues (i.e., one nation receiving the benefits of collective action to combat global warming, but not participating in helping to pay for that action). One of the key issues, if not *the* key issue, to be faced in forthcoming international negotiations on climate-change agreements will be to get all key countries concerned to sign the agreement. Some lessons can be learned from our own previous analyses of existing international environmental agreements (OECD, 1991), including the following:

1. It is unlikely that an agreement will be reached over uniform abatement obligations. This, in fact, is already happening. To date, even the industrialized (OECD) countries have failed to agree on specific international targets for reductions of CO_2.
2. Side payments will be needed to induce reticent countries to sign.
3. As mentioned earlier, if international convention allowed tradable emission-reduction permits, many countries would be unwilling to sign an agreement that allocated permits on the basis of ad hoc rules, such as population or GNP.
4. If an agreement is reached, a mechanism will have to be put in place to ensure compliance with the agreement. This means that suitable administrative rules would have to be developed to detect and punish "cheating."

All of these issues (and others) will need to be addressed in preparation for the global-warming convention foreseen for 1992.

Future Priorities for the Agriculture and Forestry Sectors

With respect to the agriculture and forestry sectors specifically, our recommendation for further research is mainly based on the work carried out a few months ago within the OECD Environment Directorate on the possible impacts of climate change on agriculture, from both national and international perspectives.

Although much effort has been devoted to quantifying the potential impact of climate change on agricultural production, this work has mainly focused on physical aspects, such as plant productivity, and less on the links between the agricultural economic system and these physical changes. Therefore, more attention should be given in future to research on (1) the development of models to link physical and microeconomic variables more directly; (2) the dynamic aspects in those links, such as employment changes over time, the evolution in the number or structure of farms in the agriculture sector, or tree-species "migrations"; and (3) the wider use of scenario analysis, incorporating both stochastic and nonstochastic variations.

The work we have been carrying out with the help of Steven Sonka on evaluating the impacts of climate change on the agriculture sector (OECD, 1990) suggests that future research should be focused on the following top priorities:

1. Linking physical and socioeconomic variables more closely. Population levels and resource availabilities (such as water stocks) are just as important in determining economic impacts as are changes in temperature or precipitation. These types of social and economic changes should be taken into account as fully as possible, including the full range of behavioral responses that might result from them. For example, researchers should look at the various management practices that producers could use; at changes in food-consumption patterns as consumers respond to differing food-supply patterns; and at efforts to create and implement technological improvements that might mitigate or exploit alternative climatic patterns.

2. Addressing, through scenario analysis, the dynamic aspects of both climate change and economic responses, although this is a very difficult task. Research in this area should look not only at the situation as it exists, but also at the time path of these variables. In this context, variability (i.e., the potential for greater frequency of extreme climate phenomena) should be addressed, together with its implications for food security and the economics of the global food system (i.e., agricultural trade).

3. Broadening the scope of economic analysis of the climate-change problem in all economic sectors, including the agriculture sector. Within the agricultural sector, nearly all studies carried out so far have focused on cereal crops in developed countries. Although the economic importance of these activities is significant, they do not represent the majority of the world's agricultural systems. Therefore, there is a need to broaden the scope of analysis in terms of both the geographic areas of analysis and the agricultural activities reviewed.

4. Considering the trade impacts as the scope of this analysis is broadened. To this effect, OECD recently undertook, with the help of Stan Johnson, a review of the potential to use various existing

agricultural models for assessing the agricultural trade impacts of global climate change (OECD, 1990). This review showed that all of the models reviewed had both advantages and disadvantages and that some type of integrated approach would likely prove to be the most informative. The best approach here would be to rely on a mixture of different types of models (i.e., micro, or farm-level; macro; sector-level; and country-specific) in order to link farm-level discussions and production functions with plant-growth simulations, as well as to provide information on the robustness of the micro-model results from the macroeconomic modeling perspective. However, before devoting a significant amount of resources to modeling activities, more work should be done to determine the key parameters on which to focus these modeling exercises. In the case of trade, what is the most important question? Is it the relative cost of agricultural production and trade levels among alternative climatic regions? Is it the location of the agricultural production? The factor-use levels? The distribution of agricultural production? The interaction between climate-change policy and trade? Something else? This point is very important, because the tendency of modelers in the past has too often been to generate research outputs without first defining the questions being researched or thinking about the potential users (customers) of this research information.

5. Focusing on opportunities to use a structural reform in the agricultural sector for achieving climate-change objectives. Here, the idea would be to assess how a progressive phasing out of subsidies, leading to decreases in the use of energy-intensive farming practices, might ultimately contribute positively to the goal of reducing greenhouse-gas emissions.

References

Intergovernmental Panel on Climate Change (IPCC). 1990. *Potential Impacts of Climate Change: Report Prepared for IPCC by Working Group II.* New York: World Meteorological Organization and United Nations Environmental Programme.

Nordhaus, W. 1990. "To Slow or Not To Slow: The Economics of the Greenhouse Effect." New Haven, Conn.: Yale University. (Discussion paper.)

Organization for Economic Cooperation and Development. 1990. A Survey of Studies of the Costs of Reducing Greenhouse Gas Emissions. Paris: OECD. (Working Paper 89.)

Organization for Economic Cooperation and Development. 1991. *Responding to Climate Change: Selected Economic Issues.* Paris: OECD.

Index

Printed in the United States
by Baker & Taylor Publisher Services

Printed in the United States
by Baker & Taylor Publisher Services